Workshop on Rebuilding Abalone Stocks in British Columbia

Inquiries: Monograph Publishing Program, NRC Research Press, National Research Council of Canada, Ottawa, Ontario K1A 0R6, Canada. Web site: www.monographs.nrc.ca

Correct citation for this publication: Campbell, A. (*Editor*). 2000. *Workshop on Rebuilding Abalone Stocks in British Columbia*. Can. Spec. Publ. Fish. Aquat. Sci. 130. 158 p.

Canadian Special Publication of Fisheries and Aquatic Sciences 130

Workshop on Rebuilding Abalone Stocks in British Columbia

Edited by

Alan Campbell

Fisheries and Oceans Canada
Pacific Biological Station
Nanaimo, BC V9R 5K6, Canada

NRC·CNRC
NRC Research Press
Ottawa 2000

Printed in Canada on acid-free paper. ♾

ISBN 0-660-18292-0
ISSN 0706-6481
NRC No. 42841

Canadian Cataloguing in Publication Data

Workshop on Rebuilding Abalone Stocks in British Columbia
(1999: Nanaimo, B.C.)

Workshop on rebuilding abalone stocks in British Columbia

(Canadian Special Publication of Fisheries and Aquatic Sciences 130)
Includes an abstract in French.
"An international Workshop on Rebuilding Abalone Stocks in British Columbia was held during February 23–26, 1999, in Nanaimo, British Columbia, Canada." — Abstract
Includes bibliographical references.
ISBN 0-660-18292-0

1. Abalone fisheries — British Columbia — Congresses.
2. Abalone culture — British Columbia — Congresses.
3. Abalone — British Columbia — Congresses.
4. Abalone fisheries — Congresses.
I. Campbell, Alan, 1944– .
II. National Research Council Canada.
III. Title.
IV. Series.

SH371.5C3W77 2000 639.4832 C00-980408-0

Contents

Abstract / Résumé

An international Workshop on Rebuilding Abalone Stocks in British Columbia was held during February 23–26, 1999, in Nanaimo, British Columbia, Canada. The main goal of the workshop was to develop a realistic strategy to rehabilitate depleted northern (pinto) abalone, *Haliotis kamtschatkana*, stocks in British Columbia. The workshop was also meant to clarify the roles, expectations, and shared interests of many of the interest groups in British Columbia, emphasizing the key role in the rebuilding plan of local communities, including First Nations and non-native communities. The workshop was attended by approximately 120 participants representing First Nations, local communities, recreational and commercial interests including aquaculture, and international (Australia, Canada, Japan, South Africa, and U.S.A.) abalone research and management. Presentations at the workshop were centered around the following five topics: (*i*) introducing local perspectives; (*ii*) status of abalone fisheries in Alaska, California, British Columbia, and South Africa; (*iii*) experiences in rebuilding abalone stocks using abalone aquaculture, larval and juvenile seed, and brood stock aggregations; (*iv*) additional pieces of the rebuilding puzzle, including abalone genetics and disease, sea otters, and the use of marine protected areas and refuges; and (*v*) summarizing discussions on a strategy for rebuilding abalone stocks in British Columbia. There are 13 peer reviewed papers in the Proceedings.

Un Atelier international sur la reconstitution des stocks d'ormeau en Colombie-Britannique s'est tenu du 23 au 26 février 1999 à Nanaimo, Colombie-Britannique (Canada). Le principal objectif visé par l'atelier était d'élaborer une stratégie réaliste de reconstitution des stocks décimés d'ormeau nordique, *Haliotis kamtschatkana*, en Colombie-Britannique. L'atelier visait également à éclaircir les rôles, les attentes et les intérêts communs de nombreux groupes d'intérêt de la Colombie-Britannique en mettant l'accent sur le rôle clé des plans de reconstitution des collectivités locales, y compris les collectivités autochtones et non autochtones. Environ 120 personnes ont assisté à l'atelier, dont des représentants des Premières Nations, des collectivités locales, des intérêts commerciaux et de loisirs, y compris de l'aquaculture ainsi que de la recherche et de la gestion internationale relativement aux stocks d'ormeau (Australie, Canada, Japon, Afrique du Sud et États-Unis). Les exposés présentés à l'atelier ont porté sur les cinq sujets suivants : (*i*) l'introduction de perspectives locales; (*ii*) la situation de la pêche à l'ormeau en Alaska, en Californie, en Colombie-Britannique et en Afrique du Sud; (*iii*) les expériences de la reconstitution des stocks d'ormeau au moyen de l'aquaculture, des larves et des reproducteurs juvéniles ainsi que des concentrations de stock de géniteurs; (*iv*) les autres éléments de la problématique de la reconstitution, c.-à-d. la génétique et les maladies de l'ormeau, les loutres de mer et le recours aux zones marines protégées et aux refuges; et (*v*) l'aperçu des discussions entourant la stratégie de reconstitution des stocks d'ormeau en Colombie-Britannique. Le compte rendu de l'atelier comprend 13 articles évalués par des pairs.

List of participants

Adkins, B., Fisheries and Oceans Canada, Pacific Biological Station, Nanaimo, BC V9R 5K6, Canada.

Alabi, Y., Island Scallops, 5552 West Island Highway, Qualicum Beach, BC V9K 2C8, Canada.

Alfred, H., Kwakuitl Territories Fisheries, Box 255, Alert Bay, BC V0N 1A0, Canada.

Alfred, P., Kwakuitl Territories Fisheries, 1441 B. NorthIsland Hwy, Campbell River, BC V9W 2E4, Canada.

Antilla, E., Fisheries and Oceans Canada, AFS, 417 2nd Ave. West, Prince Rupert, BC V8J 1G8, Canada.

Ashworth, C., West Coast Abalone, PO Box 422, Whistler, BC V0N 1B0, Canada.

Attridge, E., Haisla Fisheries Commission, PO Box 1028, Kitamaat Village, BC V0T 2B0, Canada.

Bath, M., 1248 Charman, Victoria, BC V8V 2T8, Canada.

Beggs, C., Heiltsuk Tribal Council, Box 880, Waglisla, BC V0T 1Z0, Canada.

Blackbourn, J., 1335 Sherwood, Nanaimo, BC V9T 1G8, Canada.

Boutillier, J., Fisheries and Oceans Canada, Pacific Biological Station, Nanaimo, BC V9R 5K6, Canada.

Bower, S., Fisheries and Oceans Canada, Pacific Biological Station, Nanaimo, BC V9R 5K6, Canada.

Bowler, M., Haisla Fisheries Commission, PO Box 1028, Kitamaat Village, BC V0T 2B0, Canada.

Brouwer, D., Fisheries and Oceans Canada, Pacific Biological Station, Nanaimo, BC V9R 5K6, Canada.

Bureau, D., Pacific Urchin Harvesters Association, PO Box 2122, Station A, Nanaimo, BC V9R 6X9, Canada.

Campagna, S., Pacific Sea Cucumber Harvesters Association, 364 Vancouver Ave., Nanaimo, BC V9S 4G2, Canada.

Campbell, A., Fisheries and Oceans Canada, Pacific Biological Station, Nanaimo, BC V9R 5K6, Canada.

Campbell, D., Ahousat Fisheries, Box 1383, Port Alberni, BC V9Y 7M2, Canada.

Carpenter, F., Heiltsuk Tribal Council, Box 880, Waglisla, BC V0T 1Z0, Canada.

Carter, L., United Holdings Ltd., 6970 Flemming Street, Vancouver, BC V5P 3H7, Canada.

Castledine, A., BC Ministry of Fisheries, PO Box 9359, Stn Prov Govt, Victoria, BC V8W 9M2, Canada.

Clapp, B., Underwater Harvesters Association, 559 Rothdale Place, Ladysmith, BC V0S 1N0, Canada.

Cross, S., Aquametrix, 204–2527 Beacon Avenue, Sidney, BC V8L 1Y1, Canada.

Davis, G., Channel Islands National Park California, 1901 Spinnaker Drive, Ventura, CA, 93001–4354, USA.

Dragseth, C., Fisheries Management, Pacific Regional Headquarters, Fisheries and Oceans Canada, 300–555 West Hastings Street, Vancouver, BC V6B 5G3, Canada.

Duffy, D., 312–1947 West 7th Avenue, Vancouver, BC V6J 1T2, Canada.

Dunlop, R., Nuu Chah Nulth Tribal Council, Box 1383, Port Alberni, BC V9Y 7M2, Canada.

Eccleston, V., Malaspina University-College, 206–3185 Barons Rd, Nanaimo, BC V9T 5T3, Canada.

Edwards, D., West Coast Sustainability Association, Box 77, Ucluelet, BC V0R 3A0, Canada.

Fraser, M., Fisheries and Oceans Canada, Station 13029–200 Kent Street, Ottawa, ON, K1A 0E6, Canada.

Gardner, J., Dovetail Consulting, 105–2590 Granville Street, Vancouver, BC V6H 3H1, Canada.

Ginetz, R., DFO Regional Aquaculture Coordinator, 555 W. Hastings St., Vancouver, BC V6B 5G3, Canada.

Gladstone, D., Heiltsuk Tribal Council, Box 880, Waglisla, BC V0T 1Z0, Canada.

Gladstone, W., Heiltsuk Tribal Council, Box 880, Waglisla, BC V0T 1Z0, Canada.

Good, M., Nanaimo First Nation, 1939 Prince Charles, Nanaimo, BC V9X 1M4, Canada.

Griggs, J., Dovetail Consulting, 105–2590 Granville Street, Vancouver, BC V6H 3H1, Canada.

Gueret, D., Fisheries and Oceans Canada, Operations Branch, 417–2nd Ave. West, Prince Rupert, BC V8J 1G8, Canada.

Haaker, P., California Department of Fish and Game, 330 Golden Shore, Suite 50, Long Beach, CA, 90802, USA.

Hand, C., Fisheries and Oceans Canada, Pacific Biological Station, Nanaimo, BC V9R 5K6, Canada.

Harbo, R., Fisheries and Oceans Canada, 3225 Stephenson Point Road, Nanaimo, BC V9T 1K3, Canada.

Harper, K., Fisheries and Oceans Canada, C & P, 100 Annacis Parkway, Unit 3, Delta, BC V3M 6A2, Canada.

Harper, T., Unique Sea Farms, 3145 Headland Road, Nanaimo, BC V9X 1N8, Canada.

Harry, F., Klahoose First Nation, Box 9, Squirrel Cove, Cortez Island, BC V0T 1K0, Canada.

Hashimoto, L., West Coast Abalone Ltd., PO Box 422, Whistler, BC V0N 1B0, Canada.

Heath, B., BC Ministry of Fisheries, BC Access Centre, 2500 Cliffe Avenue, Courtenay, BC V9N 5M6, Canada.

Hietcamp, F., Fisheries and Oceans Canada, 440–555 W. Hastings, Vancouver, BC V6B 5G3, Canada.

Horne, G., University of Victoria, 205–1400 Beach Drive, Victoria, BC V8S 2N7, Canada.

Howe, P., Praxis Pacific, 3848 St. Georges Avenue, North Vancouver, BC V7N 1W5, Canada.

Innes, K., Kitkatla Band, General Delivery, Kitkatla, BC V0V 1C0, Canada.

Jamieson, G., Fisheries and Oceans Canada, Science Branch, Pacific Biological Station, Nanaimo, BC V9R 5K6, Canada.

Johnny, J., Nanaimo First Nation, 1145 Totem Road, Nanaimo, BC V9R 1H1, Canada.

Johnson, D., Ethelda Bay Ventures, 5991 Trafalgar Street, Vancouver, BC V6N 1C5, Canada.

Johnstone, D., Malcolm Island Abalone Coop., Box 182, Sointula, BC V0N 3E0, Canada.

Jones, R., Council of Haida Nations, PO Box 98, Skidegate, BC V0T 1S1, Canada.

Jones, T., NewGen Resource Consultants, 5277 East Road, Denman Island, BC V0R 1T0, Canada.

Jubinville, B., Fisheries and Oceans Canada, 60 Front Street, Nanaimo, BC V9R 5H7, Canada.

Karpov, K., California Department of Fish and Game, 19160 So. Harbor Drive, Fort Bragg, CA, 95437, USA.

Klassen, T., Seafood Development, 100 First Avenue East, Prince Rupert, BC V8J 1A6, Canada.

Knight, D., Devon Knight Events, 1337 Mill Street, North Vancouver, BC V7K 1V5, Canada.

Landry, W., Sea Star Holdings Ltd., 4261 Rockbank Place, West Vancouver, BC V7W 1A8, Canada.

Lane, D., Malaspina University College, 900 Fifth Street, Nanaimo, BC V9R 5S5, Canada.

Lane, J., Nuu Chah Nulth Tribal Council, Box 1383, Port Alberni, BC V9Y 7M2, Canada.

Lee, L., M.T.E. Inc., PO Box 74, Tllel, BC V0T 1Y0, Canada.

Lewington, L., Aquaculture Technician, 2124 Cinnabar Drive, Nanaimo, BC V9X 1B2, Canada.

Lucas, B., Lucas Research, 6745 Rhodonite Drive, Sooke, BC V0S 1N0, Canada.

Maaraty, D., Fisheries and Oceans Canada, 417 2nd Ave West, Prince Rupert, BC V8J 1G8, Canada.

Mann, I., Fisheries and Oceans Canada, 12551 No. 1 Road, Richmond, BC V7E 1T7, Canada.

Marshall, A., Fisheries and Oceans Canada, Pacific Biological Station, Nanaimo, BC V9R 5K6, Canada.

Mason, E., Kitasoo Band Council, General Delivery, Klemtu, BC V0T 1L0, Canada.

McCormick, T., Proteus Sea Farms International, Inc., 323 East Matilija Street, #110–131, Ojai, CA, 92023, USA.

McKnight, D., BC Shellfish Growers Association, Box 116, 1601 Broadlands Road, Errington, BC V0R 1V0, Canada.

McPhee, B., Kitkatla Band, Tsimshian Tribal Council, 138 1st Ave. West, Prince Rupert, BC V8J 1A8, Canada.

Meneice, R.W., Samson Fishing Lodge, 4020 Magellan Place, Richmond, BC V7C 4M8, Canada.

Michael, F., PO Box 121, Zeballos, BC V0P 2A0, Canada.

Mitchell, D., Box 329, Errington, BC V0R 1V0, Canada.

Morrin, S., Fisheries and Oceans Canada, Treaty Negotiations, 460–555 West Hastings Street, Vancouver, BC V6B 5G3, Canada.

Needham, K., SALASAN Associates Inc., 102–9717 3rd Street, Sidney, BC V8L 3A3, Canada.

Newman, C., Heiltsuk Tribal Council, Box 880, Waglisla, BC V0T 1Z0, Canada.

Nookemus, R., Nuu Chah Nulth Tribal Council, Box 1383, Port Alberni, BC V9Y 7M2, Canada.

Norgard, T., Kitasoo Band Council, General Delivery, Klemtu, BC V0T 1L0, Canada.

Osborne, J., Nuu Chah Nulth Tribal Council, Box 1383, Port Alberni, BC V9Y 7M2, Canada.

Page, W., Sr., Cowichan First Nation, 5850 River Road, Qualicum Beach, BC V9K 1Z5, Canada.

Paul, A., BC Ministry of Environment, CCLRMP, 2080 Labieux Rd., Nanaimo, BC V9T 6J9, Canada.

Pennel, W., Malaspina University-College, 900 Fifth Street, Nanaimo, BC V9R 5S5, Canada.

Perkovitch, M., Kwakuitl Territories Fisheries Commission, 1441 B. North Island Hwy, Campbell River, BC V9W 2E4, Canada.

Perry, I., Fisheries and Oceans Canada, Pacific Biological Station, Nanaimo, BC V9R 5K6, Canada.

Quinn, S., Sechelt First Nation, Box 740, Sechelt, BC V0N 3A0, Canada.

Rankin, C., SALASAN Associates Inc., 102–9717 3rd Street, Sidney, BC V8L 3A3, Canada.

Redekopp, H., Fisheries and Oceans Canada, C & P, 12551 No 1 Road, Richmond, BC V7E 1T7, Canada.

Reece, G., Tsimshian Tribal Council, 138 1st Ave. West, Prince Rupert, BC V8J 1A8, Canada.

Reid, L., Heiltsuk Tribal Council, Box 880, Waglisla, BC V0T 1Z0, Canada.

Rogers, J., Fisheries and Oceans Canada, Operations, 417–2nd Ave. West, Prince Rupert, BC V8J 1G8, Canada.

Sam, D., Penelakut Tribe, Box 360, Chemainus, BC V0K 1K0, Canada.

Sam, R., Penelakut Tribe, Box 360, Chemainus, BC V0K 1K0, Canada.

Saunders, R., Island Scallops, 5552 West Island Highway, Qualicum Beach, BC V9K 2C8, Canada.

Schmidt, D., Kwakuitl Territories Fisheries Commission, 1441 B. North Island Hwy, Campbell River, BC V9W 2E4, Canada.

Seki, T., Genetics Division, National Research Institute of Aquaculture, 422-1, Nansei, Mie, 516-0193, Japan.

Sempert, C., Craig's Dive Center, Alaska, PO Box 796, Craig, Alaska, 99921, USA.

Seward, J., Penelakut Tribe, Box 360, Chemainus, BC V0K 1K0, Canada.

Sewid, A., Kwakuitl Territories Fisheries Commission, P.O. Box 120, Alert Bay, BC V0N 1A0, Canada.

Shepherd, S., SARDI, Australia, PO Box 120, Henley Beach, South Australia, Australia 5022.

Sloan, N., Gwaii Haanas Ntl Park Res., PO Box 37, Queen Charlotte City, BC V0T 1S0, Canada.

Smith, C., Tsimshian Tribal Council, 138 1st Ave. West, Prince Rupert, BC V8J 1A8, Canada.

Smith, T., Kwakuitl Territories Fisheries Commission, 1441 B. North Island Hwy, Campbell River, BC V9W 2E4, Canada.

Soto, C., Fisheries and Oceans Canada, MPAs Central Coast, 440A–555 W.Hastings Street, Vancouver, BC V6B 5G3, Canada.

Spencer, K., Fisheries and Oceans Canada, Room 315–940 Alder Street, Campbell River, BC V9W 2P8, Canada.

Symington, K., Canadian Parks and Wilderness Society, 611–207 W. Hastings Street, Vancouver, BC V6B 1H7, Canada.

Tallio, N., Nuxalk Nation, Box 65, Bella Coola, BC V0T 1C0, Canada.

Tarr, R., Marine and Coastal Management, S Private Bag X2, Rogge Bay, Cape Town 8012, South Africa.

Tegner, M., Scripps Institution of Oceanography, University of California, San Diego, La Jolla, CA 92093-0201, USA.

Thomas, J., Nanaimo First Nation, 1145 Totem Road, Nanaimo, BC V9R 1H1, Canada.

Thomas, K., Penelakut Tribe, Box 360, Chemainus, BC V0K 1K0, Canada.

Touchie, E., Ucluelet First Nation, Box 332, Ucluelet, BC V0R 3A0, Canada.

Vickers, D., Heiltsuk Tribal Council, Box 880, Waglisla, BC V0T 1Z0, Canada.

Wadhams, B., Nimgis Economic Development, c/o Namgis First Nation, Box 210, Alert Bay, BC V0N 1A0, Canada.

Watson, J., Malaspina University-College, 900 Fifth Street, Nanaimo, BC V9R 5S5, Canada.

Wells, T., Kwakuitl Territories Fisheries Commission, 1441 B. North Island Hwy, Campbell River, BC V9W 2E4, Canada.

Winther, I., Fisheries and Oceans Canada, Operations, 417–2nd Ave. West, Prince Rupert, BC V8J 1G8, Canada.

Withler, R., Fisheries and Oceans Canada, Pacific Biological Station, Nanaimo, BC V9R 5K6, Canada.

Woodby, D., Alaska Department of Fish and Game, PO Box 240020, Douglas, AK 99824-2202, USA.

Worsfold, R., Malaspina University-College, #1–2338 Rosstown Road, Nanaimo, BC V9T 2T3, Canada.

Zhang, Z., Fisheries and Oceans Canada, Pacific Biological Station, Nanaimo, BC V9R 5K6, Canada.

Acknowledgments

This workshop required the co-operation and assistance of a large number of people. Financial support was provided by Fisheries and Oceans Canada. The Organizing Committee, Bruce Adkins, Alan Campbell (Chair), Julia Gardner, Julian Griggs, Dale Gueret, Devon Knight, Barbara Lucas, and Russell Mylchreest, made a tremendous effort to ensure that the workshop was a success. The facilitators, Julia Gardner and Julian Griggs (Dovetail Consulting Inc.), Devon Knight (Devon Knight Events), Patricia Howe (Praxis Pacific), Kirsten Needham and Colin Rankin (SALASAN Associates Inc.), and Dorli Duffi helped to run the workshop smoothly and summarize the various group discussions and workshop conclusions in an excellent and timely fashion. Valerie Luedke (NEO Graphic Communications) provided the graphic designs for the workshop logo. Barbara Lucas prepared the printed workshop program for distribution to the participants. Note-takers, Virginia Eccleston, Claudia Hand, Gerry Horne, Ian Perry, Juanita Rogers, and Ryan Worsfold, recorded salient points made in the working groups. James Johnny, Chief Councillor, Nanaimo First Nation, Nanaimo, gave an introductory welcome to participants on the first day of the workshop. Doug Brouwer ensured that the audiovisual equipment ran smoothly for all the speakers.

Scientific papers in the proceedings were reviewed by at least two referees. The process of reviewing, editing, and revising manuscripts required the efforts and co-operation of authors and reviewers. The editorial assistance of Elizabeth Campbell, Nancy Daly, Sharon Humphreys, and Barbara Lucas in producing the workshop proceedings is sincerely appreciated.

Preface

Abalone has been a traditional food, a source of colorful shell jewellery, and has supported valuable fisheries in many countries. Important to many coastal communities and economies, in recent years, abalone has become vulnerable to over-exploitation due to increasing market value and low capital fishing costs. Although during the last 15 years understanding of abalone ecology, biology, fisheries management, and aquaculture has advanced considerably (e.g., Shepherd et al. 1992, 1995; Cook et al. 1998), many abalone fisheries have experienced declining abalone populations resulting in complete fishing closures. In British Columbia (B.C.), the northern abalone, *Haliotis kamtschatkana*, a traditional food of First Nations, was a target of recreational divers and a modest commercial dive fishery until 1990 when all fisheries were closed due to major stock declines and conservation concerns. The purpose of the coastwide closure was to allow depleted northern abalone populations to rebuild.

The lack of recovery of northern abalone populations in B.C., nine years after the abalone fishery closure in 1990, prompted Fisheries and Oceans Canada (DFO) to host a workshop seeking advice in developing a strategy for northern abalone stock rehabilitation. This issue of the Canadian Special Publication of Fisheries and Aquatic Sciences contains the proceedings of the Workshop on Rebuilding Abalone Stocks in British Columbia, held in Nanaimo, B.C., February 23–26, 1999. The workshop was attended by a diverse group of individuals representing First Nations, local communities, international (Australia, Canada, Japan, South Africa, and U.S.A.) abalone research and management, recreational, and commercial interests including aquaculture. The main goal of the workshop was to develop a realistic strategy to rehabilitate depleted northern abalone stocks in B.C. The workshop was also meant to clarify the roles, expectations, and shared interests of many of the interest groups in B.C., emphasizing the key role in the rebuilding plan of local communities, including First Nations and non-native communities. Reviews of stock status and management of several abalone fisheries and of the biology of abalone relevant to stock rehabilitation set the stage for the subsequent two days of facilitated discussions to draft a rebuilding strategy (see Appendix 1 for the workshop program agenda).

In Session One, Chris Dragseth gave an introductory DFO perspective. In addition, various concerns, expectations, and interests were presented on the northern abalone situation in B.C.

(*i*) *First Nations' perspectives.* Gary Reece, Tsimshian Tribal Council, and Russ Jones, Haida Nation, provided the perspective of peoples who have harvested abalone for food, social, and ceremonial purposes for centuries. Abalone meat was consumed for food, the shell was used in blanket and jewellery artwork, and both were a valuable trade commodity between tribes. Although the closed access to food, social, and ceremonial harvest has caused hardship, the need to let the stocks recover was recognized by First Nations. Previous lack of adequate consultation between DFO and First Nations prompted some participants to request DFO to listen more closely to concerns of their communities, since many have a history of being near the abalone resource. Problems of enforcement were considered to have a severe impact on abalone stock recovery.

(*ii*) *Aquaculture perspectives.* David Johnson indicated that the basic aquaculture interest in abalone was financial, with entrepreneurs willing to invest considerable amounts of money and effort in developing the abalone farming industry at many potentially excellent sites in B.C. Sufficient scientific knowledge and practical and marketing expertise are available from other parts of the world to help successfully farm abalone in B.C. However, a policy and regulatory framework is needed for abalone aquaculture. Farming abalone could contribute to rebuilding wild abalone stocks by providing public education and by reducing illegal harvest through increasing abalone supply and reducing prices.

(*iii*) *Community interests.* Non-native communities are primarily interested in economic development initiatives through organizations such as the Community Futures and Seafood Development offices.

(*iv*) *Recreational interests.* There is a large recreational interest in shellfish, including abalone. Dive-related tourism is popular in B.C., and being allowed to take abalone would be an added attraction.

(*v*) *Commercial interests.* While the commercial abalone fishery was conducted from the 1970s to 1990, this was one of the first fisheries with individual transferable quotas in B.C. Some commercial fishers felt that abalone stocks were still healthy enough to support a commercial fishery at the time of the 1990 closure.

(*vi*) *Conservation interests.* Almost all participants held an interest in the conservation of abalone stocks. The main conservation purpose expressed was a moral one reflected in Norm Sloan's statement, "We have to do right by abalone."

In Sessions Two and Three, the status of abalone stocks in California, Alaska, South Africa, and B.C. were reviewed. The management of abalone and abalone stock rebuilding is clearly a challenge around the world.

California. Kon Karpov et al. reported major declines in five abalone species in California which resulted from high fishing pressure, population expansion of a major predator of abalone, the sea otter (*Enhydra lutris*), disease, and reduced food availability due to climate factors. Serial declines occurred as fishing pressure moved from one area and species to another. Most of the abalone fisheries in California are now closed.

Alaska. Doug Woodby et al. described a boom and bust cycle of commercial northern abalone harvesting in Alaska that led to closure of the commercial fishery in 1995. Low levels of natural productivity and the reintroduction and spread of sea otters may prevent reopening of the abalone commercial fishery.

South Africa. Rob Tarr described the South African abalone fishery as a "fishery in crisis." Recruitment to the fishery was low in some areas. Recent poaching had increased dramatically with violence and organized crime being part of the illegal harvest scene. Co-management, ranching, and stock enhancement were being explored, but poaching, increased recreational catches, and ecological changes have led to a poor prognosis for the fishery.

British Columbia. Currently, no legal commercial harvest of northern (pinto) abalone is allowed throughout the northeast Pacific and abalone is currently the only shellfish species with a total harvest closure in B.C. In three separate presentations, Alan Campbell, Bruce Adkins, and Bryan Jubinville described the major decline of northern abalone, management regulations for the fishery, and enforcement efforts in response to poaching problems. Although abalone population declines are difficult to attribute to biological and (or) environmental effects quantitatively, DFO believes that illegal harvesting is a contributing factor. Some enforcement officers believe that the current illegal removal may be equal to the last commercial harvests. The biological characteristics of abalone make the species vulnerable to exploitation by poachers, in that the species is slow growing, relatively long lived, sedentary, has low or sporadic recruitment, and is easily accessible in shallow waters. The negative impacts of poaching on abalone recovery are severe because the loss of mature abalone reduces the reproductive potential as well as depressing populations. Enforcement against illegal harvesting is difficult and a number of attempts to improve enforcement are being considered.

Session Four examined the problems and experiences in rebuilding abalone stocks. There are many gaps in our knowledge about abalone biology, ecology, and habitat requirements that pose challenges to the rebuilding effort. We need to know more about spatial frequency, patch size and density of abalone, the environmental conditions required to maintain sufficient recruitment for healthy populations, and the efficacy of brood stock transplant and juvenile seed restocking methodologies. Tom McCormick generally reviewed aquaculture techniques and world aquaculture production of abalone and gave estimated costs for a proposal to rehabilitate the nearly extinct white abalone, *H. sorenseni*, in California. Mia Tegner proposed that the success of future abalone enhancement efforts in California would depend on having parasite- and pathogen-free hatcheries, establishing sufficient reserves and abalone brood stock aggregations, having effective enforcement to prevent illegal harvests, and addressing the expansion of the sea otter. Tetsuo Seki and Kazuya Taniguchi showed how the habitat type, substrates, and algae influenced the growth of transplanted hatchery-reared northern Japanese abalone, *H. discus hannai*, populations in a long-term study area in Japan. Scoresby Shepherd et al. (this issue) described experiments on abalone larval and juvenile releases at various Australian sites. They indicated that successful enhancement required choosing habitats that provide optimal carrying capacity (e.g., suitable substrates for sufficient space and food availability and few predators), minimizing handling stress in transporting juvenile seed, and optimizing release size and densities to increase survival rates of abalone. Shepherd et al. (this issue) were pessimistic about the cost effectiveness of seed abalone releases in the wild as a stock enhancement tool, except perhaps in areas where abalone stocks have disappeared or where enhancement and habitat improvement are combined with marine-farming operations.

Session Five dealt with additional pieces of the rebuilding puzzle. Ruth Withler emphasized the need to conduct genetic surveys to delineate the geographic scale of independent abalone populations, providing information for defining suitable marine reserves, rebuilding depleted populations, developing brood stocks for aquaculture, and assisting with enforcement to reduce illegal harvest of abalone. Susan Bower reviewed the infectious diseases known to affect abalone and advocated protection of wild abalone stocks and cultured abalone, by taking precautions to avoid accidentally introducing foreign pathogens during the transplanting of abalone. Jane Watson reviewed the ecological interrelation and co-existence between abalone and sea otter populations and indicated that the current collapse of abalone fisheries in Alaska, B.C., and

California could not be blamed on sea otters because of their present limited geographic range. Both Glen Jamieson and Gary Davis, in separate presentations, advocated the use of marine protected areas or refuges to restore collapsed abalone populations.

In Session Six, Julian Griggs et al. summarized the facilitated discussions on a comprehensive strategic plan for rebuilding abalone stocks in B.C. that occurred in the final two days of the workshop. The draft strategy developed at the workshop will provide an excellent initial step towards developing a comprehensive abalone stock rebuilding plan. Appendix 2 provides the verses of an abalone song which was sung at the end of the workshop reflecting the enthusiasm felt for abalone by participants at this and other abalone meetings.

The workshop was successful, judging by the lively exchange of ideas, the continuous discussions, and the enthusiasm of the participants. The workshop should have a significant impact on bringing the plight of abalone to the attention of the public, communities, and government agencies, and aid in future rehabilitation efforts.

References

Cook, P.A., Davis, G.E., Haaker, P.L., and Leighton, D.L. (*Editors*). 1998. Proceedings of the Third International Symposium of Abalone Biology, Fisheries, and Culture. 25–31 October, 1997, Monterey, California. J. Shellfish Res. **17**: 589–904.

Shepherd, S.A., Day, R.W., and Butler, A.J. (*Editors*). 1995. Progress in abalone fisheries research. Mar. Freshw. Res. **46**: v + 519–695.

Shepherd, S.A., Tegner, M.J., and Guzman del Proo, S.A. (*Editors*). 1992. Abalone of the world, biology, fisheries and culture. Fishing News Books, Blackwell Science Ltd., Oxford, U.K. 608 p.

Alan Campbell
Workshop Chair

Introductory Perspectives

Welcome and opening remarks

Chris Dragseth

On behalf of Fisheries and Oceans Canada (DFO), I welcome all of you to Nanaimo and thank you for participating in the Workshop on Rebuilding Abalone Stocks in British Columbia (B.C.).

This workshop brings together a diverse group of individuals with knowledge of and different interests in the abalone resource. Many of you are from coastal communities, representing First Nations, commercial fishers, businesses, and aquaculturists. We also have participants from abroad representing international research institutes and government organizations as far away as Alaska, California, Japan, South Africa, and Australia. The workshop will provide an excellent forum involving a multidisciplinary group to share experiences and academic and traditional knowledge from other jurisdictions, all with the intent to discuss the benefits and challenges of developing solutions to rehabilitate northern abalone stocks, which is an important natural resource in B.C.

The northern or pinto abalone, found in B.C., has been a traditional food of First Nations and a prize for recreational divers and has supported a small commercial dive fishery involving 26 licences during the 1970s and 1980s. However, due to evidence of continued abalone stock declines and the associated conservation concerns, DFO closed the northern abalone fishery to all user groups in December 1990. The objective of the closure was to allow depleted populations of northern abalone to rebuild. Although the closure was expected to last a minimum of 5 years, over 9 years after the closure, surveys have indicated no evidence of abalone stock rebuilding. Indeed, there have been further declines in abundance during that time. Consequently, there is a need to develop a comprehensive recovery plan for abalone in B.C. if harvesting by any group is to resume.

The objectives of the workshop are to develop a realistic rebuilding strategy to rehabilitate depleted northern abalone stocks and clarify the roles, expectations, and shared interests of many of the interest groups in B.C. The biology and population ecology of abalone will need to be examined and enforcement, enhancement, aquaculture, and community involvement should be considered as part of the rebuilding strategy. All communities, including First Nations and non-native communities, will need to play a key role in any rebuilding strategy. There has been and continues to be a strong linkage to the fisheries resource at the community level. Since DFO's resources are stretched to the maximum, strengthening the stewardship role that coastal communities can play in a rebuilding strategy should improve conservation efforts aimed at rebuilding the abalone resource.

The workshop will provide an update and review of the status of abalone stocks in B.C. and other parts of the world, as well as current information on abalone aquaculture and enhancement. The final two days are structured for working groups to draft a framework for a rebuilding strategy. The strategy should provide long-term benefits not only to the abalone resource but also to future resource users. The workshop is the initial step towards developing a comprehensive abalone stock rebuilding plan. Senior DFO officials in the Pacific Region, the Department's Regional Director General and ultimately the Minister look forward to the outcome of the workshop. The strategy developed through the workshop will be reviewed critically by DFO and will provide the framework for developing a rehabilitation plan. Although implementation of any plan will depend on fiscal constraints, this issue will be given a high priority in the Pacific Region.

On behalf of DFO, I wish you all the best in your deliberations over the next few days.

C. Dragseth. Director, Fisheries Management, Pacific Regional Headquarters, Fisheries and Oceans Canada, 300–555 West Hastings Street, Vancouver, BC V6B 5G3, Canada.
Correct citation: Dragseth, C. 2000. Welcome and opening remarks. *In* Workshop on Rebuilding Abalone Stocks in British Columbia. *Edited by* A. Campbell. Can. Spec. Publ. Fish. Aquat. Sci. 130. pp. 3.

Tsimshian Nation abalone issues

Garry Reece

The Tsimshian Nation traditional territories represent one third of the coast of northern British Columbia (BC) and inland to one third of the watershed of the Skeena River. We represent seven communities with a total population of roughly 10 000. The most southern community is Kitasoo and the most northern is Lax Kw'alaams. Other communities are Hartley Bay, Kitkatla, Metlakatla (BC), Kitselas, Kitsumkalum, and Metlakatla (Alaska). We are a maritime people insofar as our dependency is on sea resources as our source of food. Historically our people travelled coastwide.

First Nations have harvested northern abalone in a sustainable manner. Northern abalone was harvested at low "0" tide only and limited to the number of "0" tides in any one year. First Nations have always harvested in a sustainable manner in any one area. As a food item, northern abalone was consumed fresh, cooked, sun-dried, and jarred to preserve, and of course, with that wonderful invention, the deep freezer, they were frozen live.

Let me give you an idea of the value of this abalone resource to First Nations or at least, to the Tsimshian Nation. This resource was a high-value commodity for trade with neighboring island tribes. Not only was the meat of high value but also the shells, which were polished and shaped to be worn as adornment on a person and on ceremonial dress such as blankets, head-dress, and carved jewellery with abalone shell inlay. Abalone-shell buttons on a ceremonial blanket were a sign of wealth.

When 26 commercial licences coastwide were granted by Fisheries and Oceans Canada (DFO) in the late 1970s to fish northern abalone, none of our people were fortunate enough to receive a licence. We were told that this was a "limited entry" commercial fishery and that as First Nations our access to the resource was for the purposes of food and ceremonial use only, which is our aboriginal right. Well, we all know what happened; DFO managed northern abalone to near extinction.

The advent of dive equipment and the ease of access to any coastal area impacted heavily on our northen abalone stocks. Mismanagement and lack of consultation with various First Nations communities completely devastated our stocks.

Since the 1990 moratorium on the harvesting of northern abalone has been imposed, our people have not harvested this, what we commonly call, if you pardon the expression, "forbidden fruit." Since we believe in maintaining the high road to conservation, we have made it a point to observe and report on the "black marketers" around us.

For over a decade we have not had access to this resource. Naturally we question why these stocks have not rebuilt sufficiently. We need to examine the present-day management of our aquatic resources. Whether by design, or lack of personnel, expertise, funds, or equipment for DFO to monitor the extraction of all aquatic resources, the abalone stocks have not had a hope of rebuilding. Even as we are meeting here, someone is black marketing the abalone. Why? Simply, the demand at the market place makes this resource a high-priced commodity.

The example I always use for the abalone stocks of B.C. is this (and I mean no disrespect here for those who are legitimate harvesters of other aquatic species). If we look at the habitat of the abalone what do we have: many different species (sea urchins, sea cucumbers, crabs, bivalves, geoducks), all of which have a licensed commercial fishery. Isn't it tempting, or in our nature as human beings, that if I, as a commercial licensed fisherman of any one of these species, seeing sufficient abalone around me, would harvest abalone, especially if the price on the black market is approximately $10 000 for 45 kg of abalone? It certainly can be profitable.

The North Coast Conservation and Protection arm of DFO, Prince Rupert, was successful recently in convicting a "full-blown" operation of black-marketed abalone. Approximately 796 kg of abalone meat was seized and 6000 abalone shells that had been discarded were found. We can only guess the value, probably about $20 000. Also seized were dive equipment, a tug, and a barge. The two accused poachers had disguised their illegal activities as a logging operation.

Also recently, on the TV Outdoor Life Network, I watched an "undercover" operation in Mexico where a drug dealer decided to black market abalone because of its value, which exceeded the value of illicit drugs.

My concept of rebuilding our wild stock of northern abalone in B.C. is the following: address the black market and increase the conservation effort; pass legislation to approve heavier penalties or remove the licence of the offender granting access to the abalone environment, i.e., sea urchin, geoduck, sea cucumber, or any bivalve licences. We should view any offender of the abalone resource as a criminal, because the offender's action is outright theft of a public resource.

G. Reece. Tsimshian Tribal Council, 138 1st Ave. West, Prince Rupert, BC V8J 1A8, Canada.
Correct citation: Reece, G. 2000. Tsimshian Nation abalone issues. *In* Workshop on Rebuilding Abalone Stocks in British Columbia. *Edited by* A. Campbell. Can. Spec. Publ. Fish. Aquat. Sci. 130. pp. 4.

Haida abalone issues

Russ Jones

My name is *Hoya K'aadingaas*. My clan is *Gadasgaa K'iiguwaay*, from *Tanu* village in southern Haida Gwaii, the Queen Charlotte Islands. *Tanu* means "sea-grass town" because of the eel grass beds in front of the village. The Haida names for places have a meaning that tells you something about the place. For instance *Tlell* means fireweed because of the resource that is found there. I acknowledge the Nanaimo First Nation and thank them for welcoming us to their territory.

Today I am filling in for a Haida elder, Guujaaw. I'll read his letter (which follows) to you and then make a few comments myself.

Haida traditional harvest of northern abalone was in the intertidal or shallow subtidal zone. Usually northern abalone was taken with a two-pronged spear so abalone could be taken six feet below the tide level. I've been told that the abalone is related to the mouse. This shows that there are many ways of looking at the natural environment besides a scientific view. I think it might have to do with the shape of the abalone or color on the side of the mantle. Haida from the southern part of Haida Gwaii now live in Skidegate. They used to live in villages all around the Islands but moved to Skidegate in the 1800s after many people died from smallpox epidemics. People from Skidegate have to travel to the south or to the west coast in order to get abalone.

Not too long ago, our people would spend a lot of time in fishing camps to gather and put up foods. My aunt who passed away recently told me about spending a lot of time at Burnaby Narrows in southern Haida Gwaii. There was a clam cannery there in the 1910s and their family would dig and sell clams. They also put up k'aaw or herring spawn-on-kelp. Abalone was also dried and sold to Japanese at nearby salmon canneries. In the fall, I've also been told they would bring some of these products to the mainland for trade with the Tsimshian.

Shellfish is important to the Haida. But there are protocols between First Nations. One of our elders told me about coming from Haida Gwaii to fish in the Gulf of Georgia in the fall. I think it might have been near Qualicum. He visited the chief and asked if he could collect some oysters. The chief gave him permission and he went to the beach and got a bucketful. When he was returning on the beach he was stopped by a man who asked what he thought he was doing. The man said the oysters belonged to him because he had a lease from the government. He told him he had asked permission from the chief of the nearby village. The man said, "You've got them now anyway, so you might as well keep them." So he took them home and said he enjoyed them.

R. Jones. Haida Fisheries Program, Council of Haida Nation, PO Box 98, Skidegate, BC V0T 1S1, Canada.
Correct citation: Jones, R. 2000. Haida abalone issues. *In* Workshop on Rebuilding Abalone Stocks in British Columbia. *Edited by* A. Campbell. Can. Spec. Publ. Fish. Aquat. Sci. 130. pp. 5–6.

Haida people appreciate abalone and over the past few years of the closure I have heard elders regretfully say they might never taste abalone again in their lifetime.

We can learn a lot from the past. We should look at some of the things in fisheries management by Fisheries and Oceans Canada (DFO) that didn't help to save the abalone fishery. The abalone fishery was the first invertebrate fishery in British Columbia (B.C.) to have an individual quota. No Haida received these licenses to fish abalone. These policies determined who could fish commercially. Consequently, Haida could not make a living from abalone, which has also happened with development of many other fisheries. When the Haida and other people saw that abalone were being depleted, DFO closed a number of areas for Haida food and recreational use. When the commercial fishery was closed in 1990, the closed areas were also depleted. The area closures had not been successful.

The abalone fishery was closed in December 1990 to everybody including the First Nations in B.C. In Canada, the traditional Haida fishery for food, social, and ceremonial purposes has priority over commercial and recreational users. Although managers treat all fisheries equally during the closure, they have not treated the Haida First Nation equally or with priority when abalone were abundant, having allowed the commercial fishery to remove the majority of the available abalone resource, which I think is wrong.

In conclusion, past management of abalone has not worked. We need to look at new ways of managing abalone. Historically First Nations controlled access by others to their territory. You would have to have permission to take resources from another's territory. Maybe we should be looking at a stewardship model like this. Also, we need to recognize the different values that are at work in our society. The Haida put a high value on abalone as a traditional food, but the rest of society puts a commercial value on abalone. Whatever management system we use has to respect First Nations' values. As well, First Nations have to be involved in any management of abalone in their territory. I think this is the only way restoring abalone populations can be successful.

Letter to participants of the *Workshop on Rebuilding Abalone Stocks in British Columbia*

Good People,

It is without malice and with some hope that I prepare this paper. Please accept my words in this spirit.

In my younger years, I frequented Gwaii Haanas in the later part of winter with my uncle. At that time, we trapped river otter and marten; the price wasn't that great, about $60 per river otter and $15–20 for marten. We would take about six river otters, which would pay for our fuel. During our time in the area, however, we would also fish for clams, red snapper, ling cod, red sea urchins, halibut, and northern abalone. The trip meant so much more than that, as my uncle recounted the names and history of our people in the area. When we got back to the village we would unload our cargo onto a skiff and it would be distributed throughout the village before we tied up the boat and got home.

When the idea of a commercial northern abalone fishery was announced, our people who had traditionally gathered abalone at low water, spoke against it. One night I went to visit an older uncle. There was a stranger in the house and a bottle of whiskey on a map. The northern abalone fishery was on.

Upon returning to Gwaii Haanas in the following years, the result was obvious: abalone was diminishing. Within a few years, they were all but gone, or at least uneconomical, and certainly they were not available as a food source. My trips with my uncle ended, with one species out of the factor, it was no longer worth the trip. The same sad story was repeated all over the islands. To add insult to injury, the abalone fishermen demanded compensation. This story is to illustrate a direct effect on culture and one of the early events to strip Fisheries and Oceans Canada (DFO) of any trust.

In the years following the collapse and closure, the red sea urchins probably filled the holes left by the abalone. A few places showed promise of rebuilding abalone populations. After most of the abalone were gone, however, fishers went after the red sea urchins. During the red sea urchin fishery, the remnants of the once-abundant abalone seemed to be knocked back, time and again, by the urchin divers. The rumor was well known within the fishery, by local people who complained bitterly, and by the people managing restaurants, Japanese buyers, and DFO. In a meeting of the Haida Nation and DFO shellfish biologists, we heard directly that DFO was aware of the problem that urchin divers could be taking abalone. We were told of the difficulties of catching illegal poachers and the limited manpower and that when the abalone fishery was closed abalone were no longer a priority.

And so now abalone is a priority again. Why? Is it because we have realized a moral obligation to right a wrong or because we see another opportunity to grow a cash crop? If this conference is for real, we should be able to accomplish both. Let us protect the remnants and introduce a program to rebuild the natural stocks. This can be done with minimal intervention, starting with looking after what remains of abalone populations. The penalties and chances of getting caught have to be a deterrent. Grouping the few remaining abalone into natural areas would provide the best conditions to increase reproductive chances, one site at a time.

When this is done we will have righted a wrong, saved a species, protected a culture, and begun the process of salvaging the reputation of DFO. These things will set the stage for discussions toward growing abalone in controlled conditions, for an appreciative marketplace.

This conference can mark the beginning of the ascent from the dark ages.

Thank you for your attention.

Guujaaw
Box 1440
Skidegate, Haida Gwaii
BC V0T 1S1, Canada

Aquaculture's stake in rebuilding northern abalone wild stock in British Columbia

David Johnson

As part of a company wanting to culture northern abalone, I start my address by stating that we are here because of the money. To date, Ethelda Bay Ventures has spent over $275 000 on feasibility studies, site analysis, and provincial lease payment. We have hired foreign consultants, local consultants, visited other aquaculture projects in California and Alaska, and have worked several hundred unpaid hours trying to move our project forward. We also will have to spend at least an additional $2 million over the next 5 years before we expect to earn a single penny.

A small private company, like Ethelda Bay Ventures, doesn't make this kind of a financial, physical, and emotional commitment to abalone aquaculture without having strong concerns about the health of northern abalone stocks in British Columbia (B.C.). If we are to have viable aquaculture operations in B.C., we need a healthy northern abalone resource. We need the help and good will of scientists, government agencies, the coastal communities, and the First Nations people. We need a recognized and legitimate marketing structure, and a strong, clear, resource framework in which to operate.

I have participated in the B.C. fishing industry for over 30 years. My experience includes working as a fisherman, processor, and manufacturer of fish processing equipment, involving many commercial species (e.g., dogfish, hake, herring, salmon, geoduck, sea cucumber, and sea urchin). I have seen what happens when we strip mine our resources and what happens when we work against each other without developing a consensus, a communality of aims.

Since others at this workshop will present the history of the northern abalone stock declines in B.C., all I want to do is provide a brief overview of our company and the role we see for ourselves and the aquaculture industry in helping to rebuild abalone stocks.

In the late 1980s, realizing that high market demands and collapsing wild abalone stocks worldwide presented a remarkable opportunity for aquaculture, we set about developing a suitable aquaculture site at Ethelda Bay, Barnard Island. With local, provincial, and federal support, we created a development plan, secured a site, and set about obtaining foreshore leases, aquaculture licenses, and abalone broodstock collection permits.

D. Johnson. Ethelda Bay Ventures, 5991 Trafalgar Street, Vancouver, BC V6N 1C5, Canada.

Correct citation: Johnson, D. 2000. Aquaculture's stake in rebuilding northern abalone wildstock in British Columbia. *In* Workshop on Rebuilding Abalone Stocks in British Columbia. *Edited by* A. Campbell. Can. Spec. Publ. Fish. Aquat. Sci. 130. pp. 7–8.

Our development plan was produced with the help of I.E.C. collaborative Marine Research of Victoria, the B.C. Ministry of Agriculture, Fisheries and Food, and Fishtec Inc. of California. The sheltered moorage of Ethelda Bay, with a plentiful supply of food and water, ideal temperature conditions, and good growout sites, is an ideal location for abalone aquaculture. From what we have learned about established abalone ventures in Japan, Taiwan, China California, and Alaska, Ethelda Bay, and B.C. abalone aquaculture in general, would have competitive advantages in cultivating high quality abalone.

We propose to collect, hold, and condition wild abalone broodstock to serve as spawners at Ethelda Bay. We will periodically induce spawning and then nurture the spawn through the embryo and larval phases, growing the juveniles on diatom-covered plates. The juveniles will then be weaned from diatoms to fleshy algae before being moved to inwater growout cages where they will be raised until they reach maturity at about 5 years of age.

The biological and technical knowledge for culturing abalone, in general, already exists and is being applied in several countries. The only real significant problems for the B.C. industry are the poor state of the abalone wild stocks of B.C. and management.

I started out my presentation by saying that Ethelda Bay was here because of the money; let me elaborate. Yesterday in Vancouver you could buy a dozen frozen abalone for $168.00. As well as the natural problems that northern abalone may face from predators, climate change, and disease, a major problem is the high economic incentive for illegal exploitation of wild abalone stocks in B.C. Given the remoteness of most of the B.C. coast and an ever-increasing presense of commercial and recreational users, how can abalone wildstock be protected against illegal exploitation?

Three general areas, in which all interest groups can work towards protecting and rebuilding abalone populations, are public awareness, abalone identification, and reduced prices.

First, the public must be better educated about abalone. Given the geography, any enforcement will be futile without broad-based public knowledge and support. Many people would be interested in visiting abalone aquaculture sites, where we could provide promotional material (pamphlets, brochures, and videos) for public tours. We would share the knowledge and expertise of our scientists and technicians. The general public could be taught to value and respect wild abalone populations.

Second, easy methods are required to distinguish between wild and commercially raised abalone stock from B.C. as well as from other countries. Identification methods such as tagging, marking, and DNA fingerprinting already exist, but are generally expensive and time consuming. Abalone produced by aquaculture, however, could easily be marked by including dyes in their feed or by raising the spat on specific kinds of microalgae; the shells take on a distinct color which persists into the adult stage. The distinct color of farm-raised abalone also could be used as a marketing tool, e.g., New Zealand green mussels.

Third, the financial incentive to poach abalone must be reduced or eliminated. Stiff penalties are one approach, but that doesn't remove the enforcement problem and costs. Alternatively, by increasing abalone production, the aquaculture industry could work towards reducing abalone prices and hence reduce some of the incentive for poaching. Prices are controlled by supply and demand and if aquaculture interests can produce adequate volumes of high-quality product, a product viewed as comparable or better than wildstock, then the price of abalone would decline.

Despite all the money, time, effort, and patience that the aquaculture industry in B.C. has invested so far, not a single commercial northern abalone has yet to be produced. Nor will we be able to do so, unless a comprehensive recovery plan and management strategy for the northern abalone resource is developed. The aquaculture industry has a large stake in seeing such a strategy develop and we, at Ethelda Bay Ventures, will do everything possible to help move the process forward.

Status of Abalone Fisheries

Serial depletion and the collapse of the California abalone (*Haliotis* spp.) fishery

Konstantin A. Karpov, Peter L. Haaker, Ian K. Taniguchi, and Laura Rogers-Bennett

Abstract: Failure to manage the California abalone (*Haliotis* spp.) fishery as individual species, using spatially discrete fishing grounds and temporally discrete fishing periods, contributed to poor management decisions, ultimately leading to the 1997 closure of both the commercial fishery statewide and the southern recreational fishery. Here we examined patterns in the decline of abalone landings from 1942 to 1996 by species, catch area, and fishery period (years). There were four distinct phases in the fishery characterized by landings that increased, were apparently stable, rapidly declined, and gradually declined prior to closure. The apparently sustainable period (1952–1968) occurred as declining landings of pink abalone were replaced by increased take of red abalone. This was followed by a dramatic serial depletion of successive species during the depletion period (1969–1982). In the red, pink, and black abalone catch there was sequential spatial depletion of mainland fishing grounds, followed by smaller or nearby islands, and then finally depletion of the larger offshore islands. Fisheries independent stock survey assessments confirmed that the reduced landings resulted from population depletions. Life history differences among the species suggested that black, green, and white abalone were more vulnerable to overutilization and may be less likely to recover than red abalone. Increases in catch per unit effort and market value confounded management in the final period (1983–1996), hindering efforts to close the fishery of each abalone species until major stock declines had already occurred.

Résumé : Le fait de ne pas gérer séparément la pêche des différentes espèces d'ormeau (*Haliotis* spp.) de Californie, en établissant des aires de pêche distinctes et des périodes de pêche distinctes, a contribué à la prise de mauvaises décisions de gestion, ce qui s'est traduit finalement par la fermeture, en 1997, de la pêche commerciale dans l'ensemble de l'État et de la pêche sportive dans la partie sud. Dans le présent document, nous étudions les profils de la baisse des débarquements d'ormeau, de 1942 à 1996, par espèce, par zone de capture et par période de pêche (années). Nous avons observé quatre phases distinctes dans la pêche, à savoir : une augmentation des débarquements, une stabilité apparente des débarquements, une baisse rapide des débarquements puis une baisse progressive précédant la fermeture. La stabilité apparente (1952–1968) correspond à la période où la diminution des débarquements d'ormeau rose a été compensée par une augmentation des captures d'ormeau rouge. Cette période a été suivie d'une série spectaculaire d'effondrements des espèces, les unes après les autres, pendant la phase de baisse rapide (1969–1982). Dans le cas de la récolte des ormeaux rouges, roses et noirs, nous avons observé un épuisement séquentiel sur le plan spatial : d'abord les aires de pêche de la partie continentale, puis celles des petites îles ou des îles avoisinantes, et finalement celles des grandes îles du large. Les évaluations des stocks effectuées indépendamment du secteur de la pêche ont confirmé que la réduction des débarquements était attribuable à un affaiblissement des populations. D'après les différences du cycle vital des diverses espèces, il semble que les ormeaux noirs, verts, et blancs seraient plus vulnérables à une surexploitation, et que la probabilité de leur rétablissement serait plus faible que celle des ormeaux rouges. L'augmentation des prises par unité d'effort et de la valeur marchande ont désorganisé la gestion dans la dernière période (1983–1996), entravant les tentatives de fermeture de la pêche de chacune des espèces d'ormeau jusqu'au moment où on n'a pu que constater de graves baisses des stocks.

[Traduit par la Rédaction]

Introduction

The abalone (*Haliotis* spp.) fishery in California south of San Francisco has suffered dramatic declines over the past several decades, resulting in closure of the commercial and recreational fishery in 1997. The collapse of this major fishery, which landed more than 2000 metric tons during the 1950s and 1960s, occurred despite fishery management efforts. Today, in northern California, red abalone (*H. rufescens*) stocks continue to be abundant north of San Francisco due to important differ-

K.A. Karpov. CDFG 19160 So. Harbor Drive, Fort Bragg, CA 95437, U.S.A. e-mail: kkarpov@dfg.ca.org
P.L. Haaker and I.K. Taniguchi. CDFG 330 Golden Shore, Long Beach, CA 90802, U.S.A. e-mail: phaaker@dfg.ca.gov
L. Rogers-Bennett. CDFG and Bodega Marine Lab, PO Box 247, Bodega Bay, CA 94923, U.S.A. e-mail: rogersbennett@ucdavis.edu
Correct citation: Karpov, K.A., Haaker, P.L., Taniguchi, I.K., and Rogers-Bennett, L. 2000. Serial depletion and the collapse of the California abalone (*Haliotis* spp.) fishery. *In* Workshop on Rebuilding Abalone Stocks in British Columbia. *Edited by* A. Campbell. Can. Spec. Publ. Fish. Aquat. Sci. 130. pp. 11–24.

ences in management between the two regions (Tegner et al. 1992). The two major differences were prohibiting recreational use of SCUBA and excluding commercial take of abalone from the north (Cox 1962; Karpov and Tegner 1992; Tegner et al. 1992).

Historically, abalone have been the basis for important fisheries along the west coast of North America beginning with early aboriginal settlers (Cox 1962). Closures, however, occurred early in the commercial abalone fishery in California. During the last century, the commercial green (*H. fulgens*) and black (*H. cracherodii*) abalone fishery in the littoral zone reached its peak in 1879 (Cox 1962; Cicin- Sain et al. 1977) and by 1913 this fishery was closed (Cox 1962; Edwards 1913). More recently, in the 1980s, the once-favored white abalone, *H. sorenseni*, was fished to unsustainable levels, giving *H. sorenseni* the dubious distinction of being one of the few marine species fished to near extinction (Tegner et al. 1996; Davis et al. 1998). The commercial abalone closure was a tragedy for the divers who lost their livelihoods, the public whose natural heritage was lost, and resource managers whose fishery management efforts failed. In hindsight, the declines in the fishery might have been more easily recognized if fishery managers had examined the fishery data, taking into account individual species within the fishery as well as the spatial complexity of the stocks.

In this study, we examine abalone catch statistics from 1942 to 1996 for evidence of serial (sequential) depletion by species and by area. We separate the time series into four major fishery periods. The California abalone complex includes five major species: red, pink (*H. corrugata*), green, black, and white abalone. We examine landing trends by species and catch by area. Catch records are analyzed for patterns in spatial serial depletion at mainland, nearby island, and distant island fishing grounds. We corroborate the trends in the fishery statistics with fishery independent survey data to confirm the magnitude of the stock declines, focusing on the most recent periods of decline. We discuss additional factors, which may have contributed to management failures in this fishery, including reliance on catch per unit effort (CPUE) data, increases in abalone value, and a lack of resolve and funding for management. Finally, we examine suggestions for more conservative management which, if adopted, could help prevent future abalone fishery collapses.

Methods

Fisheries dependent and independent data were utilized in this study. Fisheries dependent data were based on landings reported by abalone divers on mandatory landing receipts. These landing data included the number of abalone landed, their value in US dollars, and location of catch by Fish and Game statistical block numbers spanning 10 minutes of latitude and longitude. Numbers of abalone landed were converted to weight using a fixed average value specific to each of the five species (Oliphant 1979).

We assigned landed weight for all five species to catch locations, i.e., specific segments of the coast and islands, based on the block number of the catch location for data from 1950 to 1996. We divided the area from San Francisco to Mexico into three areas; northern (San Francisco to Año Nuevo including the Farallon Islands), central (Año Nuevo to Point Conception, the sea otter range), and southern California (south of Point Conception) (Fig. 1). Southern California was further subdivided into three coastal segments, eight islands, and a set of combined banks. Only red abalone were fished commercially north of Point Conception. Percent of peak catch (1950–1996) by species was determined at each of the 14 locations to compare magnitude of catch declines. Our graphical analysis of area-specific catch examined locations that represent over 95% of the cumulative catch for a species.

Fisheries independent data included timed swim counts and density surveys. Timed swim counts, initiated in 1973, were used to provide an index of abundance for red, pink, and green abalone. Swims were conducted using SCUBA on appropriate habitat at the Channel Islands during 1973, 1974, 1976, 1983, and 1993–1998. Counts were depth stratified by expected species distribution in southern California, i.e., pink and red abalone from 6 to 24 m and green abalone from 3 to 6 m (Cox 1962; Tutschulte 1976).

Density surveys for red and pink abalone were noninvasive counts on 60 m^2 transects from 16 Channel Islands National Park (CINP) stations. These surveys were conducted on five northern CINP Islands, which have been surveyed annually since 1983 (Davis 1989). The methods used in our analysis of red abalone were described in Karpov et al. (1998). Here we expanded our analysis to include pink abalone at six locations on Santa Cruz Island (4), Santa Barbara Island (1), and Anacapa Island (1). Black abalone density survey data were collected on permanent intertidal quadrats at 11 locations at San Miguel Island (3), Santa Rosa Island (5), Anacapa Island (2), and Santa Barbara Island (1). At each location five quadrats from 1 to 60 m^2 were examined each year without destroying habitat or injuring abalone. Censuses were conducted during low-tide periods in the late summer and winter from 1985 through 1999.

The ex-vessel value of abalone in dollars per kilogram was computed for combined species from 1950 through 1996. Values were standardized to 1995 US dollars using the yearly Consumer Price Index for southern California. After 1982, landing data included sufficient information to calculate value by species, value of catch in dollar per vessel day, and CPUE in kilograms per diver days.

Results

Landing periods (1942–1996)

The trends in total commercial landings of abalone during 1942–1996 exhibited four stages (Fig. 2). Period A (1942–1951) was characterized by increasing landings. Period B (1952–1968) reflected apparent stability in landings. During period C (1969–1982) landings declined relatively rapidly. Finally, period D (1983–1996) reflects a gradual, but steady decline that ended in complete fishery closure in 1997.

Landings of a species complex

The patterns observed in the combined landings mask patterns of the individual species landings (Fig. 2). During period A, red abalone landings increased rapidly during the last years of World War II (1942–1945) followed by a gradual

Fig. 1. Location map of California, showing northern, central, and southern commercial abalone take areas and the range of the sea otter through 1997.

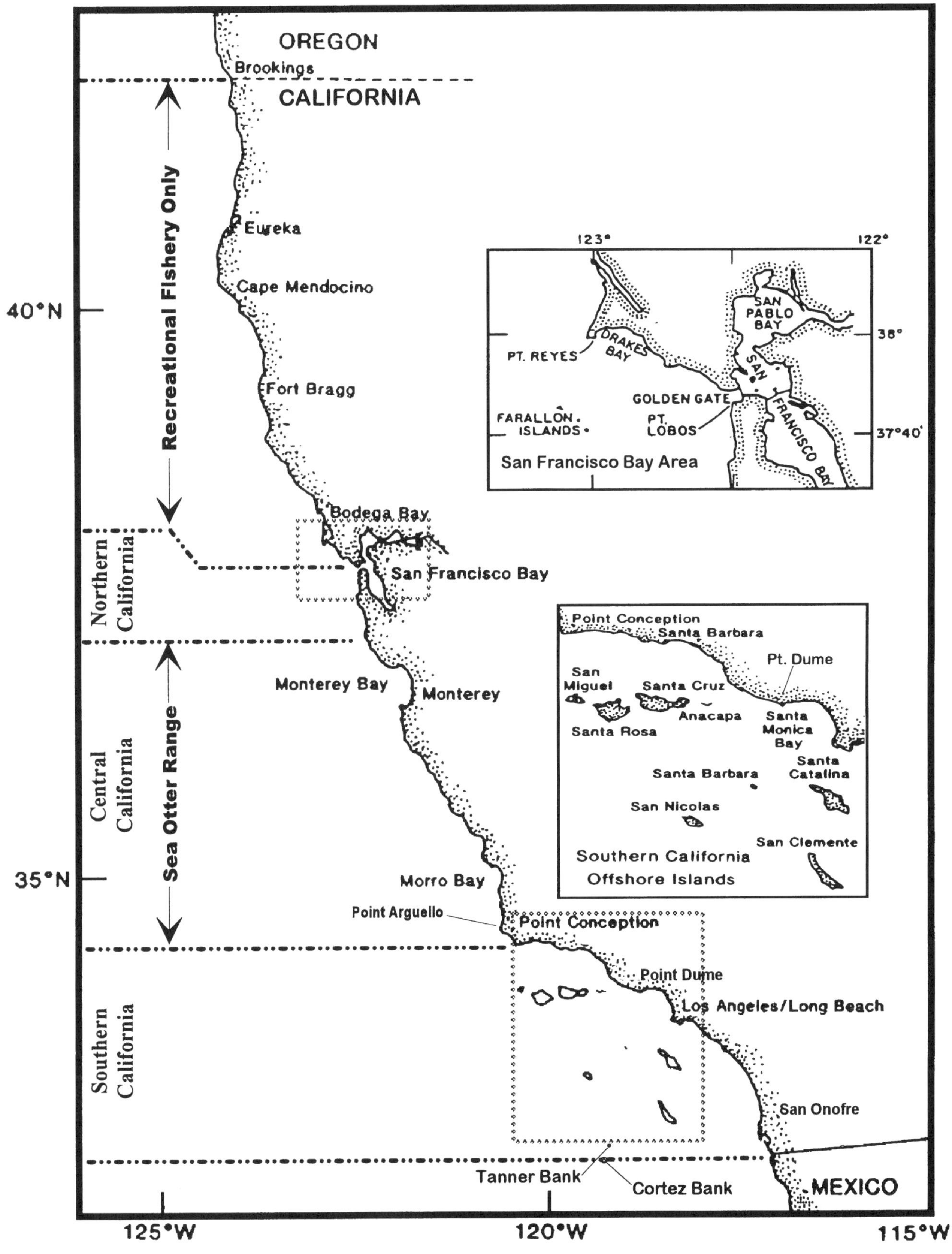

Fig. 2. Commercial landings (metric tons) of California abalone fishery (bars), combined (top), and by red, pink, green, black, and white abalone. Landings are divided into periods (A–D) by trends in the total fishery landings, with regression lines (dotted lines) for each period, (m = slope, t · year^{-1}). Regressions are provided for individual species where sufficient data exists. The average annual value of all species (US$ per kilogram) is represented by a solid line.

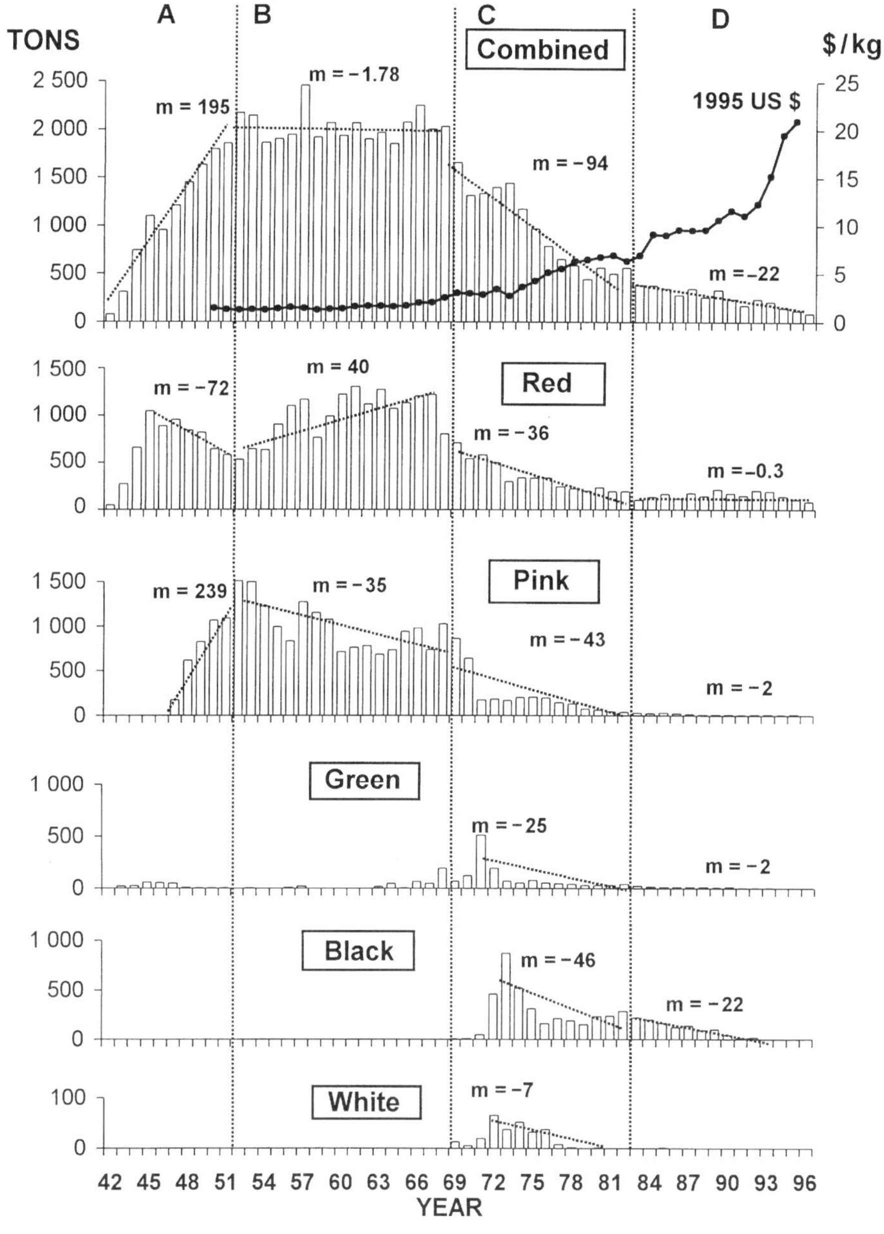

decline to 1951. Combined landings continued to increase to the end of the period as pink abalone entered the fishery from 1947 to 1951. The pink abalone fishery began after red abalone landings peaked and began to fall. The only other abalone landed during period A, green abalone, contributed little to the overall landings.

Period B was not one of stability for the component species. Red abalone landings gradually increased until 1967 followed by a long decline extending through the next period. Pink abalone landings declined throughout this period to levels lower than red abalone. Toward the end of period B, when both red and pink landings were declining, green abalone landings increased supplementing the cumulative landings.

Period C was a period of steady and rapid decline in red, pink, green, and black abalone landings, each of which declined to levels far lower than historic landing highs. The most dramatic declines of any of the abalone species was that suffered by the white abalone, which declined to 0.6% of the 1972 historic peak by 1982. None of the lesser abundant three species achieved the tonnage or apparent durability of red or pink abalone.

During period D all species, except red abalone, continued to decline until the fishery closure in 1997 (Figs. 2 and 3). Throughout this period red abalone landings appeared stable, but at far lower than historic landing levels. Black abalone were reduced to 2.0% of peak prior to the 1993 closure. The pink, green, and white abalone fisheries ended in 1995 at 0.5, 0.1, and 0.03% of their respective peak landings. During the last year of the fishery (1996), red abalone landings were 87 t, representing a 17% decline from 105 t in 1983, but almost a 90% decline from the historic peak in the early 1960s.

Catch by area and level of decline

Area specific catch from 1950 for red, pink, and black abalone revealed a spatial trend in catch with higher catches coming from mainland or nearshore islands, shifting over time to more remote areas, with catches declining at larger islands before smaller ones. Catches of green and white abalone did not show a spatial pattern in decline, but remained concentrated in a few areas, suggesting these two species were limited in their spatial distribution prior to exploitation. Declines in catch varied by species and area, but in most cases dropped two orders of magnitude from catches in peak years and areas (Table 1).

During period B, most red abalone were caught in central California, followed by southern mainland, Santa Cruz, Santa Rosa, and San Miguel islands (Fig. 4). During this period and into C, catches declined first along the central coast. Catches declined less abruptly along the southern mainland, Santa Rosa, Santa Cruz, and San Nicolas islands. During periods C and D, catches decreased off the three islands to 3, <1, and <1% of their respective peak catch (Table 1). San Miguel Island and the north coast were the exceptions to this pattern. Catches from San Miguel Island, the farthest and most northern of the Channel Islands, and the north coast (District 10) comprised 71 of the 87 t landed in 1996 (30 and 23% of peak catch, respectively) just prior to the fishery closure in 1997.

Following a similar pattern, pink abalone catches during period B increased to peak catch first along the mainland, Santa Barbara, San Clemente, and Catalina islands, followed by growth on Santa Cruz and Anacapa islands (Fig. 5). Following this initial growth phase, declines were most abrupt along the mainland and at Santa Barbara Island. During period C, San Clemente and Santa Cruz islands declined. Unlike red abalone, no remnant areas remained to sustain pink abalone landings into the final period D with all areas declining to less than 8% of peak catch (Table 1).

Green abalone catch during period B came primarily from Santa Catalina Island in 1956–1957 then later from the Southern mainland and San Clemente Island (Fig. 6). Declines during period C and into D were most pronounced off San Clemente and mainland areas with remnant catches persisting from Santa Catalina Island.

During period C, black abalone catch developed and peaked at many of the islands in 1973 with catches slowly increasing from the more remote San Nicolas Island in period D (Fig. 7). Declines in black abalone catch were first apparent at Santa Rosa Island in period C. During period D, catch continued to decline with the last landings coming from San Miguel, San Clemente, and San Nicolas islands. Before the fishery closure, catch had fallen to 4% or less of the peak catch for each of the islands (Table 1).

White abalone catch peaked and declined to <1% of the peak catch in just 14 years during period C at each of the

Fig. 3. Landings for red, pink, green, black, and white abalone, in periods C and D (bars). The ex-vessel average value (solid line), with fitted regression lines (dotted line), (m = slope, $t \cdot year^{-1}$).

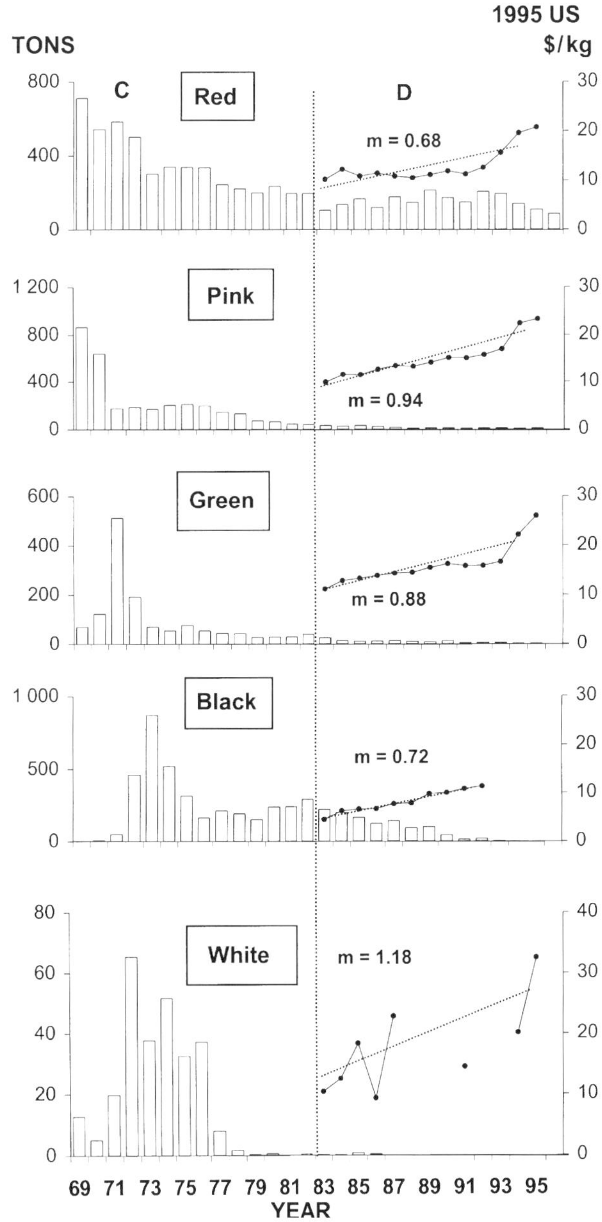

areas examined (Table 1, Fig. 8). During period D, the white abalone fishery at each of the five major locations had essentially collapsed.

Fisheries independent abundance and density data

Abundance (abalone per hour) or density (abalone per hectare) data collected during periods C and D confirmed that population estimates declined in conjunction with catch during the final stages of the abalone fishery. Abundance and density surveys revealed profound declines for most species at the locations surveyed. Many of the abundance and density figures, with few exceptions, showed dramatic declines approaching or reaching densities or counts of 0.

Table 1. Abalone catches taken one year before the fishery closures (1992–1996), expressed as a percent of peak catches that occurred in 1961 for red, 1952 for pink, 1971 for green, 1973 for black, and 1972 for white abalone in California.

	Abalone species				
Location	Red	Pink	Green	Black	White
North coast	30	—	—	—	—
Central coast	<1	—	—	—	—
Southern coast					
Point Conception to Pt. Dume	2	1	<1	—	<1
Pt. Dume to San Onofre	5	<1	<1	—	<1
San Onofre to Mexico	<1	<1	<1	—	<1
Southern islands and banks					
San Miguel Island	23	3	5	2	<1
Santa Rosa Island	3	7	<1	<1	<1
Santa Cruz Island	<1	<1	<1	<1	<1
Anacapa Island	9	<1	<1	<1	<1
San Nicolas Island	<1	6	<1	1	—
Santa Barbara Island	<1	<1	<1	2	<1
Santa Catalina Island	<1	<1	<1	3	<1
San Clemente Island	<1	1	<1	<1	<1
Tanner and Cortez banks	2	<1	<1	—	<1

Note: — Indicates area was not fished for that species.

Red abalone abundance and density trend analyses showed mixed results with declines at Santa Cruz and Santa Rosa islands, but not at San Miguel Island (Fig. 4). Santa Cruz, close to the mainland, had low initial abundance and density estimates that rapidly declined to 0. Santa Rosa Island, further offshore, had higher initial abundance and density estimates that also declined to low levels by the end of the period. At San Miguel Island abundance estimates declined, but density estimates showed no clear decline during period D.

Fishery independent pink abalone surveys yielded zeros in abundance or density at all areas except San Clemente Island by the end of period D (Fig. 5). At Santa Barbara Island densities and abundance estimates were initially low and remained low or at 0 during period D. Values fell to 0 at both Anacapa and Santa Cruz islands during period D. San Clemente Island was the only location where abundance declined but did not reach 0.

As with pink abalone, abundance surveys for green and density surveys for black abalone showed major declines through the final years of the fishery in all areas surveyed (Figs. 6 and 7). Green abalone trends in stock abundance were only monitored on Santa Catalina and San Clemente islands. In both areas final abundance declined by two orders of magnitude to timed swim counts of <1 abalone · h^{-1}.

Intertidal densities of black abalone approached or fell to 0 at all Channel Island locations monitored from 1985 through 1999 (Fig. 7). Densities declined to 0 at the islands close to mainland, first with Anacapa in 1990 and Santa Rosa Island in 1992, followed by declines at San Miguel Island (the farthest from the mainland) in 1995.

Market value, value of vessel landings, and CPUE

Market value of abalone increased at an accelerating rate during periods C through D (Fig. 2). Ex-vessel cost of abalone per kilogram in US dollars, adjusted for inflation to 1995 dollars, rose gradually during period B at \$0.06 · $year^{-1}$. During the two periods of landing decline, the combined value of abalone accelerated. The slope of adjusted cost during periods C and D was 0.37 and \$0.95 · $year^{-1}$, respectively, increasing from \$2.98 in 1969 to \$20.93 by 1995.

Value increased as landings declined during period D (Fig. 3). The value of white abalone, whose landings had essentially collapsed by 1983 at the end of period C, continued to increase at the greatest rate of all species, tripling to \$33.56 between 1983 and 1995. Red abalone, which showed no decline in landings during period D, had the lowest increase in real value of the five species, doubling to \$20.70.

Trends in CPUE did not parallel landings during period D except for red and black abalone (Figs. 2 and 9). Surprisingly, red and pink abalone CPUE was highest in later years. Green abalone CPUE remained essentially level, declining only during the last few years of the fishery in 1994 and 1995. Black abalone CPUE declined slightly while white abalone showed no definitive trend. During this same period, value per boat landing tripled for red and pink abalone, remained level for green and black abalone, and peaked in 1985 for white abalone (Fig. 9).

Discussion

Sustainable fishery?

We found no evidence for sustainable commercial catch in the California abalone fishery at any time during the fishery

Fig. 4. Red abalone catch (metric tons, bars), density (number per hectare, solid dots), and number per hour (open squares) from northern, central, and southern California mainland, and four Channel Islands. Catch is divided by landing periods. Catch areas are ordered from top to bottom by accessibility to fishery. Vertical lines around points are SE.

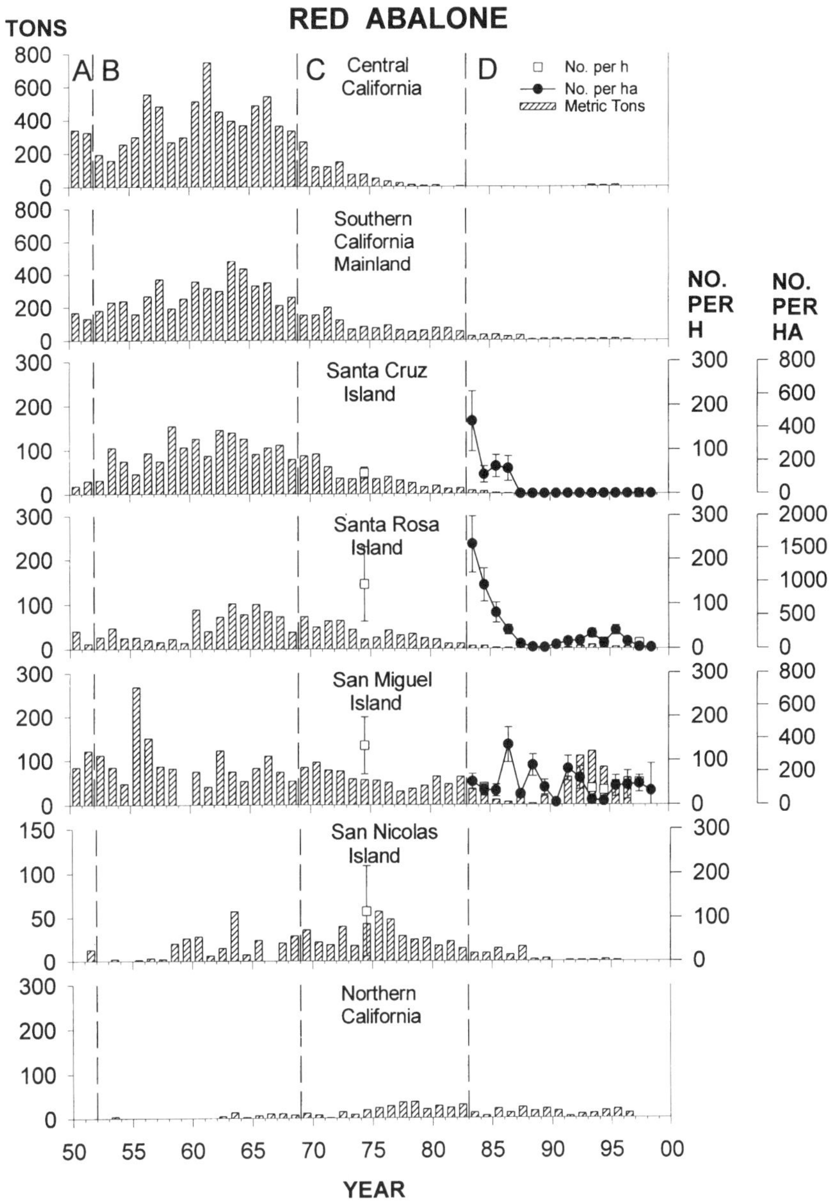

(1942–1996). Examination of combined abalone landings revealed four distinct phases in the fishery. There was a building phase (Period A), in which landings steadily increased presumably during the initial exploitation phase of the fishery. Increases in landings can occur as new areas are opened to exploitation in spatially complex fisheries (Orensanz et al. 1998) or as accumulations of old and (or) large individuals are utilized (Francis 1986). As the fishery matured, it appeared to remain stable for almost two decades (Period B), however, combined landings masked the replacement of declining pink abalone with red abalone. The fishery began a period of sharp decline in 1969 (Period C), in which a series of species were sequentially depleted. The reoccupation of central California by the southern sea otter in the 1960s decimated red abalone populations in that area and consequently commercial fishery landings (Wendell 1995). The decline in the combined landings appeared to slow as landings reached low levels (Period D); however, an examination of catch data during this period revealed a pattern of serial depletion by fishing area. Fishery independent data corroborated these declines in the fishing areas.

The one exception to this pattern of dramatic decline has been the recreational fishery for red abalone in northern California. North of San Francisco, abalone may not be taken using SCUBA or surface supplied air. This regulation results in a *de facto* reserve at depths beyond 8.5 m (Karpov et al. 1998). This has resulted in a large-scale spatial harvest refugia for the deeper portions of the population. Walters, as cited in Orensanz et al. (1998), suggested that sustainable fisheries result from spatial accidents which protect a large

Fig. 5. Pink abalone catch (metric tons, bars), density (number per hectare, solid dots), and number per hour (open squares) from southern California mainland, and five Channel Islands. Catch is divided by landing periods. Catch areas are ordered from top to bottom by accessibility to fishery. Vertical lines around points are SE.

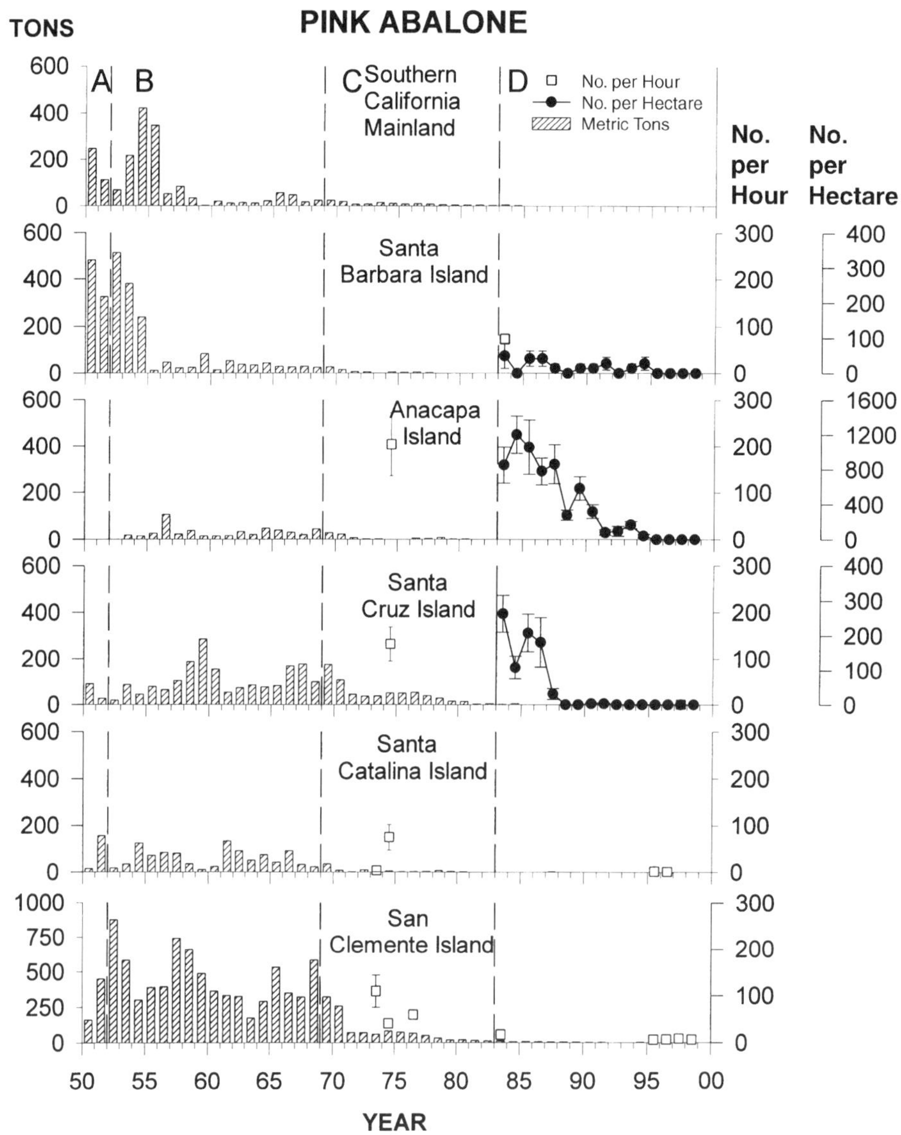

portion of the stock. This may be part of the explanation for the success of the recreational red abalone fishery in northern California. The northern fishery is also less complex to manage with only one species in a geographically simpler area with no offshore islands. Other factors that help protect these stocks include longer periods of inclement weather and large areas of coast inaccessible to recreational divers. Given the levels of decline found in southern California in both fishery-dependent and -independent data, it is perhaps surprising that stronger management action was not taken earlier to protect stocks. What masked the dramatic problems in this fishery?

Serial depletion by species

The replacement of one species or subspecies by another, which results in the appearance of stable landings, has occurred in many nearshore marine fisheries, suggesting that the management of species complexes can be problematic (Dugan and Davis 1993; Orensanz et al. 1998). Alaskan crustacean fisheries were serially depleted as a succession of species suffered dramatic declines in their landings (Orensanz et al. 1998). Similarly, the serial replacement of subspecies has been suggested as a contributor to the collapse of the eastern Atlantic cod fishery (Hutchings and Myers 1994). Replacement of one subspecies of cod by another acted to maintain the illusion of a long period of stability in the cod fishery until a rapid, unforeseen, total collapse occurred when all the subspecies had been fished out.

In the California abalone fishery south of San Francisco, apparent stability from 1952 to 1968 was an illusion produced by landings comprised of multiple species and multiple fishing areas. When the fishery complex was divided into components, a pattern of serial decline by species and area emerged. Combined landings were bolstered by increases in red abalone landings, which maintained stability in the combined landings during the decline in pink abalone landings

Fig. 6. Green abalone catch (metric tons, bars) and number per hour (open squares) from southern California mainland and two Channel Islands. Catch is divided by landing periods. Catch areas are ordered from top to bottom by accessibility to fishery. Vertical lines around points are SE.

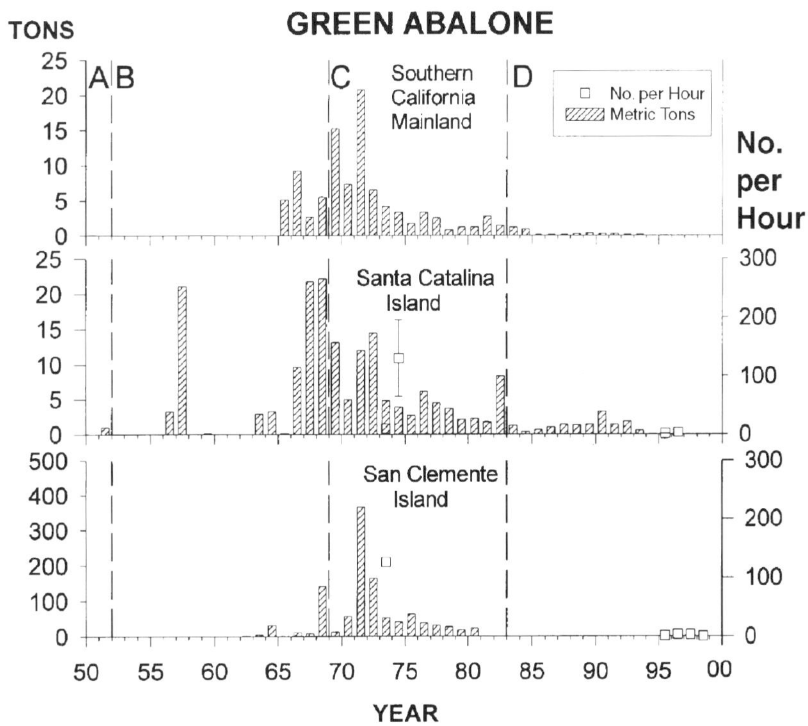

(Fig. 2). In 1971, there was an abrupt decline in pink abalone landings caused by increases in pink abalone size limits imposed by managers in an effort to protect stocks (Figs. 2 and 3). The spike in green abalone landings (1971) reflected a lowering of the green abalone size limit to placate fishermen after the pink abalone size limit increase. Red abalone also began to decline in this period, marking the start of intensive commercial fishing of green, black, and white abalone. Landings for these three species rapidly peaked and then declined. In the early 1970s, black abalone bolstered the combined landings.

During the last period of the commercial abalone fishery, many commercial divers held both abalone and sea urchin permits. As the availability of abalone decreased, efforts were shifted to the growing sea urchin fishery. Since the abalone fishery was not closed, divers searching for sea urchin continued to land abalone even with populations at extremely reduced levels. Declines in commercial abalone landings were replaced by sea urchin landings, thus maintaining the value of the combined dive fishery at $10 million from 1955 to 1985, while abalone suffered dramatic declines (Dugan and Davis 1993).

Serial depletion by area

Spatially structured populations are particularly vulnerable to depletion by area. Orensanz et al. (1998) detailed the collapse of the Alaskan crustacean fisheries due to the serial depletion of fishing areas, with areas close to port being depleted first. In the California commercial abalone fishery, an analysis of the catch by area revealed a similar pattern of spatial serial depletion. Combined landings from the different catch areas (Figs. 4–8) masked the succession of declines in fishing areas. Landings appeared stable as abalone were fished from more distant locations while nearby fishing grounds were sequentially depleted. During the apparently sustainable period B, red abalone catches in most areas peaked midperiod and then began to decline (Fig. 4). The declines in red abalone catch were masked by combined landing increases through 1967 at the end of period B (Fig. 2, slope (m) = 40 $t \cdot year^{-1}$). Likewise, apparent stability in red abalone landings during period D (Fig. 2) reflected a shift to distant fishing grounds on San Miguel Island and in northern California (Fig. 4). In the commercial fishery, distant areas (typically islands) were the last strongholds for landings with red and black abalone at San Miguel and San Nicolas islands and pink abalone at Santa Cruz and San Clemente islands (Figs. 4, 5, and 7). Another factor apparently affecting the speed of the decline of abalone was the size of the area being fished. Relatively small islands, such as Anacapa and Santa Barbara islands, and Cortez and Tanner banks, were among the first to be depleted of abalone (Figs. 5, 7, and 8).

Remnant island populations of pink abalone, as indicated by fishery independent surveys in the early 1980s, could have been protected by a pink abalone closure or by area closures, but neither of these management strategies were implemented. Pink abalone populations may have been reduced to insufficient levels to generate successful recruitment. No evidence for pink abalone recruitment was found in our fishery independent surveys. These populations eventually collapsed in the mid-1990s (Fig. 5). Periods C and D (1969–1996) would have been the time to implement pink broodstock enhancement measures (Tegner 1992; Tegner 2000). This was clearly a missed opportunity for pink abalone restoration.

Fig. 7. Black abalone catch (metric tons, bars) and density (number per hectare, solid dots) from five Channel Islands. Catch is divided by landing periods. Catch areas are ordered from top to bottom by accessibility to fishery. Vertical lines around points are SE.

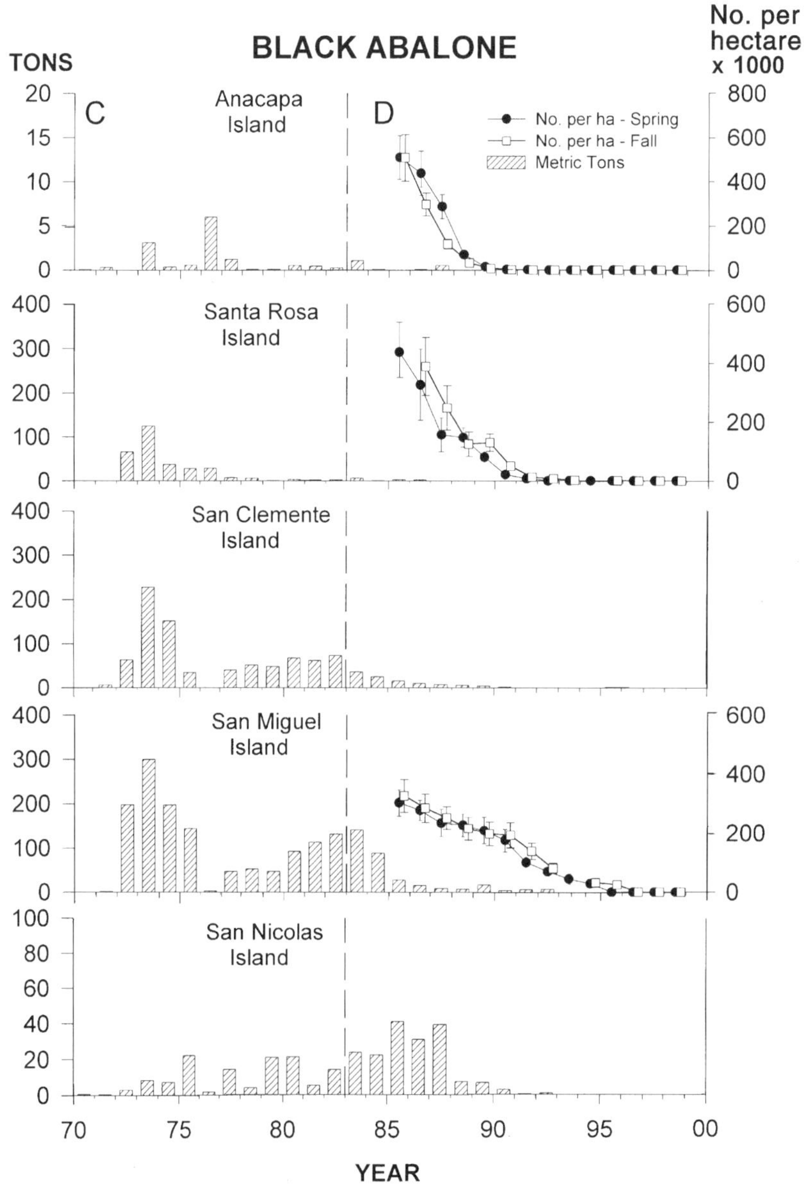

Commercial landings of red abalone in northern California and at San Miguel Island persisted in the 1990s (Figs. 4–8). Karpov et al. (1998) showed that red abalone at San Miguel Island were the last to be impacted by the commercial fishery, being furthest from the mainland. San Miguel Island is also in a more suitable water circulation pattern for this cold water species (Engle 1994; Tegner et al. 1992). However, the sustained commercial catch of red abalone in northern California south of San Francisco (Fig. 1) during period D is puzzling. Fishery independent surveys in 1993 on the fishing grounds in this area at Half Moon Bay revealed low densities averaging only 200 abalone · ha^{-1} (K. Karpov, unpublished data). One possible explanation for this disparity between fishery-dependent and -independent data is poaching from recreation-only areas north of San Francisco, which were then reported as landings from the open northern California commercial area (Daniels and Floren 1998).

Magnitude of decline

Landings in California have declined precipitously. Fisheries-independent data collected during periods C and D revealed that levels of decline observed in fishery landings presaged stock collapse. Declines in red abalone populations today are a major concern for fishery managers because this is now the only abalone in California abundant enough to support a fishery. Persistence of red abalone, as compared to some of the other species in the commercial landings, may be a reflection of differences in distribution and vulnerability to exploitation. Commercial landings of red abalone in central and southern California averaged 461 t · $year^{-1}$ from 1950 to

Fig. 8. White abalone landings (metric tons, bars) from southern California mainland, two Channel Islands, and two offshore banks. Catch is divided by landing periods. Catch areas are ordered from top to bottom by accessibility to fishery.

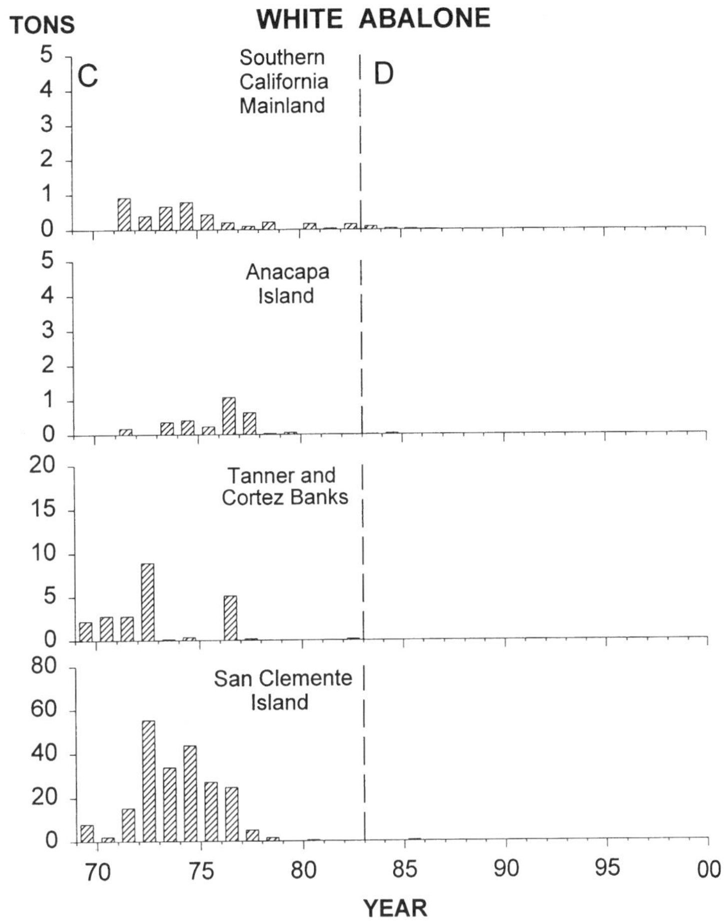

1996. Recreational catch north of San Francisco averaged 920 t · year^{-1} from 1985 to 1989 (Tegner et al. 1992). The red abalone has the broadest depth range of the abalone species in California (Tegner et al. 1992; Tutschulte 1976).

Pink abalone were the second most persistent species in catch. Pinks were the most widely distributed of the four southern species. Cox (1962) attributed a portion of the decline in pink abalone during period B to the combined effects of slow growth and starvation due to reductions in kelp biomass during the 1957–1959 El Niño.

The fisheries for the green, black, and white abalone, all with narrower depth and geographic distributions, were short-lived. Green abalone are a shallow subtidal species abundant in surf grass beds. This species was fished hard during period C and thereafter contributed little to the commercial fishery. Black abalone have a wide geographical distribution, but a narrow depth distribution restricted to the lower- and mid-intertidal zone. An added concern for black abalone is the combined effects of fishing and withering foot syndrome (Alstatt et al. 1996; Haaker et al. 1992). While landing declines occurred prior to the outbreak of this lethal disease, continued take further contributed to the potential extirpation of this species by removing remaining individuals that may have been resistant to the disease. White abalone have a narrow, deep distribution from 25 to 50 m on rocky habitat. Davis et al. (1998) reported extreme declines in white abalone abundance following commercial and sport exploitation. During extensive surveys in the 1990s at the Channel Islands, using deep diving and a manned submarine, divers found <2 white abalone · ha^{-1}, compared to 2000–10 000 · ha^{-1} in the 1970s at comparable depths (Tutschulte 1976). Davis et al. (1998) suggest that unless active restoration methods are enacted soon this species may become extinct. The white and black abalone are currently candidates for federal endangered species status.

Factors contributing to management failure

A number of factors undermined effective management action in addition to ignoring the possibility of serial depletion by species and by area. Management effort was limited to conventional strategy that primarily focused on size limits to protect stocks (Tegner et al. 1992). This perspective was based on egg-per-recruit models that assume several years of spawning success for a significant portion of the abalone population prior to reaching the minimum size for take (Tegner et al. 1989), derived from the potentially high fecundity of abalone (Giorgi and DeMartini 1977). We show that this management approach did little to ensure the sustainability of abalone resources. Focus on spawning potential ignores effects of protracted periods of recruitment failure under intense fishing

Fig. 9. CPUE (kilogram per diver landing, bars with SE) for red, pink, green, black, and white abalone and values per boat landing (solid line).

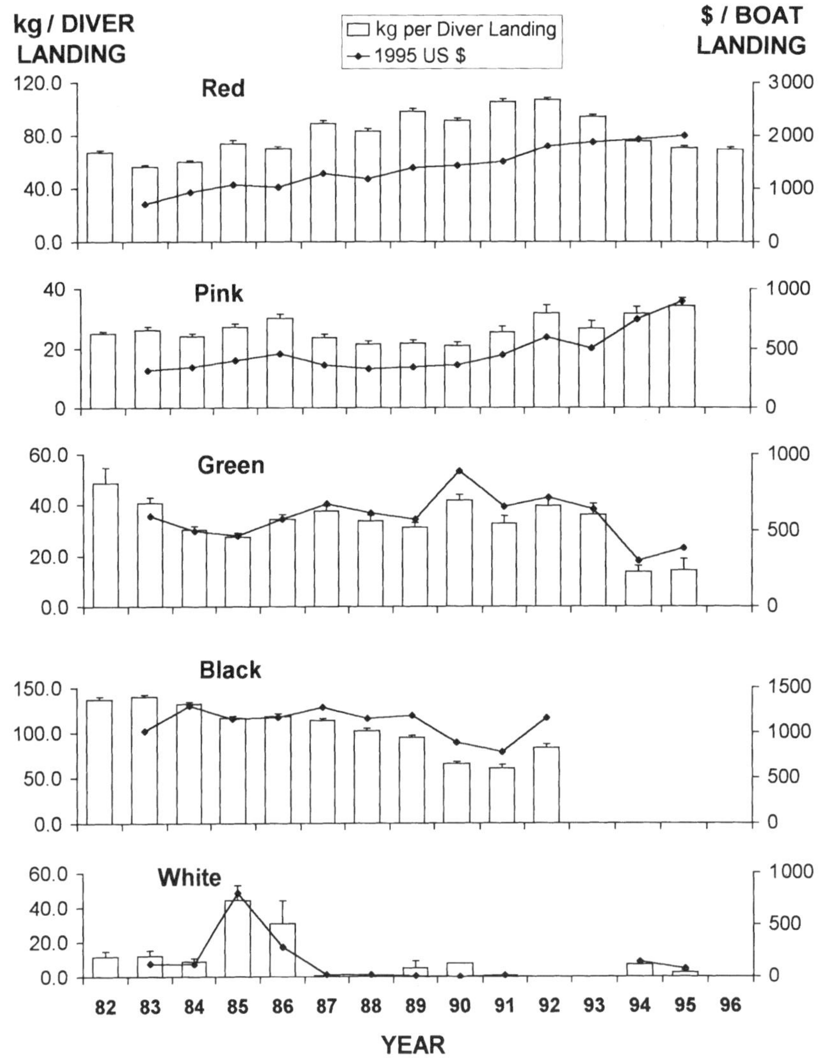

pressure (Sluczanowski 1984). Another factor may have been the loss of adult aggregations needed for spawning success (Shepherd and Brown 1993). Low densities can result in fertilization failure in free-spawning invertebrates (Pennington 1985; Levitan et al. 1992). Recruitment of young abalone may not be successful every year (Karpov et al. 1998; Rogers-Bennett and Pearse 1998). Karpov et al. (1998) reported only one red abalone recruitment event in 4 years in northern California where deep water (>8.5 m) stocks were protected. Likewise, Tegner et al. (1989) found a single recruitment peak of red abalone in a study area on Santa Rosa Island over a 5-year period.

Ultimately, during the final period of abalone decline, reliance on CPUE data further delayed closures. CPUE for abalone is a poor index of abundance, as has been shown for other spatially structured fisheries (Orensanz et al. 1998). Assumptions for use of CPUE, including the redistribution of the stock and random fishing (Gulland 1983; Ricker 1975), are violated in the abalone fishery as they are in other fisheries for sessile benthic invertebrates. CPUE for red and pink abalone (1983–1996) increased throughout much of period D, during the time when red abalone landings remained consistent and pink abalone declined (Figs. 2 and 9). Two factors during this time worked to increase CPUE: (*i*) improvements in locating sites using Loran and Global Positioning Systems (GPS), and (*ii*) increased search time of fishing grounds by abalone and urchin divers in the dual fishery.

Although the economic value of this resource was high, little funding was directed at research or assessment during the successful period of the fishery and at conservation when several species were clearly in decline. Commercial landing taxes were limited to US\$0.03 · kg^{-1} until the 1990s when an added enhancement and restoration tax of \$0.43 · kg^{-1} was established. Enhancement funds were used primarily for stocking hatchery-reared juveniles and not stock assessment. Restocking programs moved forward with abalone juvenile seed supplied by the abalone aquaculture industry, but there was little follow-up work quantifying success rates. The few studies that were conducted examining abalone juvenile restocking success suggested that survival was <1% (Tegner and Butler 1989; Rogers-Bennett and Pearse 1998). Restocking programs functioned to dissipate the resolve and funding needed for more conservative management strategies (Tegner 2000), a situation that has been repeated in other fisheries (MacCall 1989).

Increasing abalone value was another factor that further delayed conservative management action during the decline. As the landings declined the value increased exponentially, in response to demands from foreign markets and a growing population in California (Fig. 2). Increases in value intensified political pressures to continue fishing despite evidence of collapsing stocks. In this case market forces did not work to stop the fishery as the species declined and the economics of the dual fishery permitted the fishing of abalone species to near extinction.

A final factor, not considered in this paper, was the added impact of a growing SCUBA-based sport fishery in southern California (Karpov and Tegner 1992; Tegner et al. 1992). Beginning in the 1970s, this largely unmonitored fishery had a growing impact on abalone stocks on the mainland and Channel Islands. We focused on commercial landing statistics as the best fisheries-dependent record of the magnitude of decline resulting from various cumulative impacts. This paper was not intended to identify the degree of relative impacts from fishery, pollution, or the various nonanthropogenic factors.

Conclusions

In this study, we outline the case history of the failed abalone fishery in California south of San Francisco, which ended in the 1997 closure. Management strategies, including minimum size limits, season closures, and limited entry did little to prevent declines of individual species in the abalone complex or declines in catch areas. CPUE data did not forecast fishery declines and increased abalone values exacerbated fishing effort. Failure to examine landing data by species and catch by fishing grounds delayed management actions restricting fishing. Management lacked the flexibility to respond to dramatic declines in catch areas to allow for spatial closures such as closing Anacapa Island, which declined early in the fishery. In fact, the burden to prove that there were problems with the landings rested on managers rather than on fishers (Dayton 1998). Areas were not closed when they showed clear signs of decline because landings remained high elsewhere as fishers concentrated their efforts in new areas of high resource abundance. Our analyses show this was serial depletion by area. Many of the depleted areas did not recover, i.e., the mainland in southern California, despite long closures (25 years) enacted well after landings had declined. Our findings also support Dugan and Davis (1993) that component species were serially depleted. Today, several species have been reduced to such low levels that recovery without human intervention may be impossible and may be a challenge even with costly enhancement measures. The added impacts of nonfishery-related pressures such as El Niño, disease, and predation by sea otters may now drive populations to extinction.

The use of the precautionary principle in fishery management might have prevented this type of fishery disaster. This principle requires that, in the absence of information, fishery management should be conservative (Food and Agriculture Organization of the United Nations (FAO) 1996). In the case of the California abalone, localized catch declines in the absence of fishery independent information would have required earlier closures. Edwards (1913) was the first to recognize that local declines in abalone could ultimately lead to loss of the resource as a whole. He was visionary in suggesting the establishment of protected reservations to function as breeding centers for abalone at 5–10 mile intervals along the coast. This plan to establish marine protected areas (MPAs) for abalone could only be used in areas that still support minimum viable populations. Today, red abalone populations on the north coast of California are protected by a large-scale refuge by depth since fishing is restricted to breath-hold divers and landings in this fishery appear to be stable (Karpov et al. 1998). In southern California, new and creative solutions are needed to restore abalone stocks. If stocks do recover and fishing is reestablished, new and more conservative management approaches in addition to MPAs will be needed to prevent future collapse. An example might be to allow breath-hold only diving for pink abalone and other species once refuge populations in deeper water are reestablished.

Acknowledgments

We wish to thank Alan Campbell and the Canadian Department of Fisheries and Oceans for supporting our participation in this important abalone workshop. We also extend our appreciation to Gary Davis, Daniel Richards, and David Kushner, of the Channel Islands National Park, as well as the many volunteers who collected fisheries-independent data used in our assessments. Thanks are due to Frank Henry, Dave Parker, DeWayne Johnston, and Steve Nicholls of the California Department of Fish and Game for encouraging our participation. Special thanks are due Dr. Mia Tegner and Mary Patyten for help in editing this manuscript.

References

Alstatt, J.M., Ambrose, R.F., Engle, J.M., Haaker, P.L., Lafferty, K.D., and Raimondi, P.T. 1996. Recent declines of black abalone *Haliotis cracherodii* on the mainland coast of central California. Mar. Ecol. Prog. Ser. **142**: 185–192.

Cicin-Sain, B., Moore, J.E., and Wyner, A.J. 1977. Management approaches for marine fisheries: the case of the California abalone. Univ. Calif. Sea Grant College Program, Sea Grant Pub. 54. 223 p.

Cox, K.W. 1962. California abalones, Family Haliotidae. Calif. Dep. Fish Game Fish Bull. **118**: 1–133.

Daniels, R., and Floren, R. 1998. Poaching pressures on northern California's abalone fishery. J. Shellfish Res. **17**: 859–862.

Davis, G.E. 1989. Design of a long-term ecological monitoring program for Channel Islands National Park, California. Nat. Areas J. **9**: 80–89.

Davis, G.E., Haaker, P.L., and Richards, D.V. 1998. The perilous condition of white abalone *Haliotis sorenseni*, Bartsch, 1940. J. Shellfish Res. **17**: 871–876.

Dayton, P.K. 1998. Reversal of the burden of proof in fisheries management. Science (Washington, D.C.), **279**: 821–822.

Dugan, J.E., and Davis, G.E. 1993. The role of refugia in management of coastal fisheries. Can J. Fish. Aquat. Sci. **50**: 2040–2052.

Edwards, C.L. 1913. The abalone industry in California. Calif. Dep. Fish Game Fish Bull. **1**: 5–15.

Engle, J.M. 1994. Perspectives on the structure and dynamics of nearshore marine assemblages of the California Channel Islands. *In* The Fourth California Islands Symposium: Update on the Status of the Resource. *Edited by* W.L. Halvorson and G.J. Maender. Santa Barbara Museum of Natural History, Santa Barbara, Calif.

Food and Agriculture Organization of the United Nations (FAO). 1995. Precautionary approach to capture fisheries and species introductions. Elaborated by the Technical Consultation on the Precautionary Approach to Capture Fisheries (Including Species Introductions). Lysekil, Sweden, 6–13 June 1995. FAO Technical Guidelines for Responsible Fisheries. No. **2**. Web address: http://www.fao.org/icatalog/inter-e.htm. 56 p.

Francis, R.C. 1986. Two fisheries biology problems in West Coast groundfish management. North Am. J. Fish. Manage. **6**: 453–462.

Giorgi, A.E., and DeMartini, J.D. 1977. A study of the reproductive biology of the red abalone (*Haliotis rufescens*) near Mendocino, California. Calif. Fish Game, **63**: 80–94.

Gulland, J.A. 1983. Fish stock assessment: a manual of basic methods. John Wiley & Sons, Chichester, U.K. 223 p.

Haaker, P.L., Richards, D.V., Freidman, C.S, Davis, G.E., Parker, D.O., and Togstad, H.A. 1992. Mass mortality and fishery management of black abalone, (*Haliotis cracherodii*) in southern California. *In* Abalone of the world: biology, fisheries and culture. *Edited by* S.A. Shepherd, M.J. Tegner, and S. Guzmán del Próo. Blackwell Scientific, Oxford, U.K. pp. 214–224.

Hutchings, J.A., and Myers, R.A. 1994. What can be learned from the collapse of a renewable resource? Atlantic cod *Gadus morhua*, of Newfoundland and Labrador. Can J. Fish. Aquat. Sci. **51**: 2126–2146.

Karpov, K.A., and Tegner, M.J. 1992. Abalones. *In* California's living marine resources and their utilization. *Edited by* W.S. Leet, C.M Dewees, and C.W. Haugen. California Sea Grant Extension Publication, Univ. Calif. Sea Grant Exp. Prog. pp. 33–36.

Karpov, K.A., Haaker, P.L., Albin, D., Taniguchi, I.K., and Kushner, D. 1998. The red abalone, *Haliotis rufescens*, in California: importance of depth refuge to abalone management. J. Shellfish Res. **17**: 863–870.

Levitan, D.R., Sewell, M.A., and Chia, F.S. 1992. How distribution and abundance influence fertilization success in the sea urchin *Strongylocentrotus franciscanus*. Ecology, **73**: 248–254.

MacCall, A.D. 1989. Against marine fish hatcheries: ironies of fishery politics in the technological era. Calif. Coop. Oceanic Fish. Invest. Rep. **30**: 46–48.

Oliphant, M.S. 1979. California marine fish landings for 1976. Fish Bull. **170**: 1–56.

Orensanz, J.M.L., Armstrong, J., Armstrong, D., and Hilborn, R. 1998. Crustacean resources are vulnerable to serial depletion — the multifaceted decline of crab and shrimp fisheries in the Greater Gulf of Alaska. Rev. Fish Biol. Fish. **8**: 117–176.

Pennington, J.T. 1985. The ecology of fertilization of echinoid eggs: the consequences of sperm dilution, adult aggregation, and synchronous spawning. Biol. Bull. **169**: 417–430.

Ricker, W.E. 1975. Computation and interpretation of biological statistics of fish populations. Bull. Fish. Res. Board Can. 191. pp. 1–382.

Rogers-Bennett, L., and Pearse, J.S. 1998. Experimental seeding of hatchery-reared juvenile red abalone in northern California. J. Shellfish Res. **17**: 877–880.

Shepherd, S.A., and Brown, L.D. 1993. What is an abalone stock? Implications for the role of refugiain conservation. Can. J. Fish. Aquat. Sci. **50**: 2001–2009.

Sluczanowski, P.R. 1984. A management oriented model of an abalone fishery whose substocks are subject to pulse fishing. Can. J. Fish. Aquat. Sci. **41**: 1008–1014.

Tegner, M.J. 1992. Brood–stock transplants as an approach to abalone stock enhancement. *In* Abalone of the world: biology, fisheries and culture. *Edited by* S.A. Shepherd, M.J. Tegner, and S. Guzmán del Próo. Blackwell Scientific, Oxford, U.K. pp. 461–474.

Tegner, M.J. 2000. Abalone enhancement in California: what we've learned and where we go from here. *In* Workshop on rebuilding abalone stocks in British Columbia. *Edited by* A. Campbell. Can. Spec. Publ. Fish. Aquat. Sci. 130. pp. 61–71.

Tegner, M.J., and Butler, R.A. 1989. Abalone seeding. *In* Handbook of culture of abalone and other marine gastropods. *Edited by* K.O. Hahn. CRC Press, Inc., Boca Raton, Florida. pp. 157–180.

Tegner, M.J., Breen, P.A., and Lennert, C.E. 1989. Population biology of red abalone, *Haliotis rufescens*, in southern California and management of the red and pink, *H. corrugata*, abalone fisheries. Fish. Bull. **87**: 313–339.

Tegner, M.J., DeMartini, J.D., and Karpov, K.A. 1992. The California red abalone fishery: a case study in complexity. *In* Abalone of the world: biology, fisheries and culture. *Edited by* S.A. Shepherd, M.J. Tegner, and S. Guzmán del Próo. Blackwell Scientific, Oxford, U.K. pp. 370–383.

Tegner, M.J., Basch, L.V., and Dayton, P.K. 1996. Near extinction of an exploited marine invertebrate. Trends Ecol. Evol. **11**: 278–280.

Tutschulte, T.C. 1976. The comparative ecology of three sympatric abalones. Ph.D. dissertation. University of California, San Diego, Calif.

Wendell, F.E. 1995. Relationship between sea otter range expansion and red abalone abundance and size distribution in central California. Calif. Fish Game, **81**: 45–56.

Decline of the Alaska abalone (*Haliotis* spp.) fishery and prospects for rebuilding the stock

Doug Woodby, Robert Larson, and Jan Rumble

Abstract: The commercial fishery for northern abalone (*Haliotis kamtschatkana*) in Alaska has a boom and bust history and a clouded outlook for rebuilding. Unrestricted harvests in the 1970s peaked at 172 tonnes in the 1979–1980 season followed by stock collapse in the early 1980s despite restricted harvests. The fishery closed in 1995. Sea otter populations grew exponentially in the 25-year history of the fishery after otters were reintroduced to their former range in southeast Alaska. Abalone populations were decimated in areas where otters reestablished; however, these declines came after peak fishery harvests. If otters continue to expand their range, future fishing opportunities for abalone will be severely limited. Future openings also depend on improvements in understanding of abalone productivity and recruitment processes. Growth rates are estimated from historic tagging data and compared to rates estimated from shell ring counts. Those results indicated that size limits may have provided protection to individuals for at least 3 years after the average age of reproductive maturity. This protection was insufficient to prevent abalone stock collapse.

Résumé : En Alaska, la pêche commerciale à l'ormeau nordique (*Haliotis kamtschatkana*) a connu des périodes de prospérité et de débâcle, et les perspectives de reconstitution des stocks de cette espèce sont sombres. Les récoltes non réglementées des années 70 ont culminé à 172 tonnes au cours de la saison 1979–1980, puis les stocks d'ormeaux nordiques se sont effondrés au début des années 80 et ce, malgré les restrictions de récolte. On a fermé la pêche en 1995. Pendant les 25 ans d'histoire de la pêche à l'ormeau, les populations de loutres de mer ont augmenté de façon exponentielle après leur réintroduction dans leur aire de répartition d'origine dans la région du sud-est de l'Alaska. Dans les endroits où s'étaient rétablies les loutres, les populations d'ormeaux ont été décimées; toutefois, ces baisses sont survenues après des pics de récolte. Si les loutres continuent d'étendre leur aire de répartition, les possibilités d'ouvrir la pêche à l'ormeau dans l'avenir seront gravement réduites. La réouverture éventuelle de la pêche dépend également de l'approfondissement des connaissances sur les processus de productivité et de recrutement de l'ormeau. On estime les taux de croissance de cette espèce à partir des données historiques, puis on les compare aux taux estimés à partir du dénombrement des anneaux sur les coquilles. Les résultats indiquent que les restrictions portant sur les limites de taille ont peut-être protégé les individus pendant au moins trois ans après l'âge moyen de la maturité reproductive. Cette protection a cependant été insuffisante pour empêcher le déclin des stocks d'ormeaux.

[Traduit par la Rédaction]

Introduction

The commercial fishery for northern abalone (*Haliotis kamtschatkana*) in Alaska was short-lived. The 25-year history was marked by peak catches in the first decade, followed by a steep decline and a long slide towards total closure in 1995. A minimal set of harvest restrictions was imposed in a sequence that proved ineffective to alter the downward trend. The final closure was made hoping that local populations would rebuild at least to self-sustaining levels and possibly to levels adequate to provide for renewed commercial exploitation. In this paper, we recount the history of the fishery and the sequence of management actions.

D. Woodby and J. Rumble. State of Alaska Department of Fish and Game, Commercial Fisheries, P.O. Box 240020, Douglas, AK 99824-0020, U.S.A.
R. Larson. Alaska Department of Fish and Game, P.O. Box 667, Petersburg, AK 99833, U.S.A.
Correct citation: Woodby, D., Larson, R., and Rumble, J. 2000. Decline of the Alaska abalone (*Haliotis* spp.) fishery and prospects for rebuilding the stock. *In* Workshop on Rebuilding Abalone Stocks in British Columbia. *Edited by* A. Campbell. Can. Spec. Publ. Fish. Aquat. Sci. 130. pp. 25–31.

We analyse historic tagging data to estimate growth rates relative to size limits and we examine the expansion of sea otter (*Enhydra lutris*) populations in relation to the abalone fishery.

History of the commercial fishery

Commercial abalone harvests in Alaska peaked from 1978 through 1982 and declined sharply thereafter (Fig. 1a). The highest annual catch was 172 t in the 1979–1980 season (defined as October through September). Catch per unit effort rose along with the increased harvests. A clear decline in catch rates lagged the catch decline by several years, masking the downward trend in abundance until the late 1980s. Product value (ex-vessel) increased in two phases: in the late 1970s, fueling the rapid increase in harvests; and again in the late 1980s, fueling a rapid increase in the number of divers (Fig. 1b). The number of participating divers increased sharply in the early 1990s, despite decreasing daily catch rates. The fishery has been closed since October 16, 1995 to protect the remaining, reduced population.

In the 25-year history of the commercial abalone fishery in Alaska, 77% of the abalone was harvested from districts 3

Fig. 1. Catch, effort, and value in the southeast Alaska abalone fishery, 1971–1995: (a) catches (bars) and catch rate (line); (b) value (bars) and number of divers (line).

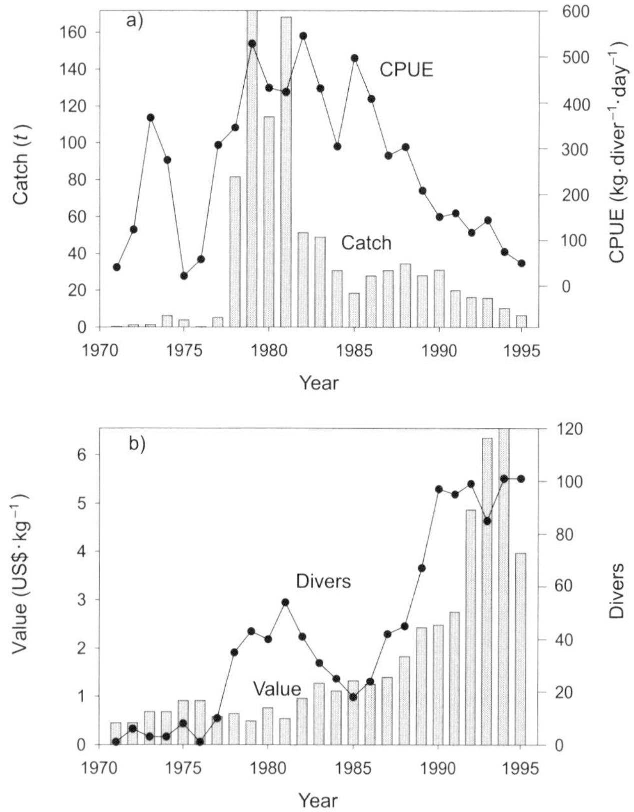

Table 1. History of abalone size limits in the commercial fishery in southeast Alaska.

Year	Minimum shell length (mm)	Fishing districts
1962	76	all
1967	102	1–4
1968	102	1–6
1976	89	9–14
1977	89	all
1979	95	all
1994	102	all

Note: The first year that the stated size limit was imposed is given for the listed districts.

and 4, at the extreme southwestern end of the Alaskan panhandle (Fig. 2). Harvests in district 13 near Sitka were 16% of the Alaska total. District 4, which produced 45% of total harvests, remains one of the areas least affected by sea otters (see below).

Size restrictions and seasonal catch limits were the primary fishery control measures. A minimum size limit of 76 mm shell length (SL) was in effect beginning in 1962 and this was modified at various times thereafter (Table 1). Size limits were generally more restrictive in fishing districts 1 through 4 (the major abalone harvest districts); however, the limit was lowered in those areas in an apparent effort to encourage the fishery in 1977, one year before the initial explosion of catch in 1978. The size limit was then raised to 95 mm SL in 1979 in an attempt to preserve reproductive potential. The limit was raised to 102 mm SL in 1994 once the stock decline was clear. At that time, many commercial divers were expressing concerns over low abundance and catch rates.

Harvest limits were first put in place at the peak of the fishery in the 1980–1981 season. These limits were reduced repeatedly in the subsequent 15 years until the final closure in 1995. The first harvest limit was 111 t, which was halved to 48 t for the 1981–1982 season. This was further reduced to 26 t for the 1985–1986 season and remained unchanged until the final year, 1995, when the limit was 7 t. In practice, total harvest was limited by adjusting fishing season length. Season length was first reduced from 12 months to 9 in 1980. Total fishing time in the final year, 1995, was reduced to 6 days.

Areas important for subsistence and personal use harvests were first closed to commercial exploitation of abalone in 1977. These closures were made because of concerns raised by subsistence harvesters and others who saw large catches being removed from traditional harvest areas. Closed areas were expanded through 1985 and remained unchanged thereafter. The total area of closed waters was small relative to that open for commercial fishing.

The number of divers participating in the fishery was not limited. The rise in participation to approximately 100 divers by the early 1990s was an important factor in the eventual closure of the fishery. The allowable catch in this decade was too small to provide for an orderly, manageable fishery with 100 or more participants under open-access regulations. Following the closure, a 4-year moratorium to entry began in 1996 for all dive fisheries in southeast Alaska.

Productivity

A tagging experiment was conducted in southern southeast Alaska from 1978 through 1981 to determine growth rates (Larson and Blankenbeckler 1980). This effort was spurred by the contemporaneous growth in commercial fishing effort and a feeling of discomfort by managers that too little was known about Alaskan abalone to justify particular minimum size limits. The first analysis of these data is presented here.

External, numbered tags were applied to abalone from 1978 through 1980 at two sites: Grant Cove (Gravina I., 55°21′13″ N 131°51′22″ W) and Ship I. (in Cordova Bay, 54°54′18″ N 132°31′45″ W) in southern southeast Alaska (Fig 1). Abalone were brought to the surface and tagged with numbered plastic disk tags and then returned to the seabed. Disk tags were attached with stainless steel wires and (or) vinyl spaghetti tags looped through two respiratory holes. The number of tags applied exceeded 1000, but is not known precisely due to lost records. Recoveries were made on multiple dates from 1979 through 1981. All periods between tagging and recovery dates at large were converted to years by dividing the number of days at large by 365.

Fig. 2. Distribution of abalone harvests (*t*) in southeast Alaska in the major fishing districts, 1971–1995. Dark shaded areas represent the location of harvests in each district or district section (D) as labeled. Two tagging sites are marked by triangles.

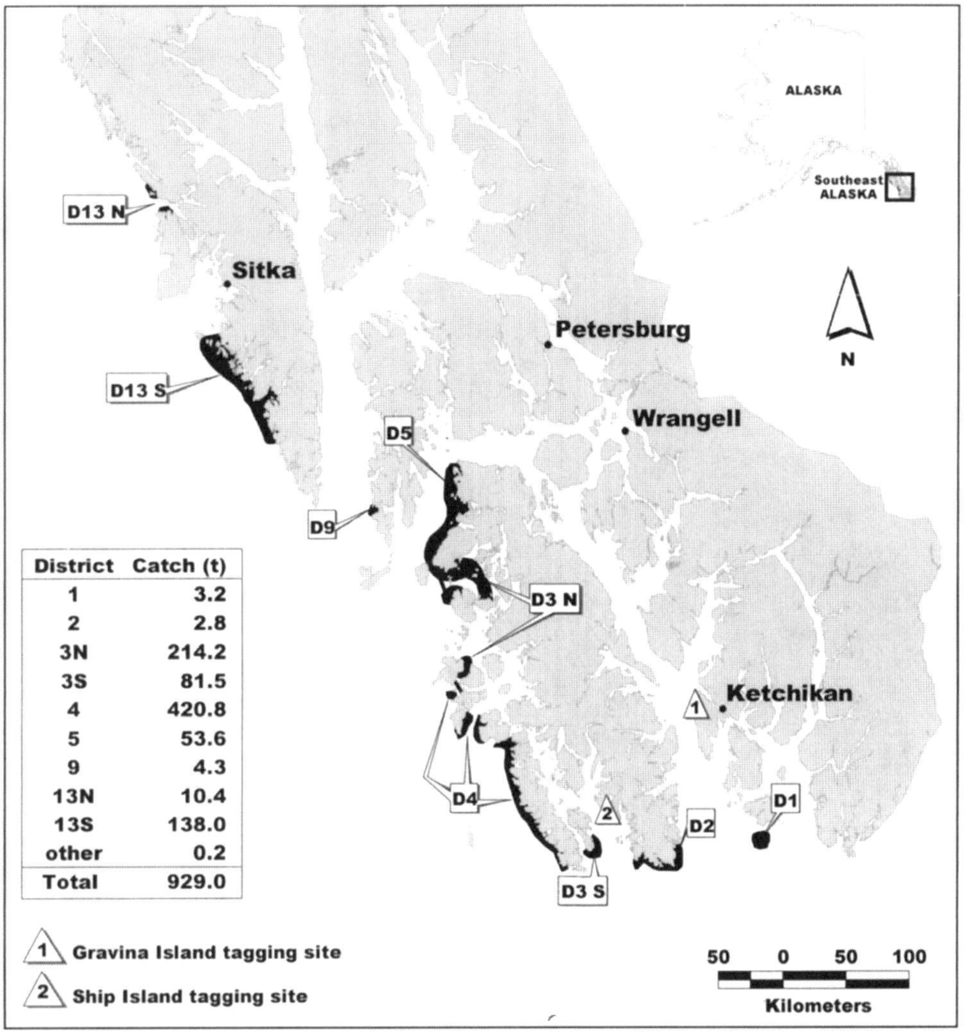

A restricted sample of shell length increments from tag recoveries was selected for analysis as follows. Recoveries were included in the growth analysis only when the period at large was within one-quarter year (91 days) of an integer year. This limit was imposed to avoid undue extrapolations of growth to the full year from seasonal periods dominated by growth or senescence. In cases where abalone were recaptured several times, only the longest period at large was used. This restriction was made to avoid correlation between increments. There were 93 tagged abalone recoveries satisfying these criteria from Gravina I. and 38 from Ship I. Tagging and recovery dates were between June 1978 and April 1984 at Gravina I. and between May 1979 and August 1981 at Ship I.

Parameters K and L_8 of a rearranged von Bertalanffy growth equation

$$l_{t+1} - l_t = (L_8 - l_t)(1 - e^{-K})$$

(Ricker 1975) were estimated with a nonlinear least-squares procedure (Solver algorithm, Microsoft 1998). Growth curves (Fig. 3) suggest a faster initial rate of growth at the Gravina I. site but a larger final size at the Ship I. site (K = 0.34 and 0.18; L_8 = 97.5 and 109.6, respectively).

Counts of internal rings in the shell can be used to age Alaskan abalone (Shepherd et al. 2000) in a method similar to that used for *H. midae* (Erasmus et al. 1994), *H. mariae* (Shepherd et al. 1995*a*), and *H. fulgens* (Shepherd et al. 1995*b*). These rings are visible under low magnification when the spire of the shell is ground flat. In particular, the number of major rings in *H. kamtschatkana* appears to be a fairly precise indicator of age in years, as corroborated by analysis of known age shells, growth checks, and electron microscopic examination of laminar width (Shepherd et al. 2000).

The use of ring counts as estimates of age in Alaskan abalone was further corroborated by comparing the number of rings at size with the growth curves predicted by the von Bertalanffy model for the tagging study reported above (Fig. 3). The range in shell lengths for each ring count interval (vertical series of dots in Fig. 3) spanned the predicted

Fig. 3. Growth rates of abalone in southeast Alaska. Vertical series of points are lengths for individual shell ring counts collected in 1998. The shaded line is a regression through the ring count data. The other lines are von Bertalanffy growth curves derived from tagging data from Gravina I. (triangles) and Ship I. (squares), 1979–1984 (see text).

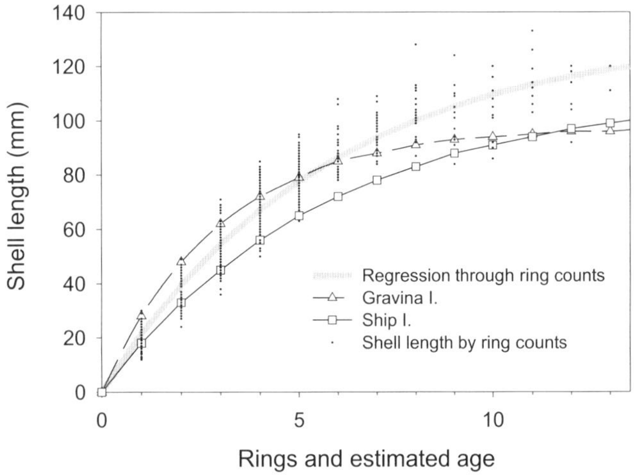

lengths at age up to age 5 for both tagging sites when the ring count was plotted as equal to the age on the x-axis.

The tagging data predicted slower growth than the ring counts at ages ≥6 years. This divergent trend was highlighted by a von Bertalanffy growth curve fitted to the ring data (Fig. 3; same fitting methods as above, assuming rings = years of age; $K = 0.18$, $L_8 = 131.9$). This discrepancy at greater sizes (ages) may be due to several sources, including (*i*) a real difference in growth between years (growth to larger sizes in the past decade than from 1978 to 1981), (*ii*) comparing results from dissimilar techniques, including the possibility that tags and tagging disturb and slow annual growth (Day and Fleming 1992), and (*iii*) an undercounting of rings at large sizes. This last possibility is unlikely given the corroboration of the ring method with known age abalone older than 6 years (Shepherd et al. 2000).

Distributions and impact of sea otters

Sea otters have recolonized much of their former range in southeast Alaska following extirpation by fur hunters prior to 1900. Between 1965 and 1969, the Alaska Department of Fish and Game (ADF&G) in cooperation with the U.S. Fish and Wildlife Service transplanted 402 otters to southeast Alaska from the Aleutian Islands and Prince William Sound (Pitcher 1989). Complete coastal surveys in 1987 and 1988 (Pitcher 1989) showed that otters had expanded their populations from the release sites to the majority of shorelines on the outer coast (Fig. 4, heavy shading). Opportunistic observations by various observers through the summer of 1998 indicated that sea otters have expanded their range further into more sheltered waters (Fig. 4, light shading).

We projected the population size of sea otters in southeast Alaska for 1998 by fitting an exponential growth function ($N_t = N_0 e^{rt}$) to the counts of otters (402 [= N_0] in 1968; 477 in 1975, 1460 in 1983, and 4521 in 1988 (Pitcher 1989)) by the method of least squares (Solver algorithm, Microsoft 1998). The best fit rate of increase over the census years is $r = 0.117$, which provides a projected population size of 13 250 in 1998 (Fig. 5). Observations during the count years suggest varying rates of increase, initially slow during the first decade of establishment, possibly due to mortality associated with transplanting, dispersal out of the area, and skewed sex and age structures (Schneider 1975). Population growth from 1975 to 1987 has been rapid, with rates as high as 23% per year for localized populations (Pitcher 1989).

The significant range expansion and rapid increase in otter populations occurred after the major harvest of abalone in the late 1970s and early 1980s. Therefore, otter predation on abalone is not considered the major factor in the decline of abalone in the 1980s. Otter predation is a major factor in the potential for further commercial abalone fishing. During the past 10 years of region-wide dive surveys (unpublished data), we observed that areas of otter recolonization lost most of the remnant populations of abalone, as well as commercially viable populations of red sea urchins (*Strongylocentrotus franciscanus*). Notable areas of incursion include northern District 3 (Sea Otter Sound), where total historic harvests exceeded 200 t, and southern District 13, with total harvests of 138 t (Fig. 2).

Pockets of abalone populations remained in former commercial harvest areas of Districts 1, 3, and 4 and near Sitka in District 13 (Fig. 2) as of 1998 (unpublished data). Densities in these pockets were judged to be considerably lower than those prior to the extensive commercial exploitation in the late 1970s and 1980s; however, there have not been quantitative abalone surveys to document density changes. The southwestern *panhandle* portion of District 4, especially the outside coast of Dall I., and passages between Dall, Suemez, Baker, and Prince of Wales islands. (Fig. 4), lack otters and have traditionally produced the largest catches of abalone (Fig. 2; 421 t over all years). For these reasons, these areas are expected to have the highest potential for commercial abalone harvests in the future.

Some abalones were seen to coexist with sea otters in the immediate vicinity of Sitka. Abalone size samples collected in 1998 on the west side of Galankin I., 2 km south of downtown Sitka, included numerous small abalone <25 mm shell length (11 of 44 shells), indicating recent settlement (past 1–2 years). This same site appeared to have recent otter predation, based on observations of red urchins existing only in cracks, where they are presumably safe from otters. Nearby areas in southern Sitka Sound, 13 km further south, were observed to have intense otter predation on red urchins in 1993, after which urchin shell fragments littered the bottom of former urchin research plots. Whether the presence of abalone near Sitka is a passing phenomenon, or if abalone will persist there, is not clear. High populations of otters may be partially excluded by local intense human activity, including boat and seaplane traffic.

Discussion

The sequence of management actions by the Alaska Department of Fish and Game and Alaska Board of Fisheries was inadequate to maintain commercially viable populations of abalone. In particular, the primary management tools,

Fig. 4. Distribution of sea otters in southeast Alaska in 1987 and 1988 (dark shading; from Pitcher 1989) and expansion though 1999 (light shading). Transplant sites from 1965 to 1969 are circled; numbers of released otters are noted. Observations since 1988 are by the authors, Tim Koeneman of the Alaska Department of Fish and Game, Brendon Kelly of the University of Alaska, and Jim Bodkin of the U.S. Fish and Wildlife Service.

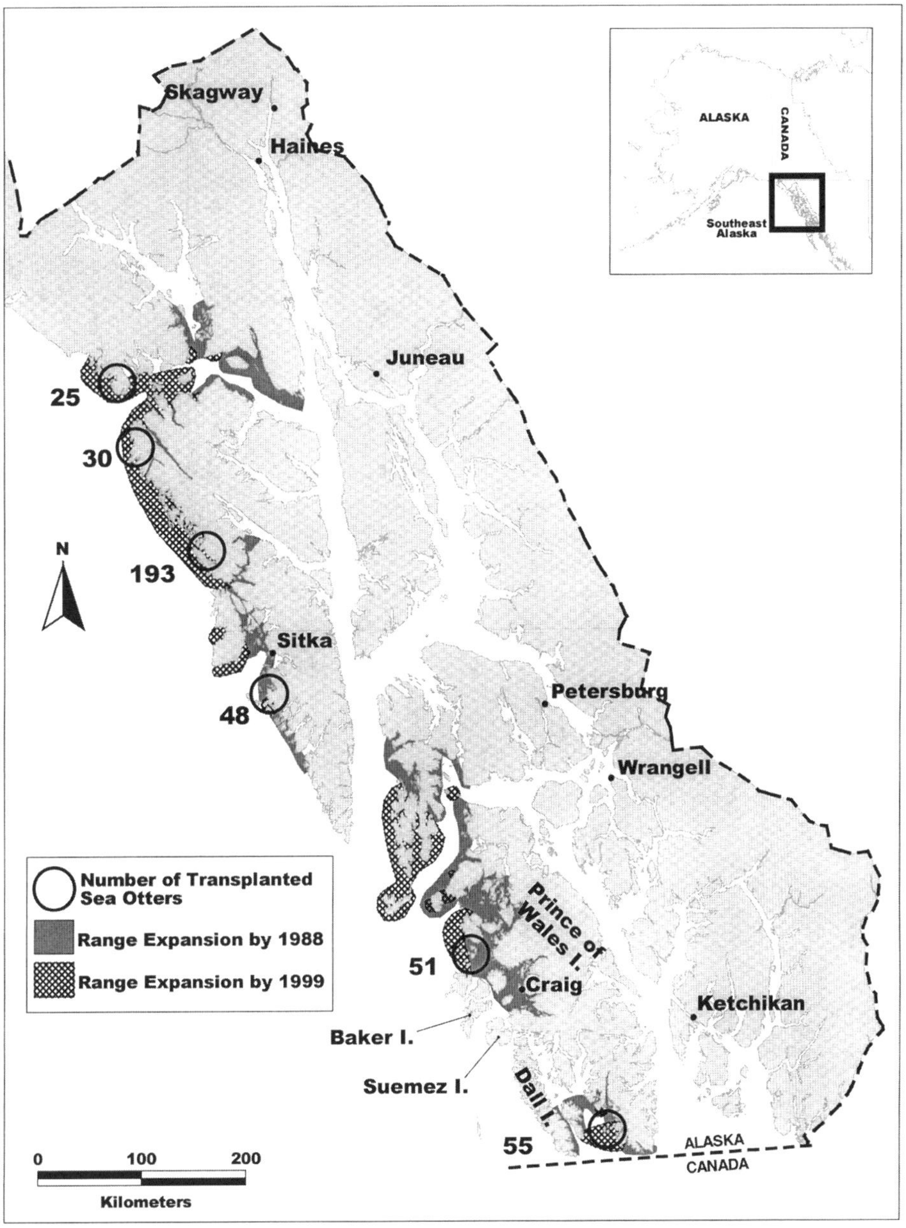

minimum size limits in combination with annual catch limits, were not sufficiently restrictive under open-access regulations. There is still too little known about population biology of northern abalone in Alaska to justify reopening a commercial fishery at this time with the existing management tools.

The reason for failure of size limits is not clear from available data, as size limits allowed at least 3 years of reproduction prior to vulnerability to harvest. This estimate is based on observations of sexual maturity at about 50 mm SL in the Sitka area (Paul and Paul 1981) and 51–64 mm SL in southern Southeast Alaska (Larson and Blankenbeckler 1980), which, by inspection of Fig. 3, indicates age at maturity is 3–4 years. Also, age at recruitment to the fishery is at least 7 years (ring count data in Fig. 3) for a minimum size of 95 mm SL, the limit for nearly the entire history of the Alaska fishery after the initial boom in catches. This minimum age at recruitment is similar to estimates for British Columbia (B.C.) (Quayle 1971; Breen 1986; Sloan and Breen 1988) and an initial estimate from Alaska (Larson and Blankenbeckler 1980). Rates of growth in B.C. were frequently slower, particularly at sites exposed to high surf (Sloan and Breen 1988), and the tagging data reported here indicate much higher ages at recruitment.

Minimum legal size limits were not set based on information on fecundity at size and prerecruit abalone were proba-

Fig. 5. Sea otter population counts (circles) and projected population size (line) for southeast Alaska with intrinsic rate of increase (r). Count data are from Pitcher (1989). See text for fitting method.

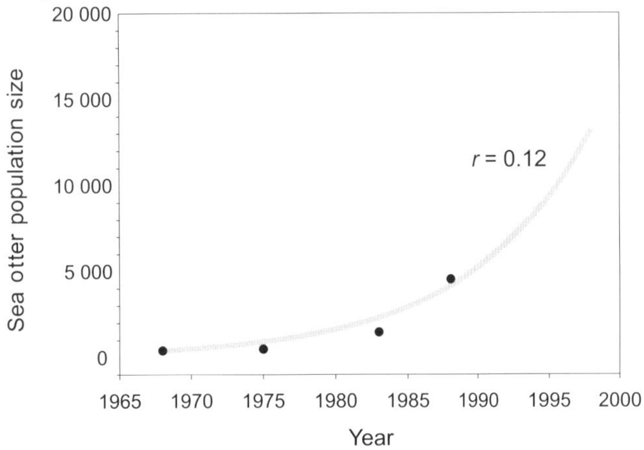

bly not sufficiently fecund to sustain a heavily harvested population. Fecundity in abalone probably increases linearly with weight (Mottet 1978), hence, geometrically with length, as has been shown for various species (e.g., Sainsbury 1982 for *H. iris*). Fecundity may also vary across habitats in *H. kamtschatkana* (Breen and Adkins 1979; Sloan and Breen 1988). Size specific fecundity has not been estimated in Alaska abalone. An egg per recruit analysis by Breen (1986) suggested that size limits for abalone in B.C. should be >100 mm SL, possibly 120 mm, and fishing mortality should be <0.40. An egg-per-recruit analysis for *H. rubra* in Tasmania suggested egg production was adequately conserved by size limits and that factors other than egg production were responsible for stock declines (Nash 1992).

Studies of failing populations suggested that a critical threshold density exists below which the risk of collapse of the fishery was high. In a South Australian population this was 0.2 abalone $\cdot$ m^{-2} (Shepherd and Partington 1995). Breen (1986) implied that the risk of collapse in northern abalone was high when densities fell below about 0.5 abalone $\cdot$ m^{-2}. A potential mechanism for this Allee effect is low fertilization success when densities of sperm and eggs are too low, as reported for urchins (Levitan et al. 1992; review in Levitan and Sewell 1998).

Larval dispersal in abalone appears to be local, implying that abalone populations may be composed of many local stocks subject to localized depletion (review by McShane 1992). The history of exploitation in Alaska suggested that fishers removed large numbers of abalone from localized aggregations before moving to more distant and less productive sites. In this case, recovery of highly exploited populations would be slow if dependent on locally produced larvae. Cool Alaskan coastal waters may exacerbate the problem in that the pelagic larval period may be prolonged and larvae are more likely to drift away from sites favorable to settlement, which appear to be patchily distributed.

The prospects for further commercial abalone fishing in Alaska are poor under current management regulations. Commercial fishing will not be allowed until a comprehensive management plan is developed that specifies a research program for determining the potential productivity of the population. The research focus is to include an estimate of size-related fecundity for the purpose of egg-per-recruit analysis, a mapping of the metapopulation distribution of productive areas, and a program to assess adult density and patterns of recruitment.

Further expansion of sea otter populations is also a serious deterrent to commercial fishing opportunities. The experience in southeast Alaska indicates that coexistence of commercially viable populations of abalone and otters is not possible and any resumption of commercial fishing should be directed to areas free of otters.

The commercial dive fishery industry in Alaska has expressed interest in rebuilding the abalone population, possibly through seeding. This and other enhancement activities are dependent on improving our understanding of productivity, recruitment processes, and spatial population structure. More effective enforcement against poaching, which is occurring to an unknown extent both for commercial and noncommercial purposes, is important.

Acknowledgments

The authors wish to thank present and former divers with the Alaska Department of Fish and Game who assisted in tagging studies. We thank Tim Koeneman, Jim Bodkin, and Brendon Kelly for their assistance in mapping the recent distribution of sea otters and we thank Scoresby Shepherd for his insights and recommendations for research on Alaskan abalone, including the hypothesis relating cold temperatures to prolonged larval periods. This is contribution PP-189 of the Alaska Department of Fish and Game, Division of Commercial Fisheries, Juneau, Alaska, U.S.A.

References

Breen, P.A. 1986. Management of the British Columbia fishery for northern abalone (*Haliotis kamtschatkana*). *In* North Pacific Workshop on Stock Assessment and Management of Invertebrates. *Edited by* G.S. Jamieson and N. Bourne. Can. Spec. Publ. Fish. Aquat. Sci. 92. pp. 300–312.

Breen, P.A., and Adkins, B.E. 1979. A survey of abalone populations on the east coast of the Queen Charlotte Islands, August 1978. Fish. Mar. Serv. MS Rep. 1490. 125 p.

Day, R.W., and Fleming A.E. 1992. The determinants and measurement of abalone growth. *In* Abalone of the world: biology, fisheries and culture. *Edited by* S.A. Shepherd, M.J. Tegner, and S.A. Guzmán del Próo. Fishing News Books, Blackwell Science Ltd., Oxford, U.K. pp. 141–168.

Erasmus, J., Cook, P.A., and Sweijd, N. 1994. The internal shell structure and growth rings in the shell of the abalone, *Haliotis midae*. J. Shellfish Res. **13**: 493–501.

Larson, R, and Blankenbeckler, D. 1980. Abalone research. Alaska Department of Fish and Game, Douglas, Alaska. 21 p.

Levitan, D.R., and Sewell, M.A. 1998. Fertilization success in free-spawning marine invertebrates: review of the evidence and fisheries implications. *In* Proceedings of the North Pacific Symposium on Invertebrate Stock Assessment and Management. *Edited by* G.S. Jamieson and A. Campbell. Can. Spec. Publ. Fish. Aquat. Sci. 125. pp. 159–164.

Levitan, D.R., Sewell, M.A., and Chia, F.S. 1992. How distribution and abundance influences fertilization success in the sea urchin *Strongylocentrotus franciscanus*. Ecology, **73**: 248–254.

McShane, P.E. 1992. Early life history of abalone: a review. *In* Abalone of the world: biology, fisheries and culture. *Edited by* S.A. Shepherd, M.J. Tegner, and S.A. Guzmán del Próo. Fishing News Books, Blackwell Science Ltd., Oxford, U.K. pp. 120–138.

Microsoft Corporation. 1998. Excel version 7. Bellevue, Wash.

Mottet, M.G. 1978. A review of the fishery biology of abalones. Wash. State Dep. Fish. Tech. Rep. No. 37.

Nash, W.J. 1992. An evaluation of egg-per-recruit analysis as a means of assessing size limits for blacklip abalone (*Haliotis rubra*) in Tasmania. *In* Abalone of the world: biology, fisheries and culture. *Edited by* S.A. Shepherd, M.J. Tegner, and S.A. Guzmán del Próo. Fishing News Books, Blackwell Science Ltd., Oxford, U.K. pp. 318–340.

Paul, A.J., and Paul, J.M. 1981. Temperature and growth of maturing *Haliotis kamtschatkana* Jonas. Veliger, **23**: 321–324.

Pitcher, K.W. 1989. Studies of southeastern Alaskan sea otter populations: distribution, abundance, structure, range expansion, and potential conflicts with shell fisheries. U.S. Fish and Wildlife Service Cooperative Agreement No. 14-16-0009-954. Alaska Department of Fish and Game, Anchorage, Alaska.

Quayle, D.B. 1971. Growth, morphometry and breeding in the British Columbia abalone (*Haliotis kamtschatkana* Jonas) Fish. Res. Board Can. Tech. Rep. 279. 84 p.

Ricker, W.E. 1975. Computation and interpretation of biological statistics of fish populations. Bull. Fish. Res. Board Can. 191.

Sainsbury, K.J. 1982. Population dynamics and fishery management in the paua, *Haliotis iris*. 1. Population structure, growth, reproduction, and mortality. N.Z. J. Mar. Freshwater Res. **16**: 147–161.

Schneider, K.B. 1975. Survey of transplanted sea otter populations in southeast Alaska, April 30 – May 16, 1975. Alaska Department of Fish and Game, Anchorage, Alaska. 30 p.

Shepherd, S.A., and Partington, D. 1995. Studies on southern Australian abalone (Genus *Haliotis*). XVI. Recruitment, habitat, and stock relations. *In* Progress in abalone fisheries research. *Edited by* S.A. Shepherd, R.W. Day, and A.J. Butler. Mar. Freshw. Res. **46**: 669–680.

Shepherd, S.A., Al-Wahaiba, D., and Al-Azri, A.R. 1995*a*. Shell growth checks and growth of the Omani abalone *Haliotis mariae*. Mar. Freshw. Res. **46**: 575–582.

Shepherd, S.A., Avalos-Borja, M., and Quintanilla, M.O. 1995*b*. Toward a chronology of *Haliotis fulgens*, with a review of abalone shell microstructure. Mar. Freshw. Res. **46**: 607–615.

Shepherd, S.A., Woodby, D., Rumble, J.M., and Avalos-Borja, M. 2000. Microstructure, chronology and growth of the pinto abalone, *Haliotis kamtschatkana*, in Alaska. J. Shellfish Res. **19**: 219–228.

Sloan, N.A., and Breen, P.A. 1988. Northern abalone, *Haliotis kamtschatkana*, in British Columbia: fisheries and synopsis of life history information. Can. Spec. Publ. Fish. Aquat. Sci. 103. 46 p.

The South African abalone (*Haliotis midae*) fishery: a decade of challenges and change

Robert J.Q. Tarr

Abstract: The South African abalone, *Haliotis midae*, fishery is now 50 years old. The commercial sector has recently been consolidated into one category "rights holders," who may both dive and market their catch. A new overall total allowable catch (TAC) of 820 t (whole mass in shell) has been initiated. This includes a commercial TAC of 523 t, a new "subsistence" TAC of 138.5 t, and a recreational TAC of 158.5 t. Poaching is a major problem with a high proportion of sublegal-sized abalone being taken. This has resulted in serial reductions in TAC in one zone from 150 to 15 t. Intensive fishery-independent diving surveys are underway. Co-management and ranching pilot projects have been initiated. Movement of rock lobster, *Jasus lalandii*, into two of the major TAC zones has interrupted the normal recruitment cycle of *H. midae* due to lobster predation on sea urchins, *Parechinus angulosus*, which juvenile abalone depend on for protection. Due to the combined effect of poaching and ecological changes, the prognosis for the resource is now poor, with future reductions in TAC likely.

Résumé : L'ormeau de Mida (*Haliotis midae*) (Afrique du Sud) est exploité depuis maintenant 50 ans. Dernièrement, les pêcheurs commerciaux ont été regroupés dans une catégorie de « titulaires de droits », qui les autorise à la fois à plonger et à vendre leurs captures. On a fixé un nouveau total autorisé des captures (TAC) global de 820 t (masse entière en coquille). Il comprend un TAC commercial de 523 t, un nouveau TAC aux fins de « subsistance » de 138,5 t et un TAC pour la pêche sportive de 158,5 t. Le braconnage est un problème important car une forte proportion d'ormeaux de taille inférieure à la taille légale sont prélevés, ce qui s'est traduit dans une zone par une série de réductions du TAC qui est passé de 150 à 15 t. Des relevés intensifs en plongée, indépendants de la pêche, sont en cours. Des projets pilotes de co-gestion et d'élevage extensif ont été mis en oeuvre. Le déplacement des langoustes (*Jasus lalandii*) vers deux des principales zones visées par un TAC a perturbé le cycle de recrutement normal de *H. midae* en raison de la prédation exercée par les langoustes sur les oursins (*Parachinus angulosus*), dont les ormeaux juvéniles dépendent pour leur protection. Présentement, en raison de l'effet combiné du braconnage et des changements écologiques, le pronostic pour la ressource est sombre, avec d'éventuelles réductions du TAC en perspective.

[Traduit par la Rédaction]

Introduction

The South African commercial abalone, *Haliotis midae*, fishery began in 1949 (Newman 1964) and it is now a mature 50-year-old fishery. Extensive research by Newman (1965, 1966, 1967, 1968, 1969) documented the early fishery and biology of the species. The commercial fishery up to the late 1980s was described by Tarr (1992) and at that stage the prognosis for the fishery and the resource was good, largely as a result of a conservative management approach and well-controlled management measures. Since then both social (Hauck and Sweijd 1999) and ecological changes (Tarr et al. 1996) have impacted the abalone resource in ways that have changed this outlook considerably. This paper describes the changes that have occurred over the past 10 years and considers the prognosis for the future.

R.J.Q. Tarr. Marine and Coastal Management, Private Bag X2, Rogge Bay, Cape Town 8012, South Africa.

Correct citation: Tarr, R.J.Q. 2000. The South African abalone *(Haliotis midae)* fishery: a decade of challenges and change. *In* Workshop on Rebuilding Abalone Stocks in British Columbia. *Edited by* A. Campbell. Can. Spec. Publ. Fish. Aquat. Sci. 130. pp. 32–40.

Commercial fishery

Management of the participants

The past decade experienced a number of minor changes (described below), until the 1998–1999 fishing season when major transformations took place. The previous commercial fishery was comprised of two components: (*i*) approximately 50 licenced divers (entitlement holders), who owned the right to dive (using hookah equipment only) and deliver a fixed percentage of the annual total allowable catch (TAC), and (*ii*) about five "packers" (quota holders), who owned the receiving, processing, and marketing rights to fixed percentages of the annual TAC. During 1991–1992, divers were permitted to employ assistant divers, enabling older divers to remain active in the fishery. This was also the first year that live abalone were exported to the far East and the amount exported has increased annually. From the 1992–1993 season, catches were sealed at the slipway, preventing possible illegal sale of the abalone before delivery to the factory. From 1993 to 1994, catch reports were completed daily instead of at month end, facilitating greater accuracy. In that

Table 1. Allocation of the new TAC for 1998–1999 was broadened to include the recreational and subsistence fisheries, previous participants, as well as new entrants.

Details	TAC (tonnes)
Recreational fishery	158.5
Subsistence fishery	158.5
New entrants, allocations by the Fisheries Transformation Council, and legal appeals	133
Previous "large" quota holders (280 t) and divers (90 t)	370
Total	820

season a 10-t community trust quota was allocated, which is still active. During 1995–1996, 10% of the TAC was allocated to 10 new entitlement holders, in amounts ranging from 2 to 10 t.

In 1998, the new Marine Living Resources Act (No. 18 of 1998), which had taken four years to develop, was gazetted. A cornerstone of the new act was, where possible, to allow more equitable access to marine resources for all South Africans. In terms of this Act, major changes were initiated for the abalone fishery in 1999 (as well as other South African fisheries).

The 1998–1999 allocation of TAC was broadened to cater to all participants in the fishery. Previously, TAC was applicable only to the commercial sector, whereas now the subsistence and recreational sectors are also included. The TAC for 1998–1999 totalled 820 t, which was calculated as the sum of a "traditional" commercial TAC of 515 t (a reduction of 15 t from the previous season), a recreational take of 220 t (equal to the 1997–1998 take), and a subsistence allotment of 85 t. The subsequent apportionment of the 820 t by the Minister had two objectives. First, the consolidation of the previous two-part commercial fishery structure (entitlement holders and quota holders) into one category, the rights holders, who now hold the right to dive, process, and market their percentage of the annual TAC. They may now employ any qualified divers to catch their abalone. Second, the introduction of new entrants to the fishery from previously politically and socio-economically disadvantaged backgrounds. The final allocations for 1998–1999 are shown in Table 1.

Reallocation of the 370 t between the divers and the previous "large" quota holders (Table 1) was achieved on the basis of a 3:1 ratio in published product prices (Stuttaford 1997) obtained by the quota holders relative to that paid to the divers. Due to the difficulties of administering these changes, the commercial season did not commence until January 1999.

Management of the resource

Four major TAC zones, A–D, on the southwest coast (including the partially protected area of Dyer Island), have consistently yielded over 90% of the TAC (Fig. 1). Poaching escalated in zone C after 1994, leading to serial reductions in the TAC from 130 t in 1995–1996 to 15 t in 1998–1999 (Table 2). Zone D was subject to a series of boundary changes, with appropriate changes to the TAC. The west

Fig. 1. The commercial *Haliotis midae* fishing zones A–G on the southwestern Cape coast.

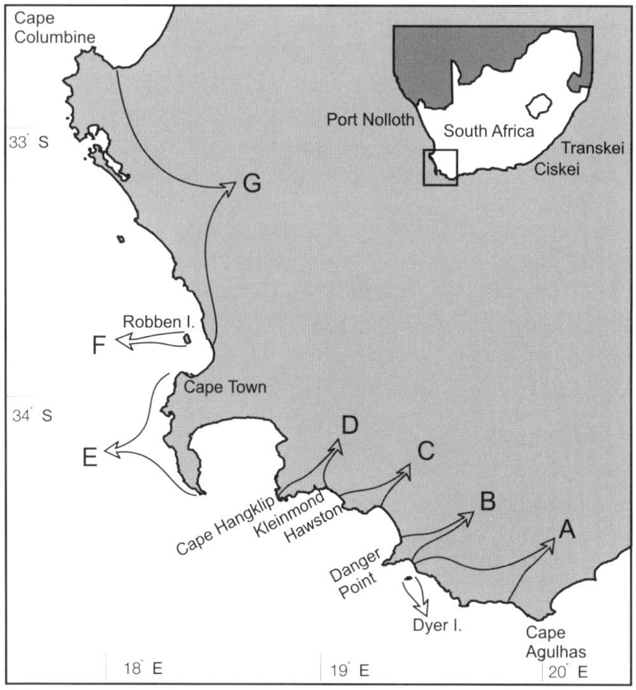

coast zones E and G supported small TACs for a few years until rapidly declining catch rates indicated an unsustainable level of harvest and, subsequently, fishing was stopped. The cause of the low productivity of these zones was probably irregular recruitment. Successful recruitment only occurred intermittently on the west coast in contrast to the regular annual recruitment cohorts found on the southwest Cape coast (Tarr 1994). However, from 1997, after requests by the fishers and quota holders, new attempts were made to fish these zones and low experimental TACs were allowed, subject to the supply of detailed fishery data from the divers. Robben Island (zone F), initially supported a dense unfished abalone population, which has been gradually fished down since 1985, with appropriate reductions in TAC aimed at arriving at a sustainable level.

Resource assessment and management

Commercial fishing of the resource has been managed on the basis of annually revised, area-specific TACs since the 1985–1986 season. Recommendations on the TAC were made by a Working Group of eight or more persons, including scientists from Marine and Coastal Management (formerly Sea Fisheries) and invited members from outside research establishments. Up until 1997, the TACs for all zones were determined on the basis of a set of decision rules. The recommendations are presented to the Consultative Advisory Forum, who in turn advise the Minister of Environmental Affairs and Tourism.

Trends in catch-per-diver hour (CPUE) indices per zone have been used as one indicator of resource status in the South African fishery since the late 1970s (Tarr 1992) and these appeared to have adequately reflected known trends in the resource. CPUE indices are traditionally accepted as being vulnerable to bias from such factors as aggregation

Table 2. Annual total allowable catch (TAC) in tonnes for the commercial zones A–G including Dyer Island.

Fishing season (years)	Zone A	B	DI	C	D	E	F	G	Total
1985–1986	—	—	—	—	—	—	50	—	—
1986–1987	180	160	0	160	40	20	50	30	640
1987–1988	190	140	0	160	50	20	50	30	640
1988–1989	190	140	0	160	50	20	50	30	640
1989–1990	195	140	0	170	50	20	50	0	625
1990–1991	195	140	0	170	50	10	30	0	595
1991–1992	195	145	25	150	55	0	30	0	600
1992–1993	195	150	25	150	55	0	30	0	605
1993–1994	195	150	25	140	90	0	15	0	615
1994–1995	205	150	25	130	90	0	15	0	615
1995–1996	205	150	25	130	90	0	15	0	615
1996–1997	205	150	25	65	90	0	15	0	550
1997–1998	185	150	25	30	105	5	15	15	530
1998–1999	185	150	25	15	105	5	15	15	515

effects, between-diver variability, and management changes (McShane 1994; Dichmont et al. 2000; Breen 1992). However, the long time series (22+ years) of detailed CPUE data (which now show a declining trend) and the absence of large aggregations of abalone (that can be serially depleted) in the mature South African abalone fishery are considered sufficient grounds for cautious use of the data. The absence of an alternate time series (such as long-term fishery independent diving survey data) and the presence of marked changes in the CPUE data in zone C (which facilitates model fitting), further support this approach. Attempting to understand and account for the factors causing bias in CPUE data is considered preferable to rejecting an apparently valuable and informative long time series of data. These CPUE data are therefore currently being subjected to generalized linear modeling (GLM) analyses to remove the effects of factors (such as between-diver variability, season effects, and between-area variability) known to affect effort (C.L. Moloney, Marine and Coastal Management, Cape Town, personal communication; E. Plaganyi, University of Cape Town, personal communication).

Length–frequency data have provided another useful indicator of resource status. However, in the last decade a change occurred in the interpretation of these data. Average size and the proportion of the catch estimated to comprise annual growth recruitment into the fishery ("percent recruit fishing": derived from the length frequency of sampled commercial catches) were used as resource indices on the assumption of constant recruitment into the fishery (Tarr 1992). A decreasing trend in recruit fishing (usually associated with an increasing mean size of catch) was taken as an indication of improving resource strength. Since the mid-1990s, the assumption of constant recruitment is no longer considered valid for two reasons. First, poaching has resulted in the removal of large volumes of sublegal-sized abalone (Marine and Coastal Management, Cape Town, unpublished data). Second, the collapse of urchin populations, probably caused by predation from lobsters in certain management zones, has resulted in high mortality to juvenile abalone (Tarr et al. 1996; Day and Branch 2000). Therefore, decreasing recruit fishing (and increasing mean size) is now interpreted as indicating potential recruitment failure and resource weakness in management zones where these effects are known to have occurred.

Diving surveys, stratified mainly in two depth ranges (0–5 and 5–10 m), were carried out in the 1980s. In each zone, an average of ten 200 × 3 m strip transects were placed within the major kelp (*Ecklonia maxima*) beds (Tarr 1993), providing data on abalone density and size distribution. Infrared aerial photographic surveys of the kelp together with depth contouring were used to produce maps of the kelp, from which total habitat area and biomass were calculated, given that over 90% of the abalone are found within the kelp forests. Results showed that more than 70% of the resource occurred within the 0–5 m depth range. However, high variances limited the direct applicability of the results. The causes of these are discussed in Tarr (1993). This necessitated a revised approach for direct surveys.

New fishery-independent abalone surveys (FIAS) were initiated in 1995 to provide an index of relative abundance (Dichmont et al. 2000). Density data from the 200-m transects were used to calculate expected coefficients of variance over a range of combinations of transect numbers versus transect length. Given constraints over available diving time and reasonable coefficients of variance, a survey design was chosen, which required twenty evenly spaced GPS-located diving stations to be surveyed per fishing zone annually, stratified in a 4–6 m depth range. Each station comprises a 30 × 2 m transect oriented perpendicular to the coastline and the expected CV is 25%. After six years the data are expected to provide an accurate estimate of abalone abundance trends (E. Plaganyi, University of Cape Town, personal communication). The FIAS data were used for the first time as inputs in resource assessment models in late 1999.

Since 1997, age-structured models have been used to reconstruct the history of resource dynamics. Growth curves are used to split the sizes into ages and the models are being fitted to CPUE data. The general conclusions from these studies are that the resource had been overexploited in the 1960s but that substantially reduced catches from 1970 onwards allowed the resource to recover. However, in recent years ecological effects, the increase in poaching, as well as the increased recreational catches, have contributed to a resource decline in some areas. The reductions in TAC for zone C for 1997–1998 and 1998–1999 were based on model results indicating that the resource was overexploited (Marine and Coastal Management, Cape Town, unpublished data). However, these results were probably optimistic because they did not include the observed poor recruitment in zones C and D, which resulted from increased predation by rock lobsters (see later in this paper; Tarr et al. 1996; Day and Branch 2000).

Recreational fishery

The recreational sector has probably been active for longer than the commercial sector, but only in the last decade

Fig. 2. Estimated recreational fishery landings of *Haliotis midae* derived from permit sale figures and reported landings derived from telephonic interviews.

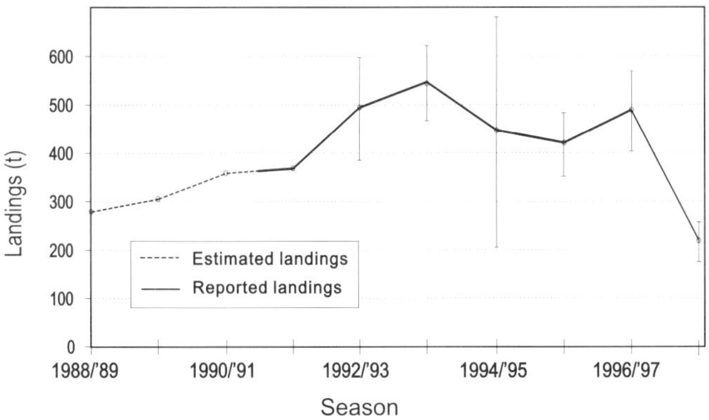

has the recreational fishery become an important management issue. The recreational fishery was not referred to in Tarr (1992). Recreational fishers may only dive from shore and only with snorkelling equipment. This fishery started as open access with the only limitation on annual take being the daily bag limit of 5 abalone per person (reduced to 4 abalone per person in 1991) and the size limit of 114 mm shell breadth (138 mm shell length, SL). In 1985, a 3-month closed season was introduced. The requirement to purchase abalone permits was introduced in 1983 but administrative problems resulted in this being stopped after 3 years. From 1988 the permit system was reintroduced and reliable data were obtained on numbers of permits being sold. Annual permit sales increased steadily, indicating an expanding effort base. However, estimates of annual take were unreliable. In 1991, permits were redesigned to incorporate the applicant's telephone number, providing a database of telephone numbers of permit holders. A professional survey company was contracted to carry out annual telephone surveys of permit holders to determine annual landings of this sector as well as permit holder demographics and other information.

Results (Fig. 2, Marine and Coastal Management, Cape Town, unpublished data) confirmed that the recreational fishery expanded initially and this growth continued to a high of 548 t in 1993–1994, which was 89% of the current commercial total allowable catch. Thereafter, landings fluctuated around 480 t caught by 34 000 permit holders. In 1997–1998, when one third of the season had elapsed, the Minister, due to concern over the effects of poaching on the resource, changed the fishing time to weekends (and public holidays) only and stopped the sale of further recreational permits. However the weekends-only ruling was soon overturned in court. The stoppage of permit sales remained, with only 64% of the permit numbers of the previous season being sold. This resulted in a large drop in estimated recreational landings to 221 t for 1997–1998. In terms of the Marine Living Resources Act of 1998, which seeks to redress past imbalances in access to marine resources by the previously disadvantaged sectors of the population, the Minister, for the 1998–1999 season, again changed the season to weekends only and reduced the season length to four months. This was expected to reduce the recreational take to about 160 t, considerably less than that of previous years, and the expected "saving" of tonnage was to be reallocated to other sectors of the fishery.

Subsistence fishery

The Marine Living Resources Act defined a new category, subsistence fishers, and these were included in the abalone fishery by the Minister for the 1997–1998 season. Noncommercial fishers can apply for a subsistence permit, in terms of which they can sell their daily bag limit of four abalone to end users only (not to the industry). A tag system was developed to prevent multiple sales of daily bags. To date, little fishery data have been provided by this sector. Nonetheless, almost 140 t was set aside by the Minister for allocation to subsistence fishers in 1998–1999.

Poaching

While illegal fishing has always been a factor in management of this resource, the increase in poaching from 1994 was unexpected and was the major cause of the change in the prognosis for the fishery from positive in 1989 to negative in 1999. A number of factors were responsible for the escalation in poaching.

(*i*) The primary cause for the upsurge in poaching appears to have been related to the political changes in South Africa. With the establishment of a new government in April 1994, there was a new emphasis on the constitutional rights of every individual. This resulted in raised expectations among the residents of previously disadvantaged coastal communities, who demanded the right to fish the abalone resource previously denied them. Inevitably, the process of transforming the country's fisheries was too slow and abalone poaching boomed.

(*ii*) A sharply declining South African Rand : US Dollar exchange rate made the export price of abalone attractive, fueling demand. Press reports, citing high overseas prices, probably raised awareness of the potential gains from abalone poaching. This resulted in the formation of sophisticated marketing networks, reportedly with connections to the drug trade and Chinese triads (Hauck and Sweijd 1999).

(*iii*) Changing priorities within the new government resulted in cuts in funding for the enforcement services of Marine and Coastal Management. This in turn lead to reduced morale among enforcement staff, with allegations of corruption further weakening the effectiveness and morale of this branch.

(*iv*) Political opportunism resulted in the legitimacy of the regulations being questioned publicly causing a degree of acceptance and support for poaching within the coastal communities.

(*v*) Many coastal communities exist in conditions of high unemployment and extreme poverty predisposing them to the benefits of short-term gains when combined with the above factors.

Zone C, historically one of the highest yielding fishery zones, became the center of the poaching expansion. Zone C housed a community with strong links to the traditional abalone fishery and the coastline in this area was difficult to access by enforcement staff. An underfunded control branch

Fig. 3. Samples of shucked (flesh mass) *Haliotis midae* from confiscated illegal product show a high percentage to be below the calculated minimum legal size (MLS).

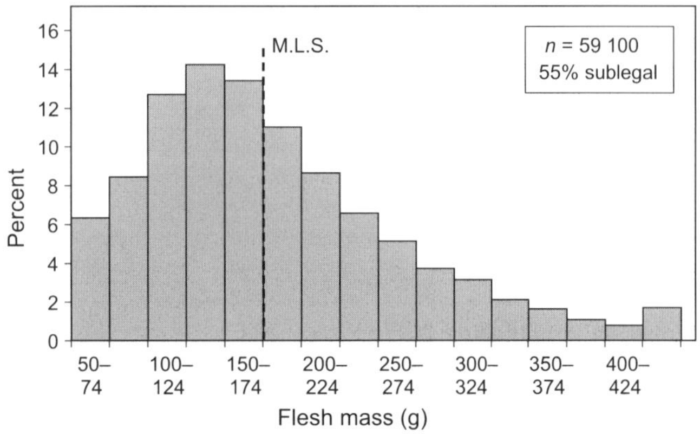

was unable to contain the poaching and on occasions when reinforcements from the police service were called in, violent confrontations frequently resulted, with firearms used on both sides. Continued lack of financial support for the enforcement branch permitted poaching to continue, expand, and consolidate; this in turn resulted in a degree of acceptance of poaching (frequently referred to as an "informal fishery"), further hampering effective control. The presence of gangs of brazen poachers became a daily phenomenon, as witnessed regularly by Marine and Coastal Management research staff, in the Hawston area of zone C.

Unfortunately, the poaching take was not confined to abalone above the fishery size limit. Larvae of *H. midae* settle primarily in the shallows (<4 m depth) and a gradual offshore movement of young animals occurs over the years with mean size increasing with depth. The shallows frequently support exposed aggregations of animals that are mature yet are below the minimum legal size of 114 mm shell breadth (138 mm SL). These are easily collected by indiscriminate shore-based poachers.

Despite being badly understaffed, control staff, assisted by police, have made many arrests since 1994 and (more frequently) confiscations of poached abalone. A sampling programme was established to record the sizes and (where possible) the origin of confiscated abalone. Because most of the confiscated abalone were shucked (out of the shell), a morphometric study (Marine and Coastal Management, Cape Town, unpublished data) was carried out so that the original shell dimensions and proportion below the minimum legal size could be estimated from drained flesh mass. Since 1994, over 59 000 confiscated abalone were measured (Fig. 3) of which 55% were below the equivalent estimated mean minimum legal size of 175 g flesh mass.

Eastern Cape resource

The traditional commercially harvested populations of *H. midae* occur in the southwestern Cape, but the species is also found east of Cape Agulhas in a discontinuous distribution pattern that extends as far as the former Transkei region (Fig. 1). This resource has never been commercially fished, yet there has always been strong recreational activity in the area, with estimated average annual landings of 13 $t \cdot year^{-1}$ (Marine and Coastal Management, Cape Town, unpublished data). In recent years interest has been expressed in commercial harvesting of this resource. Because of the extensive nature of the coastline, diving surveys were impractical and an experimental fishery was proposed, based on published (Wood 1994; Fielding 1995) and unpublished data on densities and coastal topography. One small (3 t) experimental fishery was initiated in the former Ciskei in 1992, at a reduced size limit of 100 mm shell breadth (122 mm SL) and this amount has been successfully landed annually since then. This experimental fishery is unique in South Africa in that fishing is carried out by a combination of unskilled shore pickers from the local community as well as commercial snorkel divers. The abalone occur close inshore in the surf zone and are accessible only during spring low tides or extremely calm sea conditions.

Additional experimental quotas were proposed for the entire eastern Cape coast, totaling 30 t, and applications were invited. Much interest was shown in the prospect of commercial fishing in the area, however, in the period prior and subsequent to the proposal of experimental fishing, poaching escalated in this area on a scale similar to that experienced in zone C. Anecdotal reports, as well as the volume of confiscations of illegal product from this area, suggest that the Eastern Cape abalone resource may now be denuded to the point where neither experimental nor commercial fishing will prove viable. The fishery therefore was not implemented, mainly because of a lack of personnel for enforcement purposes.

Co-management

Co-management, whereby the community is actively involved in the ownership, management, and enforcement of sustainable resource utilization, has been proposed as an alternative to top-down central government management of the abalone resource (e.g., Hauck and Sweijd 1999). An attempt to initiate a co-management project in the Hawston community failed because of disagreements within the community and an inability to find universally acceptable community leaders (Hauck and Sweijd 1999). To date, no effective co-management formula has been developed for South African abalone. However, a pilot abalone and lobster co-management project is currently being developed within the Kleinmond community (within zone D) and may yield valuable guidelines for future ventures of this nature (M. Hauck, University of Cape Town, personal communication). Zone D has been identified as being suitable, because the abalone and lobster resources have not yet been significantly denuded (as in zone C) and zone D contains a valuable Marine Protected Area that is becoming a target for poaching activities.

Ecological changes

Coincident with the increase in poaching, there was an ecological change occurring that was first noted in 1994 (Tarr et al. 1996). A large-scale movement of rock lobster, *Jasus lalandii*, into the kelp forests of zones C and D coincided with the disappearance of the entire population of sea urchins, *Parechinus angulosus*. This disappearance was probably the result of direct predation by the lobster population (Mayfield 1998). However, experimental work is planned to test whether the disappearance of *P. angulosus* could have been caused by other environmental factors.

Juvenile *H. midae* have an unusual relationship with sea urchins, deriving protection from predation by actively sheltering under their spines. This relationship permits abalone juveniles to survive in areas with few or no natural protective crevices and boulders, which are the usual shelters used by juvenile haliotids. The disappearance of the urchin population has produced a concomitant disappearance of the majority of juvenile abalone from most areas of zones C and D since 1994 (R.J.Q. Tarr, personal observation; Tarr et al. 1996), probably from direct lobster predation. The reduction of most natural recruitment into the population has not yet been directly reflected in declining fishery catch rates, but this probably will soon become apparent, with serious negative effects on the resource and fishery in the affected areas (zones C and D).

Ranching

Abalone aquaculture in South Africa has expanded in the past decade and there are now at least nine shore-based farms in various stages of production (Cook 1998). Several of these produce abalone seed for ongrowing and (or) sale to other abalone farms. In light of the declines in the natural abalone resource, stock enhancement or ranching (the release of cultured abalone into the sea) would appear an obvious route to explore. To date, however, no comprehensive stock enhancement or ranching research has been initiated within the existing commercial fishing grounds. The reasons for this are varied. (*i*) The question of ownership of seeded abalone as well as identification of seed versus natural recruitment has not been resolved. (*ii*) No studies of seeded abalone growth and survival rates have been carried out to establish these vital parameters. (*iii*) The major obstacle at present, however, is the encroachment of lobsters into two of the main abalone fishing grounds. Given the high predation levels attributed to lobsters, mortality of seeded abalone is likely to be close to 100%.

There does, however, appear to be considerable potential for establishing new populations of abalone in an area of coastline not presently inhabited by *H. midae*, e.g., the northwest coast of South Africa south of Port Nolloth (Fig. 1) (Sweijd et al. 1999). This coastline supports kelp beds similar to those in the abalone grounds of the southwest coast, yet the nearest abalone populations are 400 km to the south. Preliminary trials have shown growth rates comparable with those from natural populations, six month survival rates of up to 40%, and an economic assessment appears favourable (Sweijd et al. 1999). Further studies and an expansion of the ranching area are underway.

Discussion

Abalone fetch the highest unit prices of any South African fishery resource. Therefore, it is essential to have effective management measures in place that will minimize illegal harvesting practices (Martin 1994). Up to the early 1990s, these measures appeared to have been successful, i.e., divers reported improved catch rates, there was a strong possibility of increased TACs in the future, and new divers were buying into the fishery. However, the combined effects of poaching and ecological changes have altered this to the extent that the fishery now appears certain to face severe TAC reductions in the future.

With the change towards live marketing, there has been an increased emphasis on product quality and divers are encouraged to deliver smaller quantities spread over a longer period. This could be facilitated by the removal of the closed season and reversion to a 12-month calendar year season. South African abalone do not appear to aggregate seasonally for spawning and, in the absence of any other behavioral changes or increased vulnerability associated with the reproductive cycle, there is no scientific reason to retain a closed commercial season, given that annual commercial take is limited by a TAC.

The change in structure of the fishery to one class of licenced participant: the *Rights Holder*, should provide a number of benefits. (*i*) Greater stability, given that there is no longer conflict of interest between divers and "producers." All *rights holders* who do not harvest their own abalone can employ qualified divers of their choice to do so. (*ii*) With the admission of many new entrants, including groups representative of past "informal fishers," arguments previously used by poachers to justify their actions (i.e., historical denial of access to the resource) become invalid. (*iii*) Existing management measures should regain general acceptability within the coastal communities. (*iv*) Enforcement staff should regain lost morale because they can no longer be accused of supporting/enforcing immoral regulations. (*v*) Central government support for enforcement of the management measures should increase, because these have been derived after extensive consultation with the relevant fishery and community role players.

In the 1980s, the annual allocation of TACs was directed at maintaining stability within the fishery. In practice this meant that, during the recovery period of the late 1980s, when divers were achieving steadily higher catch rates, TACs were increased cautiously. This may have resulted in an accumulation of biomass, enabling the resource to withstand for a longer period the rapidly increased harvests, which resulted from the expansion of the recreational fishery and poaching in the 1990s.

The recent changes in fishery management practices (i.e., inclusion of many new entrants, live marketing, use of assistant divers, and boundary changes) require that CPUE data are closely examined. In addition, fishery-independent abalone surveys must continue and the data collected should be directly incorporated in modeling of the resource dynamics.

The redistribution of TAC and the consolidation of the commercial fishery into one category of participant are likely to prove beneficial in the long term. However this was achieved at the expense of the recreational sector, the previous large quota holders, and the divers. For the 1997–1998 season, the latter two had a total of 482 t in quota, whereas for 1998–1999, they had only 280 t, a loss of 42%. However, they have gained the right to employ divers of their choice, on their own terms. Similarly, the divers now have sole rights to 90 t, which they can harvest and market.

The reasons for the initial growth in the recreational fishery are difficult to determine. The expansion began before the increase in poaching; therefore, the two would not appear to be linked. Nonetheless, a significant proportion of recreational fishers were probably selling their daily bag limit of four abalone, and with the expansion of poaching in general, more markets would have been available to the "recreational poacher," possibly sustaining some of the increased effort. Measures aimed at reducing the recreational take by shortening season length were largely ineffective because most of the reductions were directed at nonpeak periods and the prime summer holiday period was left intact. To date, the most effective curbing measure was probably the change in fishing season to weekends and public holidays only. This not only markedly reduced the access time for those recreational fishers who were regularly selling their daily bag limit but also reduced the monitoring burden of enforcement personnel. The important development, however, is that for the first time the recreational fishery was allocated a "TAC" that was defined on the basis of expected catch from an undefined and unlimited number of participants. The target TAC is based on past telephone surveys, which will have to be continued in order to determine the efficacy of the changed season length. This is an important step towards more equitable management of the resource as a whole, however, because a defined limit was placed on the annual recreational take.

The newly created category of subsistence fisher is a potentially valuable one that can satisfy the needs of many coastal inhabitants, who might otherwise have started or continued poaching. However, the subsistence fishery must be effectively managed. The future of this sector will therefore depend on the successful implementation of measures that ensure that useable fishery data are obtained. In addition, the effectiveness of the tag control system must be determined in preventing the sale of multiple daily bag limits. If effective measures such as these can not be implemented, the future of the subsistence fishery should be questioned.

Poaching has the potential to cause total collapse of the commercial and recreational fisheries and is a problem faced by abalone fisheries worldwide (Brown 1986; Ward 1986; Jamieson 1989; Clavier 1992; Farlinger and Campbell 1992). Should effective management measures (be they traditional or co-management options) continue to fail, the South African abalone resource will inevitably face commercial extinction. Estimations of the total volume removed illegally are unreliable, but could range from 25 to 100% of the commercial TAC. There are two direct results of illegal harvests. (*i*) Large volumes of abalone are removed from the resource just before attaining the minimum legal size, therefore, recruitment into the fishery is reduced within a time scale of a few years. Given constant natural and fishing mortality, this will result in a depleted fishable stock, negatively affecting both egg production and catch rates. (*ii*) Removal of the shallow water aggregations of sublegal-sized yet mature abalone could have serious consequences. Because of the more aggregated nature of these populations in comparison with the generally dispersed nature of deeper water abalone populations, the fertilization success of the shallow aggregations was likely to be high. Hence, a potentially major source of egg production is being removed by the poachers. As a result of nearly five years of intensive poaching, the resource in zone C has been so severely denuded that the poachers have moved their efforts farther east to zone A. This is, like most of zone C, also a remote area of coastline, with few residential properties, favoring the undetected operations of poachers. Unless firm measures are adopted to contain this redirection of effort, zone A may suffer the same fate as zone C. However, with the recent allocation of fishing rights to many new entrants, particularly previously disadvantaged communities, a firmer approach from Marine and Coastal Management and the government has already been implemented, including an intensive six month combined exercise involving the police. The continuation of such measures to reduce poaching is imperative if the abalone resource is to remain sustainable.

Co-management of abalone resources in South Africa already has been found to be fraught with difficulty (Hauck and Sweijd 1999) and an effective formula has yet to be devised. With a high value product such as abalone, there is no doubt that enforcement of regulations will have to remain a cornerstone of management success. Ideally, this enforcement should come from within the community. However, there will always be outside groups wishing to gain access to the resources and this will place a severe strain on the cohesion and effectiveness of any community's resources.

South Africa's history of commercial access rights is different from other abalone fisheries such as Canada, New Zealand, and Australia. In those countries, aboriginal fisheries existed, which were superceded by modern commercial fishing rights. This resulted in legal claims for restitution of aboriginal fishing rights. In South Africa, although there was a measure of aboriginal fishing (Tarr 1989), this was apparently not a strong cultural or food component and hence there have been no claims for access based on historic rights. What was denied in the past were some of the rights of commercial access to the fishery during the early stages (1950s–1970s), although this only occurred with regard to obtaining processing marketing quotas. Diving rights were freely available to all races, the only criterion being that the participant had to be a *bona fide* fisher, with no additional source of income.

The ecological relationship between *H. midae* and sea urchins, whereby the former derive protection by sheltering under the spines of the urchins, may well have been an important factor behind the sustained production since inception of the abalone fishery in the 1950s. This relationship probably permitted a higher survival of abalone juveniles in areas with little or no natural protective crevices and boulders. This protection may have resulted in higher abalone survival rates during the vulnerable size range of 5–30 mm SL. In addition, the grazing behavior of urchins on the benthos may be an important factor in maintaining the coverage of healthy crustose coralline algae, which is a key element in larval settlement (Day and Branch 2000). The newly resident lobster population has also serially depleted most other edible small molluscs such as *Turbo cidaris* and *Oxystele spp.* (Marine and Coastal Management, Cape Town, South Africa, unpublished data). We have also observed lobster attempting to prey on large (110–150 mm SL) abalone by grazing on the exposed epipodial fringe. Experimental work is still planned to test the possibility that the disappearance of the urchins is an environmental or disease effect, although this seems unlikely. A few juveniles of this species can still be found under boulders. Small pockets of adults still exist in rough water shallow environments (unfavorable for lobsters) and virtually all other small (edible) benthic molluscs have disappeared. However, those with a refuge in size, such as *T. sarmaticus*, are still abundant.

Ranching with *H. midae* is still untested in South Africa, although a promising pilot project is underway on the west coast south of Port Nolloth. The advantage of this area is that the absence of any natural abalone populations simplifies the issue of ownership of harvestable abalone. Clearly, the abalone species must be absent for a reason, but if that is associated with the reproductive cycle or larval distribution and survival, then artificial placement of competent juveniles into the environment may be successful. The abalone will be harvested below the traditional minimum legal size and differentiating wild caught animals from this region from those from the traditional southwest Cape coast will be difficult. A condition of the experiment, therefore, is that the ranched abalone will be for export only; hence, they will not appear in South African retail outlets. Until further research has clarified the question of survival of transplanted cultured juveniles, ranching cannot be seen as a solution to the problems facing the present fishery.

Haliotid fisheries are acknowledged to be highly vulnerable to collapse through overfishing, because of the species' slow growth rates, aggregation effects, and vulnerable reproductive strategy. For this reason, they are challenging fisheries to manage (Farlinger and Thomas 1993; Tegner et al. 1992; Prince and Shepherd 1992; Guzmán del Próo 1992; Davis et al. 1992; Breen 1992). The South African abalone resource has, in the past decade, faced the most testing set of challenges in its 50-year history. In the future, the abalone resource may collapse to the point of commercial extinction; even if the abalone resource recovers, populations may not regain the levels of the early 1990s within the next 50 years. Unfortunately, the unique socio-political and ecological factors that may lead to the collapse of this valuable South African resource have been largely beyond the control of fishery managers.

If continued survival of this abalone fishery is to be achieved, the solution lies not only in granting access rights to resident coastal communities (with a co-management component combined with education programmes) but also in effective enforcement of management measures. These requirements are not yet adequately in place, despite the restructuring described in this paper, but this situation may be remedied in the future.

References

Breen, P.A. 1992. A review of models used for stock assessment in abalone fisheries. *In* Abalone of the world: biology, fisheries and culture. *Edited by* S.A. Shepherd, M.J. Tegner, and S.A. Guzmán del Próo. Fishing News Books, Blackwell Science Ltd., Oxford, U.K. pp. 253–275.

Brown, C.H. 1986. Conservation of the paua resource. Catch **13(3)**: 3.

Clavier, J. 1992. The ormer (*Haliotis tuberculata*) fishery of France and the Channel Islands. *In* Abalone of the world: biology, fisheries and culture. *Edited by* S.A. Shepherd, M.J. Tegner, and S.A. Guzmán del Próo. Fishing News Books, Blackwell Science Ltd., Oxford, U.K. pp. 454–457.

Cook, P. 1998. The current status of abalone farming in South Africa. J. Shellfish Res. **17**: 601–602.

Davis, G.E., Richards, D.V., Haaker, P.L., and Parker, D.O. 1992. Abalone population declines and fishery management in southern California. *In* Abalone of the world: biology, fisheries and culture. *Edited by* S.A. Shepherd, M.J. Tegner, and S.A. Guzmán del Próo. Fishing News Books, Blackwell Science Ltd, Oxford, U.K. pp. 237–249.

Day, E., and Branch, G.M. 2000. Evidence for a positive relationship between juvenile abalone *Haliotis midae* and the sea urchin *Parechinus angulosus* in the South-western Cape, South Africa. S. Afr. J. Mar. Sci. **22**. (In press).

Dichmont, C.M., Butterworth, D.S., and Cochrane, K.L. 2000. Towards adaptive management of the South African abalone *Haliotis midae* fishery. S. Afr. J. Mar. Sci. **22**. (In press).

Farlinger, R.S., and Campbell, A. 1992. Fisheries management and biology of northern abalone, *Haliotis kamtschatkana*, in the northeast Pacific. *In* Abalone of the world: biology, fisheries and culture. *Edited by* S.A. Shepherd, M.J. Tegner, and S.A. Guzmán del Próo. Fishing News Books, Blackwell Science Ltd, Oxford, U.K. pp. 395–406.

Farlinger, R.S., and Thomas, G. 1993. Management of the British Columbia abalone fishery: a square peg in a round hole. J. Shellfish Res. **12**: 144. [Abstract].

Fielding, P.J. 1995. A preliminary investigation of abalone *Haliotis midae* resources along the Transkei coast, South Africa. S. Afr. J. Mar. Sci. **15**: 253–261.

Guzmán del Próo, S.A. 1992. A review of the biology of abalone and its fishery in Mexico. *In* Abalone of the world: biology, fisheries and culture. *Edited by* S.A. Shepherd, M.J. Tegner, and S.A. Guzmán del Próo. Fishing News Books, Blackwell Science Ltd, Oxford, U.K. pp. 341–360.

Hauck, M., and Sweijd, N. 1999. A case study of abalone poaching in South Africa and its impact on fisheries management. ICES Marine Science Symposium, Cape Town, South Africa, 16–19 Nov. 1998. IECS J. Mar. Sci. **56**: 1024–1032.

Jamieson, G.S. 1989. Status of the northern abalone, *Haliotis kamtschatkana*, in Canada. Can. Field-Nat. **103**: 153–158.

Martin, R.E. 1994. South African fisheries management and abalone exploitation. Sea Changes Newsletter, **16**: 86–116.

Mayfield, S. 1998. Assessment of predation by the West Coast rock lobster (*Jasus lalandii*): relationships among growth rates, diet and benthic community composition, with implications for the survival of juvenile abalone (*Haliotis midae*). Ph.D. thesis, University of Cape Town, Cape Town, South Africa. 213 p.

McShane, P.E. 1994. Estimating the abundance of abalone (*Haliotis* spp.) stocks—examples from Victoria and southern New Zealand. Fish. Res. **19**: 379–394.

Newman, G.G. 1964. Abalone (perlemoen) research and its application to an industry. Commer. Ind. **23**: 76–78.

Newman, G.G. 1965. Abalone research in South Africa. S. Afr. Shipp. News Fish. Ind. Rev. **20**(5): 93–101.

Newman, G.G. 1966. Movements of the South African abalone *Haliotis midae*. S. Afr. Div. Sea Fish. Invest. Rep. **56**. 19 p.

Newman, G.G. 1967. Reproduction of the South African abalone *Haliotis midae*. S. Afr. Div. Sea Fish. Invest. Rep. **64**. 24 p.

Newman, G.G. 1968. Growth of the South African abalone. S. Afr. Div. Sea Fish. Invest. Rep. **67**. 24 p.

Newman, G.G. 1969. Distribution of the abalone (*Haliotis midae*) and the effect of temperature on productivity. S. Afr. Div. Sea Fish. Invest. Rep. **74**. 7 p.

Prince, J.D., and Shepherd, S.A. 1992. Australian abalone fisheries and their management. *In* Abalone of the world: biology, fisheries and culture. *Edited by* S.A. Shepherd, M.J. Tegner, and S.A. Guzmán del Próo. Fishing News Books, Blackwell Science Ltd, Oxford, U.K. pp. 407–426.

Stuttaford, M. (*Editor*). 1997. Fishing industry handbook. 25th edition. South Africa, Namibia, and Mozambique. Marine Information cc, Stellenbosch, South Africa. 415 p.

Sweijd, N., Snethlage, Q., Harvey, D., and Cook, P. 1999. Experimental abalone (*Haliotis midae*) seeding in South Africa. J. Shellfish Res. **17**: 897–904.

Tarr, R.J.Q. 1989. Abalone. *In* Oceans of life off Southern Africa. *Edited by* A.I.L. Payne and R.J.M. Crawford. Vlaeberg, Cape Town, South Africa. pp. 62–69.

Tarr, R.J.Q. 1992. The abalone fishery of South Africa. *In* Abalone of the world: biology, fisheries and culture. *Edited by* S.A. Shepherd, M.J. Tegner, and S.A. Guzmán del Próo. Fishing News Books, Blackwell Science Ltd, Oxford, U.K. pp. 438–447.

Tarr, R.J.Q. 1993. Stock assessment, and aspects of the biology of the South African abalone, *Haliotis midae*. M.Sc. thesis, University of Cape Town, Cape Town, South Africa. 156 p.

Tarr, R.J.Q. 1994. Differences in recruitment patterns and fishery productivity in two areas: a link? *In* Second International Symposium on Abalone Biology, Fisheries and Culture, 7–11 Feb., 1994, Hobart, Tasmania, Australia.

Tarr, R.J.Q., Williams, P.V.G., and Mackenzie, A.J. 1996. Abalone, sea urchins and rock lobster: a possible ecological shift may affect traditional fisheries. S. Afr. J. Mar. Sci. **17**: 311–315.

Tegner, M.J., DeMartini, J.D., and Karpov, K.A. 1992. The California red abalone fishery: a case study in complexity. *In* Abalone of the world: biology, fisheries and culture. *Edited by* S.A. Shepherd, M.J. Tegner, and S.A. Guzmán del Próo. Fishing News Books, Blackwell Science Ltd, Oxford, U.K. pp 370–383.

Ward, T.J. 1986. Abalone biology and fisheries in Australia: a review of research. Fisheries Paper No. 86-12. Australian Fisheries Service, June 1986, Canberra, Australia. 36 p.

Wood, A.D. 1994. The South African abalone *Haliotis midae*—aspects of ecology, biology. The Naturalist, **38**: 33–40.

Review of northern abalone, *Haliotis kamtschatkana*, stock status in British Columbia

Alan Campbell

Abstract: Relevant fishery and biological characteristics of the northern abalone, *Haliotis kamtschatkana*, in British Columbia (B.C.), are reviewed to provide possible directions for stock rehabilitation. After closure of the northern abalone fishery in B.C. in 1990 due to conservation concerns, surveys at indicator sites during 1978–1998 showed a continued general decline of abalone populations in the central coast of B.C. and the Queen Charlotte Islands. The northern abalone is vulnerable to overexploitation because this species has a short larval period, is slow-growing, relatively long-lived and sedentary, and has low or sporadic recruitment. Mature individuals, which tend to accumulate in shallow water, are easily accessible to harvesters. High market value, reduced availability, and difficulty of enforcing the fishery closure in a large, mostly uninhabited coastal area has encouraged illegal harvesting of northern abalone populations. Samples from abalone illegally harvested during 1995–1998 suggested that poachers indiscriminately removed mostly large mature abalone with no regard for the legal size limit of 100 mm shell length. Without size and fishing rate controls, illegal harvesting not only depletes already depressed abalone stocks, but also reduces their reproductive potential by the removal of large mature abalone, and hinders attempts to rehabilitate abalone populations through fishery closure. Without reductions in illegal harvest, protection of brood stock, a continued closure of the fishery, and other effective rehabilitation methods, abalone population abundance may remain low or continue to decline in many areas of B.C. Research, in well-monitored experimental areas, is required to determine the optimal distribution and density of abalone spawners and recruits. The efficacy of brood stock transplant and juvenile seed restocking methodologies for northern abalone as part of a rehabilitation program requires evaluation in B.C.

Résumé : Les caractéristiques halieutiques et biologiques propres à l'ormeau nordique (Haliotis kamtschatkana), en Colombie-Britannique, sont examinées afin de fournir des orientations possibles en vue du rétablissement des stocks. Après la fermeture de la pêche de cette espèce en Colombie-Britannique, en 1990, aux fins de conservation, les relevés menés de 1978 à 1998, à des sites indicateurs, ont montré une baisse générale et constante des populations d'ormeau dans la partie centrale de la côte de la C.-B. et dans les îles Reine-Charlotte. L'ormeau nordique est vulnérable à une surexploitation à cause de son stade larvaire de courte durée, de sa croissance lente, de sa longévité assez grande et de sa nature sédentaire; son recrutement est faible ou sporadique. Les individus matures, qui ont tendance à s'accumuler en eau peu profonde, sont des proies faciles pour les pêcheurs. Leur valeur marchande élevée, leur disponibilité réduite, et la difficulté de faire respecter l'interdiction de récolte dans une vaste zone côtière pratiquement inhabitée, favorisent la capture illégale de l'ormeau nordique. D'après les échantillons d'ormeaux prélevés illégalement entre 1995 et 1998, les braconniers auraient récolté au hasard des ormeaux matures, principalement de grande taille, sans tenir compte de la limite de taille légale de 100 mm de longueur de la coquille. En l'absence de contrôles sur la taille et le taux de capture, la récolte illégale épuise non seulement des stocks déjà affaiblis, mais elle réduit leur potentiel reproducteur en prélevant des ormeaux matures de grande taille, et entrave les tentatives de rétablissement des populations par la fermeture de la pêche. Si l'on n'arrive pas à réduire le nombre de captures illégales, à protéger le cheptel reproducteur, à maintenir la fermeture de la pêche ou à appliquer d'autres méthodes efficaces de rétablissement, l'abondance de la population d'ormeau risque de demeurer faible ou de continuer à décliner dans de nombreuses régions de la C.-B. Il faut effectuer des recherches dans des zones expérimentales bien surveillées pour déterminer la répartition optimale et la densité des géniteurs et des recrues. Il faut évaluer dans cette province l'efficacité des méthodes de transplantation des reproducteurs et d'ensemencement de juvéniles dans le cadre d'un programme de rétablissement de l'ormeau nordique.

[Traduit par la rédaction]

Introduction

Abalone (archaeogastropod molluscs of the Haliotidae family) have been exploited by humans for thousands of years for the mollusc's large edible foot and colourful bowl-shaped shell. Abalone are found mostly on shallow (intertidal to 30 m) shorelines of temperate and tropical seas (Mottet 1978) and are easily accessible to harvest. Important to many coastal communities and economies, abalone, in recent years, have become vulnerable to overexploitation due

A. Campbell. Shellfish Section, Stock Assessment Division, Fisheries and Oceans Canada, Science Branch, Pacific Biological Station, Nanaimo, BC V9R 5K6, Canada.
Correct citation: Campbell, A. 2000. Review of northern abalone, *Haliotis kamtschatkana*, stock status in British Columbia. *In* Workshop on Rebuilding Abalone Stocks in British Columbia. *Edited by* A. Campbell. Can. Spec. Publ. Fish. Aquat. Sci. 130. pp. 41–50.

Fig. 1. The general locations of *H. kamtschatkana* survey areas discussed in text: (A) south east Queen Charlotte Islands, (B) north central coast, (C) south central coast of British Columbia.

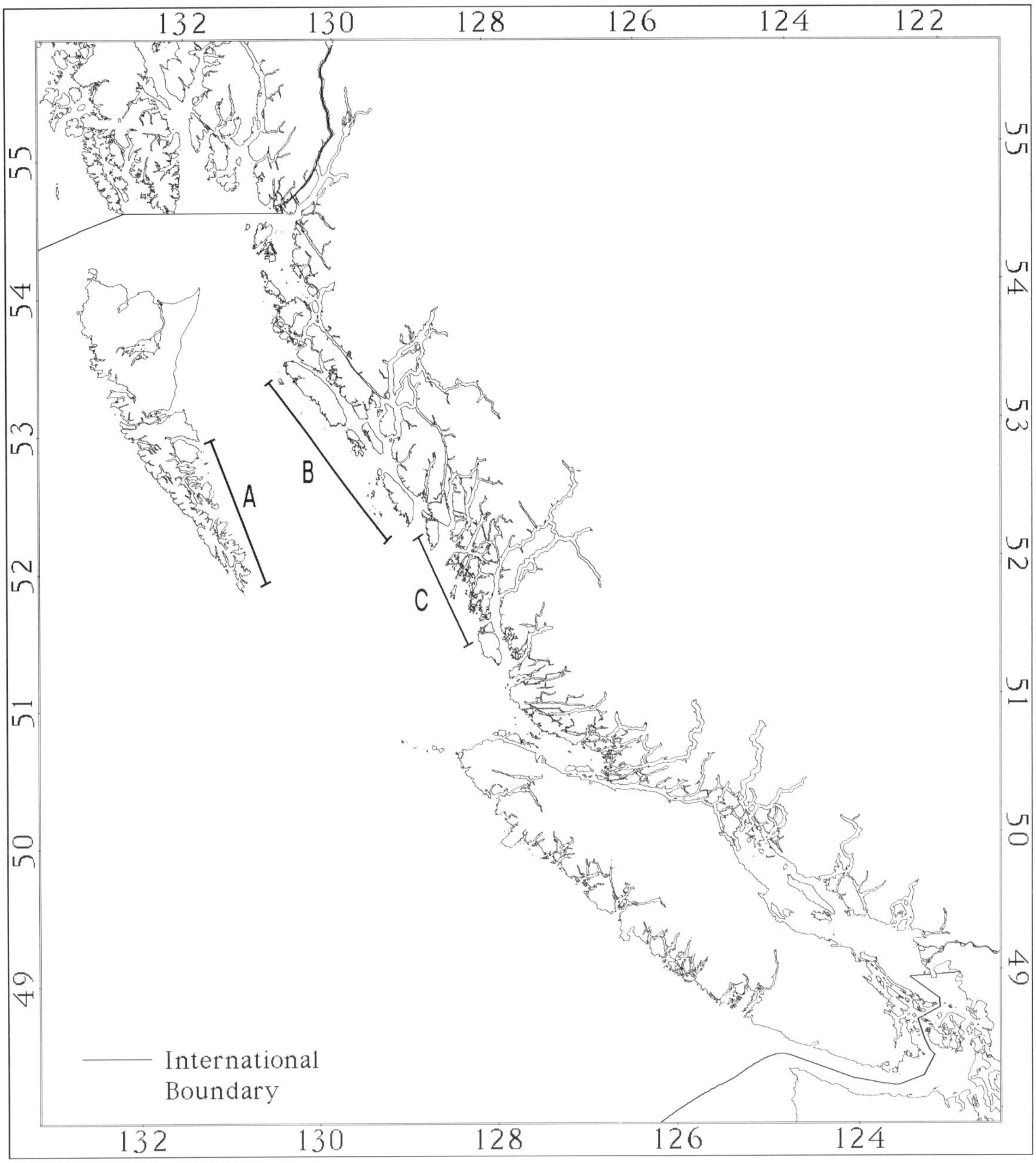

to the low capital cost of fishing and high market value. Most abalone fisheries throughout the world have experienced increased landings and subsequent declines; many resulted in complete closure to fishing and in one case the prospect of species extinction (e.g., Clavier 1992; Davis et al. 1998; Tarr 2000; Karpov et al. 2000).

The northern or pinto abalone, *Haliotis kamtschatkana*, found from Sitka Island, Alaska, to Baja California, generally occurs in patchy distribution on exposed and semi-exposed coasts in British Columbia (B.C.). The biology of *H. kamtschatkana* was reviewed by Sloan and Breen (1988). A traditional food of First Nations, the northern abalone was a target of recreational divers and a modest commercial dive fishery in B.C. during 1950–1990. Surveys at index sites in the south east Queen Charlotte Islands and the north central coast of B.C. (Fig. 1) indicated that the abundance of northern abalone had declined more than 75% during the period of 1978–1984 and remained low until 1998 (Winther et al. 1995; Thomas and Campbell 1996; Campbell et al. 1998, 2000). Faced with stock depletion and the possibility of abalone population collapse, the northern abalone fishery was closed in December 1990 to First Nations, recreational divers, and commercial fishers due to conservation concerns. The purpose of the coastwide closure was to allow depleted populations of northern abalone to rebuild.

The objectives of this paper are to review northern abalone stock status in B.C. before and after the closure and to

Fig. 2. Landed weight, total value, and unit price from the *H. kamtschatkana* fishery in British Columbia.

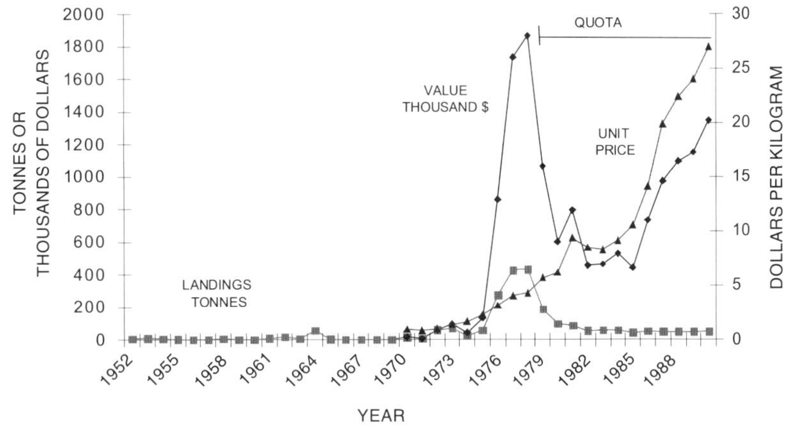

Fig. 3. Landings and catch per unit effort from the *H. kamtschatkana* fishery in British Columbia.

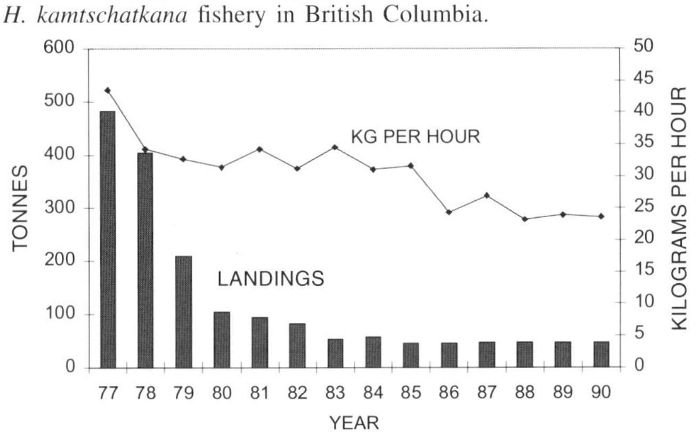

review relevant biological and fishery characteristics to suggest possible directions for stock rehabilitation.

Summary of harvest and management history

Traditional harvesting of abalone by First Nations along the north west coast of North America occurred at low tide where exposed abalone were picked by hand and subtidally where abalone were harvested with a long pole terminating in a hook (Holm 1965). The first nonaboriginal harvest of abalone began in the early 1900s using long poles armed with hooks or by diving (Thompson 1914). By the 1950s, the introduction of SCUBA allowed small recreational and commercial fisheries to begin. Little was known about the extent of First Nations and recreational harvesting, although the level was probably well below the commercial harvest during the last four decades (Sloan and Breen 1988). The commercial northern abalone fishery and management in B.C. was well documented (e.g., Breen 1986; Sloan and Breen 1988; Farlinger 1990; Farlinger and Campbell 1992; Harbo 1997). Commercial landings were low prior to the early 1970s after which total annual landings increased dramatically to a maximum of 481.4 t in 1977 (Fig. 2) due to increases in market demand and technological improvements enabling exploitation in remote areas (Fedorenko and Sprout 1982; Farlinger 1990). Thereafter, with evidence of resource depletion, landings were reduced under a quota management system to 47 t in 1985–1990 (Fig. 2). A minimum size limit (started as 101.6 mm shell length (SL) in 1908 and changed to 100 mm SL during 1981–1990) was enforced for all three fishery categories from 1990 (Farlinger and Campbell 1992). Harvests for all three fishery categories were officially closed in 1990 and have remained closed due to conservation concerns in B.C.

Landings and effort (dive hours) were recorded on fisher logbooks during the last 14 years (1977–1990) of the northern abalone fishery in B.C. Overall catch per unit effort (CPUE) declined by about 46% for the whole fishery during this period (Fig. 3). In contrast, density index surveys indicated >70% decline of northern abalone (see surveys section). Use of CPUE data as indices of stock abundance were considered inaccurate, especially in abalone fisheries (Breen 1992; Prince 1992). Serial depletion of northern abalone in many areas of B.C. could be explained by the fishing strategy of exploiting the easily seen aggregations of emergent or exposed abalone, especially during spawning periods (Breen and Adkins 1980*a*).

Surveys

Fishery-independent surveys have collected a time series of northern abalone densities and size frequencies from the south east Queen Charlotte Islands (QCI) and the Central Coast of B.C. (CC) every 3–5 years during 1978–1998 (Fig. 1) (Adkins and Stefanson 1979; Boutillier et al. 1984, 1985; Campbell et al. 1998, 2000; Carolsfeld et al. 1988; Breen and Adkins 1979, 1980*b*, 1981, 1982; Breen et al. 1978, 1982; Farlinger and Bates 1986; Farlinger et al. 1991; Thomas and Campbell 1996; Thomas et al. 1990; Winther et al. 1995). The general survey method, which consistently used the standard 16 one-m^2 quadrat survey method developed by Breen and Adkins (1979) at indicator index sites, provided a time series of northern abalone abundance indices in the QCI and CC. The general aggregated distribution of northern abalone and the time consuming dive surveying logistics made biomass estimation of this species throughout B.C. impossible on an annual basis. Although there were a few surveys of southern B.C. (Quayle 1971; Breen et al. 1978; Adkins 1996; Wallace 1999), they did not provide as extensive coverage as the surveys in north B.C. Most surveys were conducted in north B.C. where historically

Fig. 4. Mean density of legal (≥100 mm SL) northern abalone from surveys in the Queen Charlotte Islands (closed circles) and Central Coast of B.C. (crosses, north; open circle, south). All sites surveyed are included. Vertical lines are ±2 SE and numbers near dots indicate the number of sample sites.

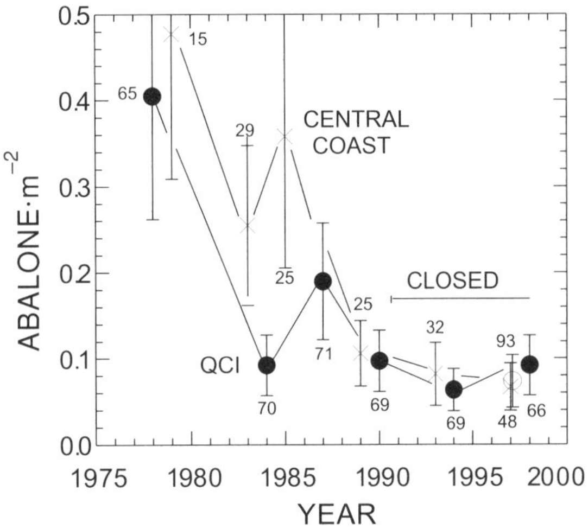

Fig. 5. Size frequency distribution of all sizes of northern abalone surveyed in (A) south east Queen Charlotte Islands, May, 1998, and (B) Central Coast of B.C. May, 1997.

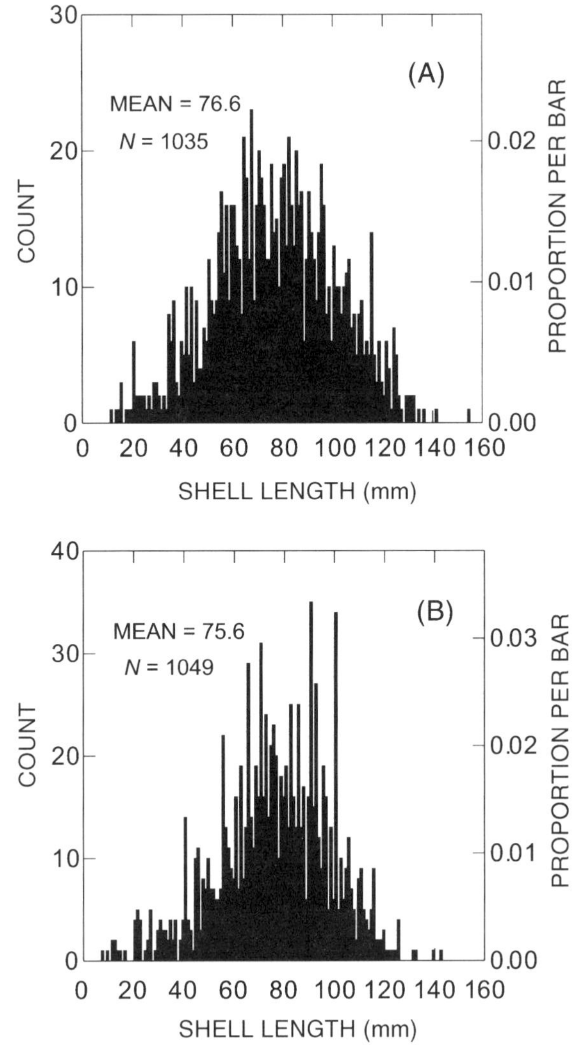

abalone, for the most part, were harvested commercially and were considered most abundant (Sloan and Breen 1988). Consequently, only the results from surveys at index sites in north B.C. were considered and summarized (additional index sites for surveys in south CC during 1997 were included (Campbell et al. 1998)) in the present paper. Only the emergent or exposed abalone (visible on rocks) were used to estimate abalone density in this paper. The majority of northern abalone ≥100 mm SL were found to be exposed in surveys, whereas immature abalone <70 mm SL (Campbell et al. 1992) tended to be more cryptic than adults (Campbell 1996). Immature abalone were not included in density estimates (except as part of the total of all sizes measured) because of the difficulty of finding small, immature abalone.

The mean density of northern abalone generally declined (except for a minor increase during 1985) for the legal (≥100 mm SL) sizes in the surveys of the CC during 1978–1997 (Fig. 4). A similar declining trend in mean density occurred for abalone in the QCI, except for minor increases during the 1987 and 1998 surveys (Fig. 4). The general decrease in mean density was reduced after 1990, suggesting that the gross decline was reduced, but little or no density rebuilding was observed. In recent years, several transect surveys in south CC have confirmed the low densities of northern abalone (Cripps and Campbell 1998; Campbell and Cripps 1998; Lucas et al. 1999).

The size frequencies of all abalone measured from the CC during 1997 and in the QCI during 1998 were similar (Fig. 5). No clear trend in changes of mean size for the total abalone measured has been observed for any of the surveys conducted in the CC and QCI during the 1980s and 1990s (Campbell 1997).

The frequency of samples (a sample normally consisted of 16 one-m^2 quadrats per site) indicated fewer sites with more than 1 abalone·m^{-2} in the QCI between 1978 and 1998 (Fig. 6). The index sites with no legal abalone (≥100 mm SL) generally increased from zero and 25% to 55% and over 65% for the CC and QCI, respectively, in the late 1970s to the late 1990s (Fig. 7). There were fewer sites during 1997–1998 than 1978–1979 with densities of ≥0.25 legal abalone·m^{-2} (i.e., ≥4 northern abalone·16 m^{-2}·site^{-1}) (Fig. 8). The decrease in density and decline in the number of sites with northern abalone suggested serial depletion of large legal-sized abalone with little recruitment in most of the 1990s.

Evidence of poaching

Poaching of the northern abalone has probably occurred during the last two decades and has been difficult to quantify and control. The illegal harvest of this high market value (ranging from Can$45 to $100 kg^{-1}) shellfish has made northern abalone a lucrative product. When the commercial fishery was open in B.C., how much the poaching of northern abalone affected overall exploitation and landings was unknown. Since the fishery closure in 1990, there probably has been a significant illegal harvest of northern abalone. Campbell (1997) estimated that *F* (instantaneous fishing mortality) was at least 0.20 for the south east QCI and from 0.14 to 0.70 in some areas in the north CC during the post-

Fig. 6. Frequency distributions of different densities (number per square metre) of total northern abalone (all sizes) from all sample sites of south east Queen Charlotte Islands during 1978 and 1998 surveys.

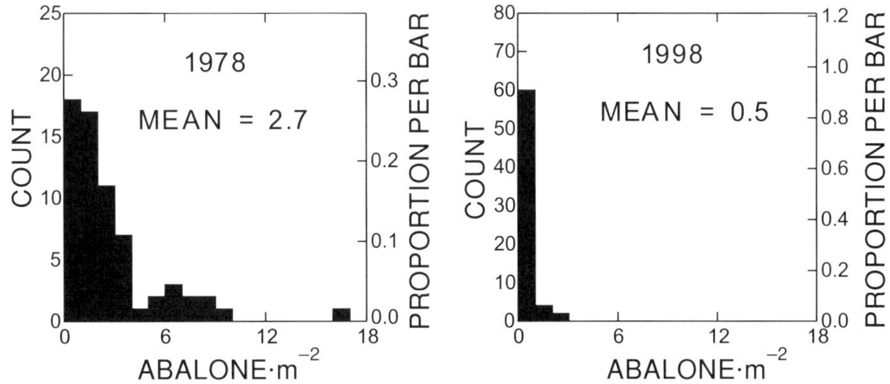

Fig. 7. Percent of index sites that did not have any northern abalone ≥100 mm SL from surveys in south east Queen Charlotte Islands (closed circles) and the central coast of B.C. (crosses, north; asterisk, south). Numbers are sample sizes.

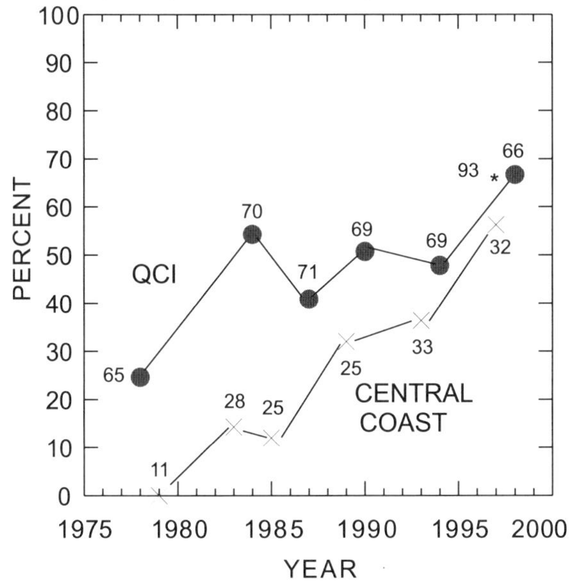

Fig. 8. Percent of index sites that had densities ≥0.25 northern abalone (≥100 mm SL) · m^{-2} from surveys in south east Queen Charlotte Islands (closed circles) and the central coast of B.C. (crosses, north; asterisk, south). Numbers are sample sizes.

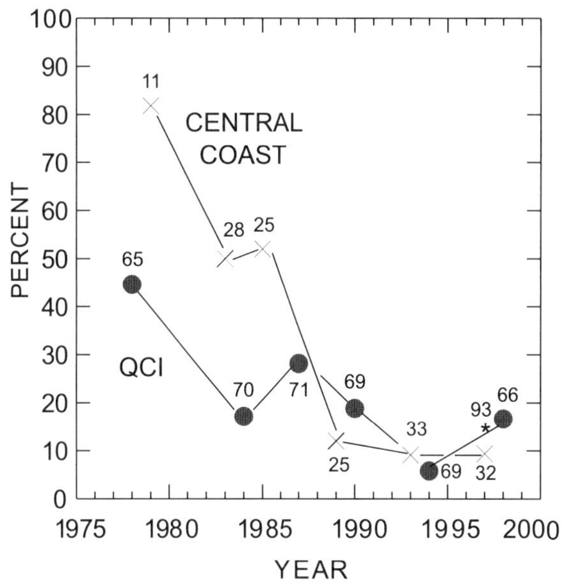

fishery closure (1993–1996) period. Evidence of poaching was provided through seizures of many thousands of poached abalone and numerous court convictions for poaching in B.C. (Jubinville 2000). Samples of abalone seized from poachers in B.C. indicated that poachers harvested mature abalone with no regard for the legal size, with 2–42% being smaller than 100 mm SL (Table 1). The negative influence that the unknown annual illegal landings have had on attempts to rehabilitate northern abalone stocks in B.C. may be substantial. Without size and fishing rate controls, the illegal harvest of mature abalone reduces the reproductive potential of brood stock (Campbell 1997). The combined high unit value and difficulty of enforcing a fishery closure in a large (mostly uninhabited) coastline suggests that poaching pressure is a serious hindrance to attempts to rehabilitate northern abalone populations through fishery closure in B.C.

Spawning and the importance of patch size

Northern abalone spawn synchronously, with groups of males and females in close vicinity to each other in shallow waters broadcasting their gametes into the water column (Breen and Adkins 1980*a*). Cues that cause mass spawning in abalone can include environmental factors such as temperature changes (Sloan and Breen 1988), minor storms, and typhoons (Sasaki and Shepherd 1995). Shepherd (1986) suggested that abalone should be protected from fishing during the spawning season because of the vulnerability of these spawning aggregations to exploitation. Recent studies in abalone (Clavier 1992; McShane 1995*a*, 1995*b*; Shepherd and Partington 1995) and sea urchins (Levitan et al. 1992) have emphasized reduced fertilization success can be caused by dilution of gametes through reduced adult spawner densities (Levitan and Sewell 1998). Because fertilization success

Table 1. Summary statistics of shell length samples recorded from thousands of abalone seized from poachers.

		Shell length (mm)				Percent of total	
General area	Date	Mean	Min.	Max.	Total no.	<100 mm SL	≥120 mm SL
Port Hardy	March 1995	125	96	163	443	2.0	64.6
Port Hardy	April 1995	109	85	150	845	10.2	15.7
Port Hardy	May 1995	108	90	138	177	18.1	13.6
Port Hardy	April 1996	103	77	140	206	41.8	6.8
S. Banks Island	Feb. 1997	109	74	137	1971	16.6	12.5
Tofino	April 1998	113	70	154	398	22.9	40.2
Chrome Island	March 1999	109	79	143	198	24.2	21.2

depends on the aggregation density of abalone, exploitation rates on abalone aggregations may be important in influencing production of larvae. After fertilization, the planktonic phase of northern abalone has a short period (about 5–6 days) (Olsen 1984). Recent studies have suggested that larval exchange in some abalone species may occur in small geographic areas (on a scale of hundreds of metres to several kilometres) (Tegner and Butler 1985*a*; Prince et al. 1987; McShane et al. 1988; McShane 1992, 1995*a*, 1995*b*). Almost nothing is known about the early juvenile stages (1–3 years) of the northern abalone in B.C. and further study is required (Sloan and Breen 1988). Estimates of the age at which northern abalone reach the recruit size of 100 mm SL are between 6 to 8 years in B.C. (Quayle 1971; Breen 1986). Surf abalone at exposed sites in B.C. may never reach the recruit size of 100 mm SL (Sloan and Breen 1988). The frequency and size of patches of northern abalone required to maintain sufficient recruitment for a healthy population requires investigation.

Natural mortality

Natural instantaneous mortality rates (*M*) for adult northern abalone were found to vary from 0.15 to 0.36 between sites (Breen 1980, 1986; Olsen 1984; Emmett and Jamieson 1988). Natural mortality may be caused by environmental factors (episodic high temperature, low salinity, storms), predators (e.g., sea otter, *Enhydra lutris*; crab, *Cancer productus*; sea stars, *Pycnopodia helianthoides*; and octopus, *Octopus dofleini*), starvation (lack of algae), competitors for space and food (e.g., red sea urchins, *Strongylocentrotus franciscanus*), parasites, and disease (Sloan and Breen 1988; Shepherd and Breen 1992; Bower 2000). Population expansion of the sea otter, a major predator of *H. kamtschatkana* in B.C. (Watson 2000), in addition to human exploitation, poses a serious potential threat to future northern abalone populations.

Stock and recruitment

Stock definition of northern abalone has been considered, in the fisheries management context, as an abalone population within arbitrarily chosen geographic or management areas. Consequently, most of the stock assessment surveys of northern abalone in B.C. have been on a broad geographic scale. Evidence from recent studies has suggested that some abalone species may be made up of many populations in which stock recruitment relations may occur in small geographic areas (on a scale of hundreds of metres to several kilometres) based on gene exchange (Brown 1991; Brown and Murray 1992*a*, 1992*b*) and larval exchange (Tegner and Butler 1985*a*, 1985*b*; Prince et al. 1987; McShane et al. 1988). Shepherd and Brown (1993) suggested that an abalone stock be defined as a metapopulation consisting of several local discrete populations that have limited larval interchange. This stock definition allows for the management of local abalone populations that may have variable demographic processes. Fishing zones subdivided into smaller management units have been advocated for Australian abalone fisheries (Prince and Shepherd 1992; Keesing and Baker 1998). Although demographic differences in growth and mortality rates of northern abalone occur between locations (Sloan and Breen 1988), research is required on the genetic variation and larval dispersal mechanisms to assist in defining northern abalone stock sizes in B.C.

Recruitment is defined as the number of abalone reaching the fishable size per year. Little is known about the process of recruitment for northern abalone in B.C. A large number of factors can influence the quantity and location of abalone recruitment (e.g., local hydrodynamics and storms may entrain or widely disperse larvae during their short larval period prior to settlement and local mortality and growth rates may differentially act on juvenile and adult abalone survival) (McShane 1992, 1995*a*, 1995*b*; Sasaki and Shepherd 1995; Shepherd 1998).

Few studies have shown strong stock recruitment relationships and the requirement of maintaining high adult abalone densities to ensure sufficient recruitment. Shepherd and Partington (1995), using a Ricker stock recruitment curve, showed that there was a critical stock density threshold (0.15 m^{-2}) for the *H. laevigata* in Waterloo Bay, South Australia, below which the risk of recruitment failure was high. Shepherd and Brown (1993) found that a minimum viable population of more than 800 individuals of *H. laevigata* was required; anything less at West Island caused recruitment failure. Shepherd and Baker (1998) suggested that recruitment to an abalone fishery could be relatively weaker and more variable in small rather than in large abalone populations, in which case, small populations would need to conserve relatively more egg production to prevent depletion. These studies supported the influence of the Allee effect or depensation (Allee et al. 1949) in which low abalone densities and aggregations reduced reproductive success due to low fertilization of gametes.

Using a stock reduction analysis, Breen (1986) concluded that recruitment failure in *H. kamtschatkana* occurred independently of fishing pressure in B.C. Breen (1992) suggested that recruitment was lower than replacement levels during 1975–1983 and that the stock would have declined by about a third for this period even without fishing. Breen (1986) and Sloan and Breen (1988) acknowledged that the causes of poor recruitment in northern abalone were unknown. Possible changes in environmental factors during 1970–1998 in relation to *H. kamtschatkana* recruitment in northern B.C. have yet to be examined.

Implications for rehabilitation

Until northern abalone stocks rebuild in B.C., the fishery should remain closed in an attempt to eliminate fishing mortality. Poaching should continue to be discouraged through enforcement efforts and public education. Developing genetic studies (Withler 2000) to help enforcement and courts identify location and species of abalone found in illegal seizures and in restaurants would also discourage illegal abalone trade. The designation by COSEWIC (Committee on the Status of Endangered Wildlife in Canada) to classify *H. kamtschatkana* as threatened (a species likely to become endangered if limiting factors are not reversed), on April 23, 1999, should help emphasize the serious problems of this species. The combination of environmental factors, predation by the expanding sea otter populations, and exploitation by humans has potentially serious future implications for the placement of northern abalone on the endangered species list in B.C.

Understanding the spatial scale of stock units or metapopulations of northern abalone to be exploited is important for developing appropriate rehabilitation and management strategies. Prince at al. (1998) suggested that a small spatial scale of monitoring and management was appropriate because of the high biological variability between abalone populations and susceptibility to sequential over-exploitation of abalone substocks. Although broad scale monitoring is useful to determine overall health of the stock and to maintain a time series of data, additional, more intensive local fine-scale surveys should be conducted to determine variability within and between areas of biologically different northern abalone populations. Appropriate survey methodology will be important to any successful application of small-scale experimental management activities. A genetic evaluation of *H. kamtschatkana* is required to help determine genetic variability and extent of genetic discreteness or spatial scale of northern abalone stocks (Withler 2000).

Well-monitored, small-scale experimental management studies in several locations could provide information on resource and rebuilding characteristics and the effectiveness of manipulation of brood stock (Tegner 1992, 2000) and juvenile seed restocking methodologies (McCormick 2000) to improve recruitment abundance. Local community participation, especially in co-operation with First Nations, should be encouraged to monitor and protect local abalone populations through experimental management areas. Prince et al. (1998) advocated Territorial User Rights Fisheries (TURF) for protecting abalone populations. A form of TURF management was recently implemented to reduce poaching in the heavily exploited gastropod Concholepas fishery in Chile (Castilla et al. 1998). The creation of more Marine Protected Areas or refuges has also been suggested (Shepherd and Brown 1993; Tegner 1993; Wallace 1999; Davis 2000; Jamieson 2000). Refuges designed to effectively assist northern abalone must consider the population biology, the local oceanographic regime, distances or spatial scales of brood stock sources, and subsequent recruitment and feasibility of monitoring populations and enforcement (Tegner 1993). In some areas, protecting the transplanted northern abalone adults and (or) juvenile seed not only from poaching, but also from sea otter predation, will be important.

Acknowledgments

I thank R.I. Perry and N.A. Sloan for reviewing this manuscript.

References

Adkins, B.E. 1996. Abalone surveys in south coast areas during 1982, 1985, and 1986. Can. Tech. Rep. Fish. Aquat. Sci. 2089. pp. 72–96.

Adkins, B.E., and Stefanson, A.P. 1979. North coast abalone survey in harvested areas, November 1978. Fish. Mar. Serv. Manuscr. Rep. 1500. 15 p.

Allee, W.C, Emerson, A.E., Park, O., Park T., and Schmidt, K.P. 1949. Principles of animal ecology. Saunders, Philadelphia. 837 p.

Boutillier, J.A., Carolsfeld, W., Breen P.A., and Bates, K. 1984. Abalone survey in the Estevan group and Aristazabal Island May, 1983. Can. Manuscr. Rep. Fish. Aquat. Sci. 1747. 60 p.

Boutillier, J.A., Carolsfeld, W., Breen, P.A., Farlinger, S., and Bates, K. 1985. Abalone resurvey in the southeast Queen Charlotte Islands, July 1984. Can. Manuscr. Rep. Fish. Aquat. Sci. 1818. 87 p.

Bower, S.M. 2000. Infectious diseases of abalone (*Haliotis* spp.) and risks associated with transplantation. *In* Workshop on Rebuilding Abalone Stocks in British Columbia. *Edited by* A. Campbell. Can. Spec. Publ. Fish. Aquat. Sci. 130. pp. 111–122.

Breen, P.A. 1980. Measuring fishing intensity and annual production in the abalone fishery of British Columbia. Can. Tech. Rep. Fish. Aquat. Sci. 947. 49 p.

Breen, P.A. 1986. Management of the British Columbia fishery for northern abalone (*Haliotis kamtschatkana*). *In* North Pacific Workshop on Stock Assessment and Management of Invertebrates. *Edited by* G.S. Jamieson and N. Bourne. Can. Spec. Publ. Fish. Aquat. Sci. 92. pp. 300–312.

Breen, P.A. 1992. A review of models used for stock assessment in abalone fisheries. *In* Abalone of the world: biology, fisheries and culture. *Edited by* S.A. Shepherd, M.J. Tegner, and S.A. Guzmán del Próo. Fishing News Books, Blackwell Science Ltd., Oxford. U.K. pp. 253–257.

Breen, P.A., and Adkins, B.E. 1979. A survey of abalone populations on the east coast of the Queen Charlotte Islands, August 1978. Fish. Mar. Serv. Manuscr. Rep. 1490. 125 p.

Breen, P.A., and Adkins, B.E. 1980*a*. Spawning in a British Columbia population of northern abalone, *Haliotis kamtschatkana*. Veliger, **23**: 177–179.

Breen, P.A., and Adkins, B.E. 1980*b*. Observations of abalone populations in Emily Carr Inlet and Lobtbiniere Bay, April 1980. Can. Manuscr. Rep. Fish. Aquat. Sci. 1576. 17 p.

Breen, P.A., and Adkins, B.E. 1981. Abalone surveys and tagging conducted during 1979. Can. Manuscr. Rep. Fish. Aquat. Sci. 1623. 88 p.

Breen, P.A., and Adkins, B.E. 1982. Observations of abalone populations on the north coast of British Columbia, July 1980. Can. Manuscr. Rep. Fish. Aquat. Sci. 1633. 55 p.

Breen, P.A., Adkins, B.E., and Heritage, G.D. 1978. Observations of abalone and subtidal communities made during a survey of the Queen Charlotte Strait and upper Johnstone Strait areas July 13–20, 1977. Fish. Mar. Serv. Tech. Rep. 789. 91 p.

Breen, P.A., Stefanson, A.P., and Adkins, B.E. 1978. North coast abalone surveys in harvested areas, spring 1978. Can. Manuscr. Rep. Fish. Aquat. Sci. 1480. 61 p.

Breen, P.A., and Adkins, B.E., and Sprout, P. 1982. Abalone populations on the west coast of Banks Island, June 1980. Can. Manuscr. Rep. Fish. Aquat. Sci. 1640. 42 p.

Brown, L.D. 1991. Genetic variation and population structure in the blacklip abalone, *Haliotis rubra*. Aust. J. Mar. Freshwater Res. **42**: 77–90.

Brown, L.D., and Murray, N.D. 1992*a*. Population genetics, gene flow, and stock structure in *Haliotis rubra* and *Haliotis laevigata*. *In* Abalone of the world: biology, fisheries and culture. *Edited by* S.A. Shepherd, M.J. Tegner, and S.A. Guzmán del Próo. Fishing News Books, Blackwell Science Ltd., Oxford. U.K. pp. 24–33.

Brown, L.D., and Murray, N.D. 1992*b*. Genetic relationships within the genus *Haliotis*. *In* Abalone of the world: biology, fisheries and culture. *Edited by* S.A. Shepherd, M.J. Tegner, and S.A. Guzmán del Próo. Fishing News Books, Blackwell Science Ltd., Oxford. U.K. pp. 19–23.

Campbell, A. 1996. An evaluation of abalone surveys off southeast Queen Charlotte Island. Can. Tech. Rep. Fish. Aquat. Sci. 2089. pp. 111–131.

Campbell, A. 1997. Possible criteria for reopening the northern abalone (*Haliotis kamtschatkana*) fishery in British Columbia. Can. Stock Assessment Secretariat Res. Doc. 97/64. Fisheries and Oceans Canada, 200 Kent Street, Ottawa, Ont. K1A 0E6, Canada. 47 p.

Campbell, A., and Cripps, K. 1998. Survey of abalone populations at Stryker Island, Tribal Group and Simonds Group, central coast of British Columbia, May, 1997. Can. Manuscr. Rep. Fish. Aquat. Sci. 245. 21 p.

Campbell, A., Manley, I., and Carolsfeld, W. 1992. Size at maturity and fecundity of the abalone, *Haliotis kamtschatkana*, in British Columbia. Can. Manuscr. Rep. Fish. Aquat. Sci. 2169. pp. 47–65.

Campbell, A., Winther, I., Adkins, B., Brouwer, D., and Miller, D. 1998. Survey of the Northern Abalone (*Haliotis kamtschatkana*) in the central coast of British Columbia, May 1997. Can. Stock Assessment Secretariat Res. Doc. 98/89. Fisheries and Oceans Canada, 200 Kent Street, Ottawa, Ont. K1A 0E6, Canada. 28 p.

Campbell, A., Brouwer, D., Rogers, J., and Miller, D. 2000. Abalone resurvey in the south east Queen Charlotte Islands, 1998. Can. Manuscr. Rep. Fish. Aquat. Sci. 2528. 30 p.

Carolsfeld, W., Farlinger, S., Kingzett, B.C., Sloan, N.A., and Thomas, G. 1988. Abalone resurvey in the southeast Queen Charlotte Islands, June 1987. Can. Manuscr. Rep. Fish. Aquat. Sci. 1966. 90 p.

Castilla, J.C., Manriquez, P., Alvarado, J., Rosson, A., Pino, C., Espoz, C., Soto, S., Oliva, D., and Defeo, O. 1998. Artisanal "caletas" as units of production and co-managers of benthic invertebrates in Chile. *In* Proceedings of the North Pacific Symposium on Invertebrate Stock Assessment and Management. *Edited by* G.S. Jamieson and A. Campbell. Can. Spec. Publ. Fish. Aquat. Sci. 125. pp. 407–413.

Clavier, J. 1992. The ormer (*Haliotis tuberculata*) fishery of France and the Channel Islands. *In* Abalone of the world: biology, fisheries and culture. *Edited by* S.A. Shepherd, M.J. Tegner, and S.A. Guzmán del Próo. Fishing News Books, Blackwell Science Ltd., Oxford. U.K. pp. 454–457.

Cripps, K., and Campbell, A. 1998. Survey of abalone populations at Dallain Point and Higgins Pass, central coast of British Columbia, 1995–96. Can. Manuscr. Rep. Fish. Aquat. Sci. 2445. 31 p.

Davis, E.G. 2000. Refugia-based strategies to restore and sustain abalone (*Haliotis* spp.) populations in Southern California. *In* Workshop on Rebuilding Abalone Stocks in British Columbia. *Edited by* A. Campbell. Can. Spec. Publ. Fish. Aquat. Sci. 130. pp. 133–138.

Davis, E.G., Haaker, P.L., and Richards, D.V. 1998. The perilous condition of the white abalone *Haliotis sorenseni*, Bartsch, 1940. J. Shellfish Res. **17**: 871–875.

Emmett, B., and Jamieson, G.S. 1988. An experimental transplant of northern abalone, *Haliotis kamtschatkana*, in Barkley Sound, British Columbia. Fish. Bull. **87**: 95–104.

Farlinger, S. 1990. Review of the biological basis for management of the British Columbia abalone fishery. Can. Manuscr. Rep. Fish. Aquat. Sci. 2099. pp. 41–65.

Farlinger, S., and Bates, K.T. 1986. Abalone survey in the Estevan group and Aristazabal Island, June 1985. Can. Manuscr. Rep. Fish. Aquat. Sci. 1896. 45 p.

Farlinger, S., and Campbell, A. 1992. Fisheries management and biology of northern abalone, *Haliotis kamtschatkana*, in the northeast Pacific. *In* Abalone of the world: biology, fisheries and culture. *Edited by* S.A. Shepherd, M.J. Tegner, and S.A. Guzmán del Próo. Fishing News Books, Blackwell Science Ltd., Oxford. U.K. pp. 395–406.

Farlinger, S., Thomas, G.A., Winther, I., and Carolsfeld, W. 1991. Abalone resurvey in the Estevan group and Aristazabal Island, June 1989. Can. Manuscr. Rep. Fish. Aquat. Sci. 2104. 39 p.

Fedorenko, A.Y., and Sprout, P.E. 1982. Abalone biology, fishery regulations, commercial catch (1952–1980), and current status of resources in British Columbia. Can. Manuscr. Rep. Fish. Aquat. Sci. 1658. 74 p.

Harbo, R. 1997. Abalone dive fishery (closed). Can. Manuscr. Rep. Fish. Aquat. Sci. 2369. pp. 86–92.

Holm, W. 1965. Northwest coast Indian art. An analysis of form. Thomas Burke Memorial Washington State Museum Monograph 1. Univ. Washington Press, Seattle, Washington. 115 p.

Jamieson, G.S. 2000. Marine protected areas and their relevance to abalone (*Haliotis kamtschatkana*) conservation in British Columbia. *In* Workshop on Rebuilding Abalone Stocks in British Columbia. *Edited by* A. Campbell. Can. Spec. Publ. Fish. Aquat. Sci. 130. pp. 139–147.

Jubinville, B. 2000. Enforcing the fishery closure for northern (pinto) abalone (*Haliotis kamtschatkana*) in British Columbia [Abstract]. *In* Workshop on Rebuilding Abalone Stocks in British Columbia. *Edited by* A. Campbell. Can. Spec. Publ. Fish. Aquat. Sci. 130. pp. 52.

Karpov, K.A., Haaker, P., Taniguchi, I., and Rogers-Bennett, L. 2000. Serial depletion and the collapse of the California abalone (*Haliotis* spp.) fishery. *In* Workshop on Rebuilding Abalone Stocks in British Columbia. *Edited by* A. Campbell. Can. Spec. Publ. Fish. Aquat. Sci. 130. pp. 11–24.

Keesing, J.K., and Baker, J.L. 1998. The benefits of catch and effort data at a fine spatial scale in the South Australian abalone (*Haliotis laevigata* and *H. rubra*) fishery. *In* Proceedings of the North Pacific Symposium on Invertebrate Stock Assessment and Management. *Edited by* G.S. Jamieson and A. Campbell. Can. Spec. Publ. Fish. Aquat. Sci. 125. pp. 179–186.

Levitan, D.R., and Sewell, M.A. 1998. Fertilization success in free-spawning marine invertebrates: review of the evidence and fisheries implications. *In* Proceedings of the North Pacific Symposium on Invertebrate Stock Assessment and Management. *Edited by* G.S. Jamieson and A. Campbell. Can. Spec. Publ. Fish. Aquat. Sci. 125. pp. 159–164.

Levitan, D.R., Sewell, M.A., and Chia, F.S. 1992. How distribution and abundance influences fertilization success in the sea urchin *Strongylocentrotus franciscanus*. Ecology, **73**: 248–254.

Lucas, B.G., Campbell, A., and Cripps, K. 1999. Resurvey of abalone populations at Tribal Group, Simonds Group and Stryker Island, central coast of British Columbia, 1998. Can. Manuscr. Rep. Fish. Aquat. Sci. 2487. iv + 18 p.

McCormick, T.B. 2000. Abalone (*Haliotis* spp.) aquaculture: farming status and stock enhancement tool. *In* Workshop on Rebuilding Abalone Stocks in British Columbia. *Edited by* A. Campbell. Can. Spec. Publ. Fish. Aquat. Sci. 130. pp. 55–60.

McShane, P.E. 1992. Early life history of abalone: a review. *In* Abalone of the world: biology, fisheries and culture. *Edited by* S.A. Shepherd, M.J. Tegner, and S.A. Guzmán del Próo. Fishing News Books, Blackwell Science Ltd., Oxford. U.K. pp. 120–138.

McShane, P.E. 1995*a*. Recruitment variation in abalone: its importance to fisheries management. Mar. Freshw. Res. **46**: 555–570.

McShane, P.E. 1995*b*. Estimating the abundance of abalone: the importance of patch size. Mar. Freshw. Res. **46**: 657–662.

McShane, P.E., Black, K.P., and Smith, M.G. 1988. Recruitment processes in *Haliotis rubra* (Mollusca: Gastropoda) and regional hydrodynamics in southeast Australia imply localised dispersal of larvae. J. Exp. Mar. Biol. Ecol. **124**: 175–203.

Mottet, M.G. 1978. A review of the fishery biology of abalones. Wash. Dep. Fish. Tech. Rep. 37. 81 p.

Olsen, S. 1984. Shellfish enhancement report. NOAA, NMFS, Final report for project 1-144-R. Wash. State Dep. Fish., Seattle, Wash. 85 p.

Prince, J.D. 1992. Using a spatial model to explore the dynamics of an exploited stock of the abalone *Haliotis rubra*. *In* Abalone of the world: biology, fisheries and culture. *Edited by* S.A. Shepherd, M.J. Tegner, and S.A. Guzmán del Próo. Fishing News Books, Blackwell Science Ltd., Oxford. U.K. pp. 305–317.

Prince, J.D., Sellers, T.L., Ford, W.B., and Talbot, S.R. 1987. Experimental evidence for limited dispersal of haliotid larvae (genus *Haliotis*; Molluscs: Gastropoda). J. Exp. Mar. Biol. Ecol. **106**: 243–263.

Prince, J.D., and Shepherd, S.A. 1992. Australian abalone fisheries and their management. *In* Abalone of the world: biology, fisheries and culture. *Edited by* S.A. Shepherd, M.J. Tegner, and S.A. Guzmán del Próo. Fishing News Books, Blackwell Science Ltd., Oxford. U.K. pp. 407–426.

Prince, J.D., Walters, C., Ruiz-Avila, R., and Sluczanowski, P. 1998. Territorial user's rights and the Australian abalone (*Haliotis* sp.) fishery. *In* Proceedings of the North Pacific Symposium on Invertebrate Stock Assessment and Management. *Edited by* G.S. Jamieson and A. Campbell. Can. Spec. Publ. Fish. Aquat. Sci. 125. pp. 367–375.

Quayle, D.B. 1971. Growth, morphometry and breeding in the British Columbia abalone (*Haliotis kamtschatkana* Jonas). Fish. Res. Board Can. Tech. Rep. 279. 84 p.

Sasaki, R., and Shepherd, S.A. 1995. Larval dispersal and recruitment of *Haliotis discus hannai* and *Tegula* spp. on Miyagi Coasts, Japan. Mar. Freshwater Res. **46**: 519–529.

Shepherd, S.A. 1986. Studies on southern Australian abalone (*Genus Haliotis*). VII. Aggregative behaviour of *H. laevigata* in relation to spawning. Mar. Biol. **90**: 231–236.

Shepherd, S.A. 1998. Studies on southern Australian abalone (Genus *Haliotis*). XIX: long-term juvenile mortality dynamics. J. Shellfish Res. **17**: 813–825.

Shepherd, S.A., and Baker, J.L.1998. Biological reference points in an abalone fishery. *In* Proceedings of the North Pacific Symposium on Invertebrate Stock Assessment and Management. *Edited by* G.S. Jamieson and A. Campbell. Can. Spec. Publ. Fish. Aquat. Sci. 125. pp. 235–245.

Shepherd, S.A., and Breen, P.A. 1992. Mortality in abalone: its estimation, variability and causes. *In* Abalone of the world: biology, fisheries and culture. *Edited by* S.A. Shepherd, M.J. Tegner, and S.A. Guzmán del Próo. Fishing News Books, Blackwell Sci. Pub. Ltd., Oxford, U.K. pp. 276–304.

Shepherd, S.A., and Brown, L.D. 1993. What is an abalone stock: implications for the role of refugia in conservation. Can. J. Fish. Aquat. Sci. **50**: 2001–2009.

Shepherd, S.A., and Partington, D. 1995. Studies on southern Australian abalone (Genus *Haliotis*). XVI. Recruitment, habitat and stock relations. Mar Freshw. Res. **46**: 669–680.

Sloan, N.A., and Breen, P.A. 1988. Northern abalone, *Haliotis kamtschatkana*, in British Columbia: fisheries and synopsis of life history information. Can Spec. Publ. Fish. Aquat. Sci. 103. 46 p.

Tarr, R.J.Q. 2000. The South African abalone (*Haliotis midae*) fishery: a decade of challenges and change. *In* Workshop on Rebuilding Abalone Stocks in British Columbia. *Edited by* A. Campbell. Can. Spec. Publ. Fish. Aquat. Sci. 130. pp. 32–40.

Tegner, M.J. 1992. Brood-stock transplants as an approach to abalone stock enhancement. *In* Abalone of the world: biology, fisheries and culture. *Edited by* S.A. Shepherd, M.J. Tegner, and S.A. Guzmán del Próo. Fishing News Books, Blackwell Science Ltd., Oxford. U.K. pp. 461–473.

Tegner, M.J. 1993. Southern California abalones: can stocks be rebuilt using marine harvest refugia. Can. J. Fish. Aquat. Sci. **50**: 2010–2018.

Tegner, M.J. 2000. Abalone (*Haliotis* spp.) enhancement in California: what we've learned and where we go from here. *In* Workshop on Rebuilding Abalone Stocks in British Columbia. *Edited by* A. Campbell. Can. Spec. Publ. Fish. Aquat. Sci. 130. pp. 61–71.

Tegner, M.J., and Butler, R.A. 1985*a*. Drift-tube study of the dispersal potential of green abalone (*Haliotis fulgens*) larvae in the southern California Bight: implications for the recovery of depleted populations. Mar. Ecol. Prog. Ser. **26**: 73–84.

Tegner, M.J., and Butler, R.A. 1985*b*. The survival and mortality of seeded and native red abalone, *Haliotis rufescens*, on the Palos Verdes Peninsula. Calif. Fish Game, **71**: 150–163.

Thomas, G., and. Campbell, A. 1996. Abalone resurvey in Aristazabal Island, the Estevan group and Banks Island, June 1993. Can. Tech. Rep. Fish. Aquat. Sci. 2089. pp. 97–109.

Thomas, G., Farlinger, S., and Carolsfeld, W. 1990. Abalone resurvey in the southeast Queen Charlotte Islands in 1990. Can. Manuscr. Rep. Fish. Aquat. Sci. 2099. pp. 66–82.

Thompson, W.F. 1914. Report on the abalones of British Columbia (*Haliotis gigantae*, Chemnitz) [= *Haliotis kamtschatkana* Jonas]. Rep. B.C. Commissioner Fish., Victoria, B.C. (1913). pp. 126–130.

Wallace, S.S. 1999. Evaluating three forms of marine reserves on Northern Abalone (*Haliotis kamtschatkana*) populations in British Columbia, Canada. Conserv. Biol. **13**: 882–887.

Watson, J.C. 2000. The effects of sea otters (*Enhydra lutris*) on abalone (*Haliotis* spp.) populations. *In* Workshop on Rebuilding Abalone Stocks in British Columbia. *Edited by* A. Campbell. Can. Spec. Publ. Fish. Aquat. Sci. 130. pp. 123–132.

Winther, I., Campbell, A., Thomas, G.A., Adkins, B.E., and Clapp, B.G. 1995. Abalone resurvey in the southeast Queen Charlotte Islands, 1994. Can Manuscr. Rep. Fish. Aquat. Sci. 2273. 46 p.

Withler, R.E. 2000. Genetic tools for identification and conservation of exploited abalone (*Haliotis* spp.) species. *In* Workshop on Rebuilding Abalone Stocks in British Columbia. *Edited by* A. Campbell. Can. Spec. Publ. Fish. Aquat. Sci. 130. pp. 101–110.

The British Columbia fishery for northern abalone, *Haliotis kamtschatkana*: management from inception to closure and beyond [Abstract]

Bruce E. Adkins

Abstract: Northern abalone (*Haliotis kamtschatkana*) have been harvested by First Nations in British Columbia (B.C.) for food, social, and ceremonial purposes since prerecorded time. While recreational and commercial abalone fisheries have occurred in B.C. since as early as 1900, these were small and largely unregulated until 1972 when a directed commercial dive fishery began.

The commercial abalone fishery, developed during the 1970s, typifies an emerging fishery. Of the total commercial abalone landings (1766 t) recorded on harvest log books during the 1977–1990 period, 42.6% came from the Queen Charlotte Islands (Pacific Fishery Management (PFM) areas 1 and 2), 46.4% from north coast (PFM areas 3–10), and 11.0% came from the south coast (PFM areas 11–29) of B.C. Total annual landings increased rapidly to a peak of 475 t in 1977, but then declined quickly to <100 t by 1981, and remained below the 47-t quota during 1985–1990. Despite the increasingly restrictive management measures (including limitation of vessels to 26, minimum size limit of 100 mm shell length, overall quotas reduced to 47 t and divided into individual quotas, seasonal fishing restrictions, and numerous local permanent area closures to the commercial fishery), in an attempt to establish sustainable harvests during this period, abalone stocks, measured at key index sites, continued to decline. Consequently, in 1990, conservation concerns led Fisheries and Oceans Canada to close the abalone fishery to harvesting by all user groups.

Aboriginal, commercial, and recreational fisheries have remained closed because assessment surveys since 1990 have shown further declines in abalone abundance at key index sites. Although this may be biologically related, illegal fishing is considered to be a contributing factor in northern abalone population declines. This case study of an emerging fishery highlights the need for a highly precautionary approach in any future abalone exploitation and the immediate requirements for a proactive rehabilitation plan as natural northern abalone stocks may take decades to rebuild in B.C.

B.E. Adkins. Fisheries and Oceans Canada, Pacific Biological Station, Nanaimo, BC V9R 5K6, Canada.
Correct citation: Adkins, B.E. 2000. The British Columbia fishery for northern abalone, *Haliotis kamtschatkana*: management from inception to closure and beyond [Abstract]. *In* Workshop on Rebuilding Abalone Stocks in British Columbia. *Edited by* A. Campbell. Can. Spec. Publ. Fish. Aquat. Sci. 130. pp. 51.

Enforcing the fishery closure for northern (pinto) abalone (*Haliotis kamtschatkana*) in British Columbia [Abstract]

Bryan Jubinville

Abstract: In December 1990, the Department of Fisheries and Oceans (DFO) closed the northern (pinto) abalone (*Haliotis kamtschatkana*) fishery to all user groups (commercial, sports, and First Nations people) for conservation reasons and to allow abalone stocks to recover to appropriate levels. At the time of the British Columbia (B.C.) closure, harvesting of pinto abalone was prohibited in the states of Washington and Oregon and Alaska was contemplating a closure of its fishery because of dwindling stocks. Currently, harvesting of pinto abalone is illegal throughout coastal B.C. and the states of Washington, Oregon, and Alaska.

Known quantities of poached abalone have ranged from <45 to 4500 kg. Large quantities of northern abalone shells have been found on some isolated island beaches and underwater. In these locations, the meats were removed and prepared for markets with no regard to abalone size. Illegally harvested abalone have been found hidden in shipments of legally caught fish and shellfish species. The price paid for black market abalone ranged from \$55 to \$100·kg^{-1} (fresh) and up to \$2600·kg^{-1} (dried meat).

The enforcement arm of the DFO, Conservation and Protection, is tasked with developing plans and instituting enforcement actions in order to curb the illegal trade of northern abalone in B.C. Enforcement actions undertaken in the past several years included regular and routine enforcement patrols by uniformed officers in different locations where illegally harvested abalone may have been harvested, transported, hidden, bartered, and (or) sold. These included plainclothes and covert enforcement actions involving the observation of suspects in their endeavors. In our attempts to address the illegal trade, we undertook several undercover operations of varying lengths of time that met with a variety of successes. Recently, a joint-forces operation was successful in apprehending several people dealing in all types of wildlife animals and (or) parts, alive or dead, including abalone. Another seizure of 4000 abalone resulted in the conviction and sentencing of several people to over \$170 000 worth in fines and equipment forfeitures.

In addition to the active enforcement work, we have undertaken to develop posters written in several different languages hoping to educate people who buy the illegal product. Also, a short film clip was developed for news and television broadcast time to educate the public and request that persons report any illegal activity that may involve the illegal harvest and (or) sale of abalone.

Although our enforcement and education efforts continue, a major part of the problem is that illegal removal of abalone is harming the fragile abalone resource.

Depending on funding, future efforts should include increased surveillance, use of dogs to detect abalone, improved public awareness, and the use of DNA identification of seized abalone meat to species and hopefully location of original removal from B.C. waters.

B. Jubinville. Fisheries and Oceans Canada, 60 Front Street, Nanaimo, BC V9R 5H7, Canada.
Correct citation: Jubinville, B. 2000. Enforcing the fishery closure for northern (pinto) abalone (*Haliotis kamtschatkana*) in British Columbia [Abstract]. *In* Workshop on Rebuilding Abalone Stocks in British Columbia. *Edited by* A. Campbell. Can. Spec. Publ. Fish. Aquat. Sci. 130. pp. 52.

Experiences in Rebuilding Abalone Stocks

Abalone (*Haliotis* spp.) aquaculture: present status and a stock enhancement tool

Thomas B. McCormick

Abstract: With the decline of wild stocks there has been increasing interest in the development of abalone (*Haliotis* spp.) aquaculture as a tool for commercial production and stock enhancement. China and Taiwan lead production with 3500 and 2500 t, respectively, of farmed abalone. Production for the rest of the world is less than 260 t. The southern hemisphere shows the greatest promise for growth with an expected production capacity in excess of 2000 t by the year 2005. In California, the white abalone (*H. sorenseni*), which was the target of commercial and sport fisheries, may be threatened with extinction. Farming methods used for other abalone in California will be modified to produce white abalone for resource enhancement. An *ad hoc* group of federal and state agencies and private organizations is now involved in the restoration process. Costs associated with hatchery production of white abalone are summarized.

Résumé : Depuis l'effondrement des stocks sauvages d'ormeaux (*Haliotis* spp.), l'intérêt pour l'aquaculture de ce gastéropode comme moyen de production commerciale et de mise en valeur ne cesse d'augmenter. La Chine et Taiwan mènent la production avec 3500 et 2500 t d'ormeaux d'élevage, respectivement. La production dans le reste du monde se chiffre à moins de 260 t. L'hémisphère Sud présente la meilleure perspective de croissance; on prévoit que sa capacité de production atteindra plus de 2 000 t d'ici l'an 2005. En Californie, l'ormeau blanc (*H. sorenseni*), victime de la pêche commerciale et sportive, est peut-être menacé d'extinction. Les méthodes d'élevage utilisées pour d'autres espèces d'ormeaux en Californie seront modifiées pour produire des ormeaux blancs afin de mettre en valeur la ressource. Un groupe spécial, formé d'organismes fédéraux, des États et du secteur privé participe maintenant au processus de restauration. La présente étude résume les coûts associés à la production d'ormeaux blancs en écloserie.

[Traduit par la Rédaction]

Introduction

World fisheries for abalone reached maximum landings of 28 000 t in the late 1960s and declined to 14 000 t in the early 1990s and to even lower levels at the end of the decade. The relatively immobile stocks of these marine snails are easily harvested by divers and susceptible to overfishing (Sloan and Breen 1988; Tegner et al. 1992). The remaining stocks are fully exploited, in decline (Vega et al. 1997), or closed following their collapse, as is the case in southern California. Shepherd and Baker (1998) noted that only those fisheries that conserved a percentage Egg-Per-Recruit (Nash 1992) level of 40–50% were sustained in the long term. A continued strong market has driven interest in commercial abalone farming throughout the world. Present production figures and five-year projections are provided for the leading abalone farming countries.

The same methodologies that have been developed to produce abalone for consumption can and have been similarly utilized for the enhancement of wild fisheries. Dramatic declines in the numbers of wild white abalone throughout their range in southern California have prompted federal consideration of this species as threatened or endangered. A restoration plan and the costs for hatchery production are described in this paper.

T.B. McCormick. Proteus SeaFarms International, Inc., PMB 131, 323 East Matilija Street #110, Ojai, CA 93023, U.S.A. e-mail: tbmccormic@aol.com
Correct citation: McCormick, T.B. 2000. Abalone (*Haliotis* spp.) aquaculture: present status and a stock enhancement tool. *In* Workshop on Rebuilding Abalone Stocks in British Columbia. *Edited by* A. Campbell. Can. Spec. Publ. Fish. Aquat. Sci. 130. pp. 55–60.

Overview of world abalone farming

Historically, the largest abalone fisheries were located in Japan, Australia, New Zealand, Mexico, the United States, and South Africa, with small landings from Canada, Europe, and the western Pacific. Collectively, these countries saw a dramatic decrease in landings, from 28 000 t in 1960s to less than 14 000 t in early 1990s. Because of their historical fisheries, it is not surprising that these are the very countries where abalone cultivation efforts have been the strongest.

In the United States and Mexico, abalone farming efforts (Table 1) have been conducted since the 1970s, with commercial sales starting in earnest in the 1980s (Oaks and Ponte 1996). A survey by McBride (1998) identified 14 farms in California and 2 in Baja California, Mexico, that produce 143 t of abalone valued at US$3.2 million. Farms range from research operations to industrial production with the bulk of production coming from one farm in California. Most farms are land-based, however, three farms in the U.S. and one in Mexico grow abalone to market size in ocean cages. Most of the production is red abalone (*H. rufescens*), which is fed kelp (*Macrocystis pyrifera*). Production in the U.S. is not expected to grow rapidly due to limited sites, rigorous permitting processes, and high operating costs.

Abalone farming is expanding more rapidly in the southern hemisphere countries of Australia, New Zealand, South Africa, and Chile (Table 1). As in the U.S., farms range from

Table 1. Current and projected production of farmed abalone.

Country	Species	Number of farms	Production (t) Current	Production (t) 5-year projection	Reference
Australia	*H. laevigata*				Allan 1999
	H. rubra	24	10	500	
Chile	*H. rufescens*	6	<5	100–500	McCormick, personal observation
	H. discus hannai				
China	*H. discus hannai*	250	2700[a]	3500[a]	Mai, personal communication
	H. diversicolor supertaxta		1800[b]	2500[b]	
Mexico	*H. rufescens*	2[c]	10	15	McBride 1998
	H. fulgens	6[d]			
New Zealand	*H. iris*	20	<5	100	Roberts, personal communication
	H. Australis				Schiel 1997
South Africa	*H. midae*	10	<100	900–1100[e]	Cook 1998; Cook, personal communication
Taiwan	*H. diversicolor supertaxta*	180	2500	3500	Chen, personal communication
United States	*H. rufescens*	14	133	150	McBride 1998
	H. fulgens				

Note: Figures for recent production based on cited reference. Five-year estimates have been made by the author.
[a]Production of *H. discus hannai.*
[b]Production of *H. diversicolor supertaxta.*
[c]Hatchery, nursery, and growout facilities.
[d]Hatchery and nursery facilities for stock enhancement.
[e]Based on Cook's production estimate of 600–700 t by the year 2001. Personal communications: J. Chen, Aquaculture Department, Taiwan; P. Cook, University of Capetown, South Africa; K. Mai, College of Fisheries, Ocean University of Qingdao, Qingdao, China; R. Roberts, Cawthron Institute, Private Bag 2, Nelson, New Zealand.

research and development operations to multi-million dollar international efforts. Most farms are in the early stages of development and production is currently small. New Zealand now has 20 farms in operation with several more in the planning stage. Most businesses are cultivating the Paua (*H. iris*) in shore-based farms, while some ocean culture is conducted in cages or barrels. Abalone are grown for meat, with some farmers producing pearls as well. Abalone are fed pelletized feeds due to restrictions on the use of wild kelp. A number of farms anticipate production of 20–50 t each within a 5-year time frame. By that time the total production from these countries could reach 1000 t.

The picture of abalone farming in Australia and South Africa is similar to that in New Zealand. Most of the farms are onshore, are in the start-up stage, and anticipate sales to begin within the next five years. There are now 24 farms in Australia raising *H. laevigata* or *H. rubra* and 10 farms in South Africa raising *H. midae.*

Efforts to cultivate abalone in Chile started with the introduction of red abalone (*H. rufescens*) in the early 1980s (Owen et al. 1984; Godoy and Jerez 1998). In 1999, there were five abalone farms, some of which were also raising Japanese Ezo (*H. discus hannai*) abalone. Abalone are fed kelp and pelletized diets. Production from farms located in these southern hemisphere countries could reach 500 t within the next 5 years.

Efforts are underway to commercially farm temperate abalone species such as *H. tuberculata*, *H. fulgens*, and *H. discus hannai* in Europe, Israel, and the Oman (Johnson et al. 1992).

The predominant producers of abalone in Asia are China and Taiwan, farming 4500 and 3500 t of abalone per year, respectively (K. Mai, College of Fisheries, Ocean University of Qingdao, Qingdao, China, personal communication; J. Chen, Aquaculture Department, Taiwan, personal communication). South Korea also produces *H. discus hannai.*

The majority of China's production is the coldwater *H. discus hannai.* In southern China and Taiwan, the fast growing *H. diversicolor supertaxta* is produced. Abalone are highly prized by the Chinese and farming efforts have greatly expanded the investment in mollusc culture in the last decade. With growing consumer demand, China has moved from an abalone exporter to an abalone importer. The pace of marine aquaculture expansion in China will decrease because production in the northern regions is limited by low seawater temperatures in winter. In south China, typhoons and increasing pollution limit farm expansion.

Japanese abalone production remains centered in government hatcheries. Up to 40 million seeds (20 mm shell length, SL) are produced each year for enhancement of coastal fisheries (Uki 1989; Seki 1997). Production costs are underwritten by the government (McCormick and Hahn 1983) and by fishing cooperatives who use abalone seed to stock their fishing concessions. Continually declining wild harvests have led Japanese researchers to consider the importance of habitat and ecological relationships when using abalone seed to enhance wild fisheries (Seki and Sano 1998).

Techniques for large-scale aquaculture

Abalone cultivation techniques have been described in Hahn (1989) and Fallu (1991). A summary of cultivation methods and techniques that are being used for red abalone is presented here. These techniques will be modified and applied to the cultivation of white abalone.

Techniques for the large-scale production of abalone were developed in Japan in the 1960s (Imai 1967; Seki 1980) where for the last 30 years they have been employed primarily for the enhancement of coastal stocks. Methods to cultivate other species have subsequently been developed in regions such as California, New Zealand, Australia, China, South Africa, and the Channel Islands where wild stocks were previously fished (Ebert 1992; Ebert and Houk 1984; Tong and Moss 1992; Zong 1992; Cook 1998; Bossy 1990) and in regions such as South America (Godoy and Jerez 1998) where wild stocks of abalone do not exist.

Cultivation methods for red abalone (*H. rufescens*) have been adapted with slight modifications for two other species in California, the green (*H. fulgens*) and pink (*H. corrugata*) abalone. These methods, with appropriate modifications, can be used for the production of white abalone. The methods are summarized here.

Collection and conditioning of broodstock are the first steps involved in the husbandry of abalone. Wild adult animals may be ready to spawn immediately, but more reliable results are obtained when the animals are conditioned for several weeks or months. Wild adult broodstock should be obtained from the area where the hatchery-raised animals will eventually be placed to avoid the mixing of unrelated stocks. Animals should be gathered from several locations because a sole source location may not provide an adequate sample of the stock's genetic diversity (Templeton 1990). The prolific nature of abalone can lead to the temptation to use a few females to produce an entire crop. For example, a single female red abalone, *Haliotis rufescens,* can produce over 10 million eggs per spawn. To preserve genetic diversity within the hatchery-raised population, a minimum of 20–30 animals should be used as a founding population (Soulé et al. 1986; Ralls and Ballou 1992).

Sexual maturation and the onset of spawning are highly synchronized for white abalone in the ocean. The result is a well-defined spawning season, from February to April, correlated with changes in water temperature (Tutschulte and Connell 1981). This and other cues such as photoperiod and food availability can be used in the hatchery to extend the normal spawning season and enhance spawning success. The effect of temperature on maturation has been documented by Kikuchi and Uki (1974*a*) whose calculations enable accurate prediction of the time required to bring *H. discus hannai* to maturation at any given water temperature. Gonadal conditioning with temperature and photoperiod is also effective for other species such as the green abalone (*H. fulgens*) but may be of less utility for others. In these cases, other conditioning methods, e.g., *ad libitum* feeding (Ault 1985) in red abalone, may be effective. Spawning readiness is determined by observing the gonadal index of the abalone (Uki and Kikuchi 1982).

Abalone spawn in response to a variety of stimuli, e.g., rapidly changing water or air temperature, changes in water chemistry, storms, or the release of gametes from other individuals. The development of a reliable spawning technique by Kikuchi and Uki (1974*b*) opened the door for large-scale cultivation. This technique, which uses ultraviolet light (UV) to produce free oxygen radicals, and a variation of this method using hydrogen peroxide (Morse et al. 1977, 1978) are effective in triggering the spawning response in sexually mature abalone. Other stimuli, e.g., a period of desiccation (Carlisle 1945), a rapid temperature increase of 3–5°C above ambient followed by gradual cooling (Ino 1952), and darkening of the spawning chamber, may be used in concert with UV or hydrogen peroxide to increase the likelihood of successful spawning. Spawning usually commences following a 2–3 h exposure to UV irradiated seawater or hydrogen peroxide. Eggs of red, green, and pink abalone have an average diameter between 220–250 μm. After ejection from the female, the eggs settle to the bottom of the spawning container and are siphoned or poured from the container through a 250- or 276-μm sieve to remove waste and break up egg clumps. Males are held in containers separate from the females so the sperm density can be determined and the timing of fertilization controlled.

Optimal sperm density varies with species (Kikuchi and Uki 1974*c*; Mill and McCormick 1992) as well as the age of the egg and sperm. Sperm (Hahn 1989) and egg densities are determined and eggs are fertilized simultaneously to insure synchronous larval development (Seki and Kan-no 1974). Following a 2–5 min fertilization period, excess sperm is washed from the egg in a 60–90 μm sieve. Hatchout and larvae cultivation can be conducted in static water cultures (Hahn 1989), although large-scale production is better facilitated in flow-through systems (Tong and Moss 1992). The rate of larval development is a function of temperature. No feeding occurs during this life stage. Characteristic development stages (Seki and Kan-no 1980) are similar between species and can be used to accurately determine developmental progress. The entire process of spawning, fertilization, hatchout, larval cultivation, and settlement are usually concluded in the course of 4–7 days, depending on water temperature and the species cultivated.

Larval abalone are settled on the mucus trails of juvenile abalone and on films of diatoms and other micro-algae. The former serves as a signal for settlement (Seki and Kan-no 1980) and the algae serve as food for the early juvenile abalone (Uki and Kikuchi 1979). They are cultivated on clear plastic plates or the walls of shallow tanks. The algal films contain diatoms, such as *Navicula* spp., *Nitzschia* spp., and *Amphora* spp., bacteria, and may include benthic monostromatic algae types, such as *Myrionema* spp. and *Uvella* spp. (Seki 1980). Preparation of the micro-algal cultures may also employ juvenile abalone to pregraze the micro-algae cultures. These grazers remove the upper layer of diatoms but leave intact the diatoms preferred by post-larvae (Matthews and Cook 1995). There are several feeding strategies in use for the cultivation of early juvenile abalone. Using the plate culture, or Japanese method (Grant 1981), abalone consume only diatoms and micro-algae for the first

Table 2. White abalone operational expense summary.

	Year									
Expense	1	2	3	4	5	6	7	8	9	10
Capital[a]	21 914	17 544	2 515	9 141	8 637	2 786	262	390	4 191	0
Salaries and taxes[b]	69 993	69 993	70 854	73 195	78 467	80 104	80 104	80 104	80 104	80 104
Utilities[c]	1 407	2 728	2 924	3 120	3 708	4 493	80 104	80t104	80 104	80 104
Lease	5 000	5 000	5 000	5 000	5 000	5 000	5 000	5 000	5 000	5 000
Total	98 314	95 265	81 293	90 456	95 812	92 383	90 055	93 183	93 984	89 793

Note: Cost (US$) for an annual production of 10 000 abalone (88 mm SL).
[a]Capital expenditures include pumps, blowers, pipes, tanks, substrate, generators, building, shade, and fence.
[b]Salaries and taxes include a hatchery manager at 25% time, an administrator 20% time, and a hatchery worker at 50% time.
[c]Utilities include electricity for pumping, telephone, and water/sewer/refuse.

six months of life. Juvenile abalone may be weaned onto macro-algae such as *Macrocystis* or *Microcladia* (Trevelyan et al. 1998) when larger than 5–12 mm SL. An alternative feeding strategy is to provide diatom cultures for only 2 or 3 months and then to introduce formulated feeds in a granular form. Survival to the weaning stage is typically quite low, averaging 2–10%.

Growout from juvenile to adult size is accomplished in shallow seawater tanks provided with a constant flow of seawater to maintain temperature, supply oxygenation, and remove waste. Supplemental aeration ensures mixing of the water column, moves food toward the abalone, and moves waste away. A number of self-cleaning tank designs are in use. Higher stocking densities lead to slower growth rates as crowding and the levels of waste increase (Sano and Maniwa 1962). Restocking once or twice per year optimizes stocking densities. When kelp is available, its low cost makes it an economical feed. Several farms in California supplement kelp diets with commercially formulated feeds that utilize vegetable and animal protein sources (Uki and Watanabe 1992) and are in commercial production in many parts of the world. For red abalone, the time required for juveniles to grow to adult sizes (70–100 mm SL) can take much longer, 40–48 months depending on water temperature, feed, and stocking densities. The green abalone, also native to southern California, can grow at slightly faster rates while the pink abalone grow at slower rates.

White abalone enhancement in southern California

In southern California, the white abalone was the target of an intense commercial and recreational fishery in the mid-1970s until densities were too low to fish economically. Densities dropped from 2 000 to10 000 ha^{-1} in the early 1970s (Tutschulte 1976) to 1.6 ha^{-1} in the 1990s (Davis et al. 1996). The white abalone fishery was officially closed in 1996. This abalone species has since been proposed as a candidate for listing as threatened or endangered under the Endangered Species Act of 1973 (Federal Register Vol. 62, 1997). Current densities are insufficient to maintain white abalone populations and evidence indicates that the remaining adults are dying of old age. Without human intervention, the species may become extinct (Tegner et al. 1996).

An *ad hoc* collaborative effort (Davis et al. 1998), involving the National Marine Fisheries Service, National Parks System, California Department of Fish and Game, University of California, Santa Barbara, and the Channel Island Marine Resource Institute (a nonprofit organization) has been undertaken to pull white abalone from the brink of extinction. The program consists of the following five steps:

(*i*) Conduct underwater surveys to locate remaining abalone. Thus far, surveys of white abalone habitat with a 3-person submarine have located in excess of 100 individuals.
(*ii*) Collect wild animals. At least 2 groups of 20 abalone will be moved from the wild into holding facilities for conditioning.
(*iii*) Acquire information on reproductive biology and ecology of the white abalone. This information will be required to rebuild the population through captive breeding and release of the offspring.
(*iv*) Operate a long-term captive breeding and rearing program. Using the procedures outlined above, a production facility will produce 10 000 abalone per year for 10 years.
(*v*) Place adult hatchery-raised abalone on reefs in protected zones. Groups of 500–1000 abalone will be placed in appropriate habitat to form the basis of new communities.

Previous enhancement efforts in southern California have failed to restore or sustain existing abalone populations (Tegner and Butler 1989). Similar efforts in other parts of the world have produced a range of results indicating that planting success is variable and affected by the condition and size of the animal at release, planting method, habitat type, predation, and food availability (McCormick et al. 1994). Larger abalone are more resistant to predation and have greater survival over long periods of time (Saito 1979; Emmett and Jamieson 1989). For this reason, white abalone will be raised in tanks for up to 4 years until they attain 80–100 mm SL. These larger, sexually mature animals should be more resistant to predation than smaller animals and will be capable of spawning immediately after being placed in the ocean.

White abalone share some characteristics with red abalone, which should facilitate ease in culture. Indeed, natural hybrids of these species occur and have been produced in the laboratory (Leighton and Lewis 1982). Throughout their range in southern California, white abalone are typically found below the 20 m thermocline where temperatures range from 12 to 18°C. Temperature preferences of the larval stages have been demonstrated by Leighton (1972). Care must be taken to ensure that cultivation temperatures do not exceed 18°C during the growout period. Like red abalone, white abalone apparently prefer brown algae such as *Macrocystis* and *Laminaria*. There are two characteristics of white abalone that will aid in the breeding program. First, white abalone show a strong degree of spawning synchrony (Tutschulte and Connell 1981). Spawning occurs as temperatures increase from annual lows of 12–13°C. Second, white abalone have a significantly higher fecundity than either pink or green abalone with each female producing 4–6 million eggs per year (Tutschulte and Connell 1981). This should result in a plentiful number of eggs being available for the hatchery program.

The costs for the abalone cultivation portion of this resource enhancement effort have been estimated. A computer model first considers the biological and environmental requirements of the abalone, then determines the required capital, energy, feed, and labor inputs (Table 2). This model assumes that 10 000 white abalone will be grown to a size of 80–100 mm SL each year for a period of 10 years. The first crop will be ready for placement in the ocean at the end of year 4. The production facility is located at an existing aquaculture site so the existing seawater system and buildings can be used.

With the decline of wild fisheries throughout the world, the farming of abalone is becoming widespread. Farmed production is expected to increase in the next decade as more producers, mostly in the southern hemisphere, bring cultivated abalone to the world market and farming technology improves.

Efforts to enhance native abalone populations in different parts of the world have been tried for several decades with varying degrees of success. Hatchery techniques are now being used to cultivate white abalone in an effort to save the species from extinction. Experience gained in this program will no doubt prove useful in other restoration efforts.

References

Allan, G.L. 1999. Aquaculture in Australia: now and in the future. J. World Aquacul. Soc. **30**: 39–54.

Ault, J. 1985. Some quantitative aspects of reproduction and growth of the red abalone, *Haliotis rufescens* Swainson. J. World Maric. Soc. **16**: 398.

Bossy, S.F. 1990. Ongrowing the ormer *Haliotis tuberculata* in Jersey, Channel Islands. Spec. Publ. Eur. Aquacult. Soc. **12**: 172–173.

Carlisle, J.G. 1945. The technique of inducing spawning in *Haliotis rufescens* Swainson. Science (Washington, D.C.), **102**: 566.

Cook, P. 1998. The current status of abalone farming in South Africa. J. Shellfish Res. **17**: 601 602.

Davis, G.E., Haaker, P.L., and Richards, D.V. 1996. Status and trends of white abalone at the California Channel Islands. Trans. Am. Fish. Soc. **125**: 42–48.

Davis, G.E., Haaker, P.L., and Richards, D.V. 1998. The perilous condition of white abalone *Haliotis sorenseni*, Bartsch 1940. J. Shellfish Res. **17**: 871–875.

Ebert, E.E. 1992. Abalone aquaculture: a North American regional review. *In* Abalone of the world: biology, fisheries and culture. Proceedings of the 1st International Symposium on Abalone, 2125 November 1989, La Paz, Mexico. *Edited by* S.A. Shepherd, M.J. Tegner, and S.A. Guzmán del Próo. Fishing News Books, Blackwell Science Ltd., Oxford, U.K. pp. 570–582.

Ebert, E.E., and Houk, J.L. 1984. Elements and innovations in the cultivation of red abalone *Haliotis rufescens*. Aquaculture, **39**: 375–392.

Emmett, B., and Jamieson, G.S. 1989. An experimental transplant of northern abalone, *Haliotis kamtschatkana*, in Barkley sound, British Columbia. Fish. Bull. **87**: 95–104.

Fallu, R. 1991. Abalone farming. Fishing News Books, Blackwell Science Ltd., Oxford, U.K. 195 p.

Godoy, C., and Jerez, G. 1998. The introduction of abalone in Chile: ten years later. J. Shellfish Res. **17**: 603–606.

Grant, J.F. 1981. Abalone culture in Japan: development and current commercial practice. Tasman. Fish. Res. **23**: 1–17.

Hahn, K.O. (*Editor*) 1989. Handbook of culture of abalone and other marine gastropods. CRC Press, Boca Raton, Florida. 348 p.

Imai, T. 1967. Mass production of molluscs by means of rearing the larvae in tanks. Venus, **25**: 159–157.

Ino, T., 1952. Biological studies on the propagation of Japanese abalone (genus *Haliotus*). Bull. Tohoku Reg. Fish. Res. Lab. **5**: 1.

Johnson, D.W., Al-Harassy, A., and Al-Harthy, M. 1992. The sultanate of Oman abalone fishery. *In* Abalone of the world: biology, fisheries and culture. Proceedings of the 1st International Symposium on Abalone, 21–25 November 1989, La Paz, Mexico. *Edited by* S.A. Shepherd, M.J. Tegner, and S.A. Guzmán del Próo. Fishing News Books, Blackwell Science Ltd., Oxford, U.K. pp. 448–453.

Kikuchi, S., and Uki, N. 1974*a*. Technical study on artificial spawning of abalone, genus *Haliotis*. V. Relation between water temperature and advancing sexual maturity of *Haliotis discus* Reeve. Bull. Tohoku Reg. Fish. Lab. **34**: 77–85.

Kikuchi, S., and Uki, N. 1974*b*. Technical study on artificial spawning of abalone, genus *Haliotis*. II. Effect of irradiated seawater with ultraviolet rays on inducing to spawn. Bull. Tohoku Reg. Fish. Lab. **33**: 80–92.

Kikuchi, S., and Uki, N. 1974*c*. Technical study on artificial spawning of abalone, genus *Haliotis*. III. Reasonable sperm density for fertilization. Bull. Tohoku Reg. Fish. Lab. **34**: 67–71.

Leighton, D.L. 1972. Laboratory observations on the early growth of the abalone *Haliotis sorenseni*, and the effect of temperature on larval development and settling success. Fish. Bull. **70**: 373–381.

Leighton, D.L., and Lewis, C.A. 1982. Experimental hybridization in abalones. Int. J. Invertebr. Reprod. **5**: 273–282.

Matthews, I., and Cook, P.A. 1995. Diatom diet of abalone post-larvae (*Halioitis midae*) and the effect of pre-grazing the diatom overstory. *In* Progress in abalone fisheries research. Proceedings of the 2nd International Symposium on Abalone Biology, Fisheries and Culture. Mar. Freshw. Res. **46**: 545–548.

McBride, S. 1998. Current status of abalone aquaculture in California. J. Shellfish Res. **17**: 593–600.

McCormick, T.B., and Hahn, K.O. 1983. Japanese abalone culture practices and estimated costs of juvenile produciton in the USA. J. World Maric. Soc. **14**: 149 161.

McCormick, T.B., Herbinson, K., Mill, T.S., and Altick, J. 1994. A review of abalone seeding, possible significance and a new seeding device. Bull. Mar. Sci. **55**: 680–693.

Mill, T.S., and McCormick, T.B. 1992. Optimal sperm density for fertilization in three species of abalone (*Haliotis*). *In* Proceedings of the 1st International Symposium on abalone, La Paz,

Mexico 21025 November 1989. Fish. Res. Pap. Dep. Fish. (S. Aust.). **24**: 42–48.

Morse, D.E., Duncan, H., Hooker, N., and Morse, A. 1977. Hydrogen peroxide induces spawning in mollusks with activation of prostaglandin endoperoxide synthesis. Science (Washington, D.C.), **196**: 298.

Morse, D.E., Hooker, N., and Morse, A. 1978. Chemical control and reproduction in bivalve and gastropod molluscs. III. An inexpensive technique for mariculture of many species. Proc. World Maric. Soc. **9**: 543.

Nash, W.J. 1992. An evaluation of egg-per-recruit analysis as a means of assessing size limits for black abalone (*Haliotis rubra*) in Tasmania. *In* Abalone of the world: biology, fisheries and culture. Proceedings of the 1st International Symposium on Abalone, 21–25 November 1989, La Paz, Mexico. *Edited by* S.A. Shepherd, M.J. Tegner, and S.A. Guzmán del Próo. Fishing News Books, Blackwell Science Ltd., Oxford, U.K. pp. 318–340.

Oaks, F.R., and Ponte, R.D. 1996. The abalone market: opportunities for cultured abalone. Aquaculture, **140**: 187–195.

Owen, B., DeSalvo, L., Ebert, E,. and Fonck, E. 1984. Culture of the California red abalone (*Haliotis rufescens*) Swainson (1822) in Chile. Veliger, **27**: 101–105.

Ralls, K., and Ballou, J.D. 1992. Managing genetic diversity in captive breeding and reintroduction programs. Trans. 57th N. Am. Wildl. Nat. Res. Conf. pp. 263–282.

Saito, K. 1979. Studies on the propagation of ezo abalone, *Haliotis discus hannai* Ino.- I. Analysis of the relationship between transplantation and catch in Funka Bay coals. Bull. Jpn. Soc. Sci. Fish. **46**: 695–704.

Sano, T., and Maniwa, R. 1962. Studies on the environmental factors having an influence on the growth of *Haliotis discus hannai*. Bull. Tohku Reg. Fish. Res. Lab. **21**: 79–86.

Schiel, D. 1997. Review of abalone culture and research in New Zealand. Moll. Res. **18**: 289–298.

Seki, T. 1980. An advanced biological engineering system for abalone seed production. *In* International Symposium on Coastal Pacific Marine Life. May 1980. Western Washington State College, Bellingham, Wash.

Seki, T. 1997. Biological studies on the seed production of the northern Japanese abalone, *Haliotis discus hannai* Ino. Bull. Tohoku Natl. Fish. Res. Inst. **59**. 71 p.

Seki, T., and Kan-no, H. 1974. Synchronized control of early life in the abalone, *Haliotis discus hannai* Ino, Haliotidae, Gastropoda. Bull. Tohoku Reg. Fish. Res. Lab. **38**: 143–153.

Seki, T., and Kan-no, H. 1980. An advanced biological engineering system for abalone seed production. *In* International. Symposium on Coastal Pacific Marine Life. October 15–16, 1979. Western Washington State College, Bellingham, Wash.

Seki, T., and Sano, M. 1998. An ecological basis for the restoration of abalone populations. Bull. Tohoku Natl. Fish. Res. Inst. **60**: 23–40.

Shepherd, S.A., and Baker, J.L. 1998. Biological reference points in an abalone (*Haliotis laevigata*) fishery. *In* Proceedings of the North Pacific Symposium on Invertebrate Stock Assessment and Management. *Edited by* J.S. Jamieson and A. Campbell. Can. Spec. Publ. Fish. Aquat. Sci. 125. pp. 235–245.

Sloan, N.A., and Breen, P.A. 1988. Northern abalone, *Haliotis kantschatkana,* in British Columbia: fisheries and synopsis of life history information. Can. Spec. Publ. Fish. Aquat. Sci. 103. 46 p.

Soulé, M.E., Gilpin, M., Conway, W., and Foose, T. 1986. The millennium ark: how long a voyage, how many staterooms, how many passengers? Zoo Biol. **5**: 101–114.

Tegner, M.J., and Butler, R.A. 1989. Abalone seeding. *In* Handbook on the culture of abalone and other marine gastropods. *Edited by* K.O. Hahn. CRC Press, Boca Raton, Florida. pp. 157–182.

Tegner, M.J., DeMartini, J.D., and Karpov, K.A. 1992. The California red abalone fishery: a case study in complexity. *In* Abalone of the world: biology, fisheries and culture. Proceedings of the 1st International Symposium on Abalone, 21–25 November 1989, La Paz, Mexico. *Edited by* S.A. Shepherd, M.J. Tegne,r and S.A. Guzmán del Próo. Fishing News Books, Blackwell Science Ltd., Oxford, U.K. pp. 370–383.

Tegner, M.J., Basch, L.V., and Dayton, P.D. 1996. Near extinction of an exploited marine invertebrate. TREE, **11**: 278–279.

Templeton, A.R. 1990. The role of genetics in captive breeding and reintroduction programs for species conservation. Endangered Species Update. **8**: 15–17.

Tong, L.J., and Moss, G.A. 1992. The New Zealand culture system for abalone. *In* Abalone of the world: biology, fisheries and culture. Proceedings of the 1st International Symposium on Abalone, 21–25 November 1989, La Paz, Mexico. *Edited by* S.A. Shepherd, M.J. Tegner, and S.A. Guzmán del Próo. Fishing News Books, Blackwell Science Ltd., Oxford, U.K. pp. 370–383.

Trevelyan, G.A., Mendoza, J.L., and Buckeley, B. 1998. Increasing yield of red abalone with the alga, Microcladia coulteri. J. Shellfish Res. **17**: 631–633.

Tutschulte, T.C. 1976. The comparative ecology of three sympatric abalones. Ph.D. dissertation. University of California, San Diego, Calif.

Tutschulte, T., and Connell, J.H. 1981. Reproductive biology of three species of abalones (*Haliotis*) in southern California. Veliger, **23**: 195–206.

Uki, N. 1989. Abalone seeding production and its theory. Intl. J. Aquacult. Fish. Tech. **1**: 3–15.

Uki, N., and Kikuchi, S. 1979. Food value of six benthic microalgae on growth of juvenile abalone, *Haliotis discus hannai.* Bull. Tohoku Reg. Fish. Lab. **40**: 47–56.

Uki, N., and Kikuchi, S. 1982. Influence of food levels on maturation and spawning of the abalone, *Haliotis discus hannai* related to effective accumulative temperature. Bull. Tohoku Reg. Fish. Lab. **45**: 45–53.

Uki, N., and Waanabe,T. 1992. Review of the nutritional requirements of abalone (*Haliotis spp.*) and development of more efficient artificial diets. *In* Abalone of the world: biology, fisheries and culture. Proceedings of the 1st International Symposium on Abalone, 21–25 November 1989, La Paz, Mexico. *Edited by* S.A. Shepherd, M.J. Tegner, and S.A. Guzmán del Próo. Fishing News Books, Blackwell Science Ltd., Oxford, U.K. pp. 504–517.

Vega, A., Lluch-Balda, D., M. Mucino, M., Leon, G., Hernandez, S., Lluch-Cota, D., Ramade, M., and Espinoza, G. 1997. Development, perspectives, and management of lobster and abalone fisheries off northwest Mexico under a limited access system. *In* Developing and sustaining world fisheries resources. *Edited by* D.A. Handcock, D.C. Smith, A. Grant, and J.P. Beumer. CISRO, Collingwood, Victoria, Australia. pp. 136–142.

Zong, Q.N. 1992. A review of abalone culture in China. *In* Abalone of the world: biology, fisheries and culture. Proceedings of the 1st International Symposium on Abalone, 21–25 November 1989, La Paz, Mexico. *Edited by* S.A. Shepherd, M.J. Tegner, and S.A. Guzmán del Próo. Fishing News Books, Blackwell Science Ltd., Oxford, U.K. pp. 592–602.

Abalone (*Haliotis* spp.) enhancement in California: what we've learned and where we go from here

Mia J. Tegner

Abstract: California's once productive commercial and recreational abalone (*Haliotis* spp.) fisheries have been declining from San Francisco to the Mexican border since the late 1960s. Enhancement experiments have included local fishery closures, outplanting of hatchery-reared juveniles, transplantation of wild adults, and larval seeding. The documented survival of seeded hatchery-reared juveniles has been poor, despite experiments with three species, a variety of seed sizes, outplanting methods, and locations. An experimental transplant of reproductively mature green abalones (*H. fulgens*) was followed by strong evidence of successful local recruitment before poaching led to loss of the brood stock. Similarly, the incipient recovery of four haliotid species in the abalone fishing closure on the Palos Verdes Peninsula near Los Angeles was apparently the victim of inadequate enforcement and poaching. Evaluation of larval seeding experiments in southern California is still ongoing. Outplanting of juveniles and larvae has been curtailed due to parasite and pathogen problems in hatchery stocks in order to protect remaining wild populations. Withering foot syndrome, apparently a rickettsial infection that has devasted wild black abalone (*H. cracherodii*) populations, has been detected in hatcheries throughout the state. An exotic sabellid polychaete also has infested stock in most culture facilities in California and has already escaped from one facility into the wild. In January 1998, the legislature enacted a moratorium on all abalone fishing for this region and one species is a candidate for the endangered species list. Funds for additional enhancement efforts are available from taxes previously levied on commercial landings and a new fee on recreational divers in the sport-only fishery north of San Francisco. The success of future enhancement efforts will depend on being able to certify hatcheries free of parasite and pathogen problems, establishing meaningful reserves and brood stock aggregations, achieving effective enforcement, and sea otter range expansion.

Résumé : De San Francisco aux frontières du Mexique, la pêche commerciale et récréative de l'ormeau (*Haliotis* spp.), naguère productive en Californie, est en déclin depuis la fin des années 1960. Des expériences de mise en valeur ont eu lieu, comportant des fermetures locales de la pêche, l'ensemencement de juvéniles d'élevage, la transplantation d'adultes sauvages et l'ensemencement de larves. L'information disponible révèle que la survie des juvéniles d'écloseries a été faible, malgré l'utilisation de trois espèces différentes dans les expériences ainsi que la diversification de la taille des semences, des méthodes et des lieux d'ensemencement. Tout indiquait qu'une expérience de transplantation d'ormeaux de l'espèce *H. fulgens* à maturité sexuelle constituait une réussite sur le plan du recrutement local, jusqu'à ce que le braconnage n'entraîne la disparition du stock de géniteurs. De même, à la suite de la fermeture de la pêche de l'ormeau de Palos Verdes Peninsula, aux environs de Los Angeles, le rétablissement à peine amorcé de quatre espèces d'haliotides a apparemment été anéanti en raison d'une application déficiente des règlements et du braconnage. L'évaluation d'expériences d'ensemencement se poursuit au sud de la Californie. L'ensemencement de juvéniles et de larves a été abrégé afin de protéger les populations sauvages restantes contre certains parasites et agents pathogènes présents dans les stocks d'écloseries. Le syndrome du flétrissement du pied de l'ormeau, une maladie qui serait causée par une rickettsie et qui a dévasté les populations sauvages de *H. cracherodii*, a été détecté dans des écloseries sur tout le territoire de l'État. Un ver polychète sabellidé exotique a également infesté les stocks de la plupart des installations d'aquaculture en Californie et s'est déjà échappé dans la nature à partir de l'une d'elles. Ainsi, en janvier 1998, un moratoire a été imposé sur la pêche de l'ormeau dans toute cette région, et une espèce pourrait même être inscrite à la liste des espèces menacées de disparition. Des fonds sont rendus disponibles pour les efforts supplémentaires de mise en valeur grâce aux taxes prélevées sur les débarquements commerciaux dans le passé et à l'imposition de nouveaux droits aux plongeurs qui pratiquent la pêche récréative au nord de San Francisco. Le succès des efforts futurs en ce sens repose sur un certain nombre de facteurs, dont la capacité d'attester que les écloseries sont exemptes de parasites et d'agents pathogènes, l'établissement de réserves valables, la constitution de concentrations de stocks de géniteurs, l'application rigoureuse des règles et l'expansion de l'aire de répartition de la loutre.

[Traduit par la Rédaction]

M.J. Tegner. Scripps Institution of Oceanography, University of California, San Diego, La Jolla, California 92093-0201, U.S.A. e-mail: mtegner@ucsd.edu

Correct citation: Tegner, M.J. 2000. Abalone (*Haliotis* spp.) enhancement in California: what we've learned and where we go from here. *In* Workshop on Rebuilding Abalone Stocks in British Columbia. *Edited by* A. Campbell. Can. Spec. Publ. Fish. Aquat. Sci. 130. pp. 61–71.

Introduction

The precipitous decline in California abalone (*Haliotis* spp.) landings (Fig. 1) that began in the late 1960s led to limited entry to the commercial fishery, other reductions in access including closure of a stretch of the southern California mainland coast (Fig. 2), and experimental studies of enhancement in the mid-1970s (Wyner et al. 1977; reviewed by Tegner 1989). Enhancement efforts, a welcomed alternative to more painful cuts in fishing pressure, were undertaken by agencies, academics, and commercial and recreational fishing interests. Two decades later, the failure of both management changes and enhancement efforts to stem the decline was apparent when all abalone fishing had to be halted in central and southern California to allow badly overexploited stocks a chance to recover. The white abalone (*Haliotis sorenseni*) is a candidate for listing under the Endangered Species Act (Davis et al. 1998) and there is concern that black abalone (*H. cracherodii*) populations may also have declined to dangerously low levels (Altstatt et al. 1996). Stocks of other southern California haliotids are in less critical condition, but recovery of fishable concentrations is threatened by the range expansion of the sea otter (*Enhydra lutris*). California Senate Bill 463 (Thompson), which went into effect in January 1998, established a moratorium on all abalone harvest south of San Francisco and a $12 stamp on recreational harvest on the north coast to support recreational abalone management, research, and restoration efforts. (The text of this bill can be viewed at http://www.sen.ca.gov). Importantly, some of the funds may also be used for enforcement.

Thus, it is an appropriate time to review what we have learned, where we stand, and to reevaluate the potential for abalone restoration in California. Recent research on proximity requirements for successful fertilization in broadcast spawners (e.g., Pennington 1985; Levitan 1995; Babcock and Keesing 1997) has suggested a major reason for stock collapse with important implications for how restoration should be carried out. Ocean climate changes in southern California have clearly affected kelp populations (Tegner et al. 1996) and may have affected abalone reproduction and recruitment as well. Declining availability of legally harvested abalones has led to severe poaching pressure (Karpov and Tegner 1992; Daniels and Floren 1998). I will review California enhancement efforts and issues, including the curtailment of outplanting activities since 1996 due to disease problems in commercial abalone aquaculture facilities. California is faced with very differ-

Fig. 1. California commercial abalone landings (metric tons, MT), 1940–1997. Black abalones were closed to commercial fishing on 2 August 1993, pink, green, and white abalones were closed on 1 March 1996, and red abalones were closed on 22 May 1997. (California Department of Fish and Game Landing Statistics.)

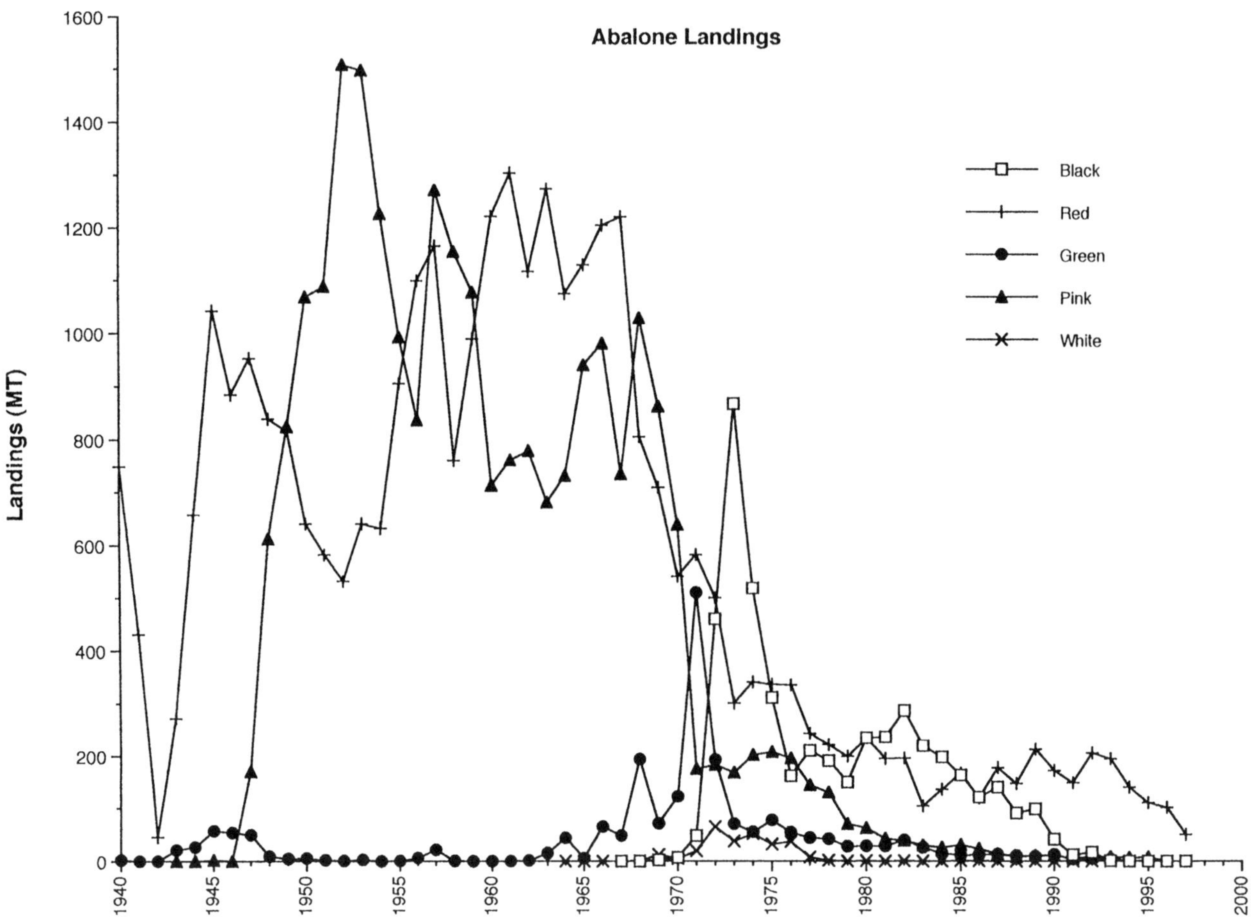

Fig. 2. Map of California illustrating regions discussed in the text.

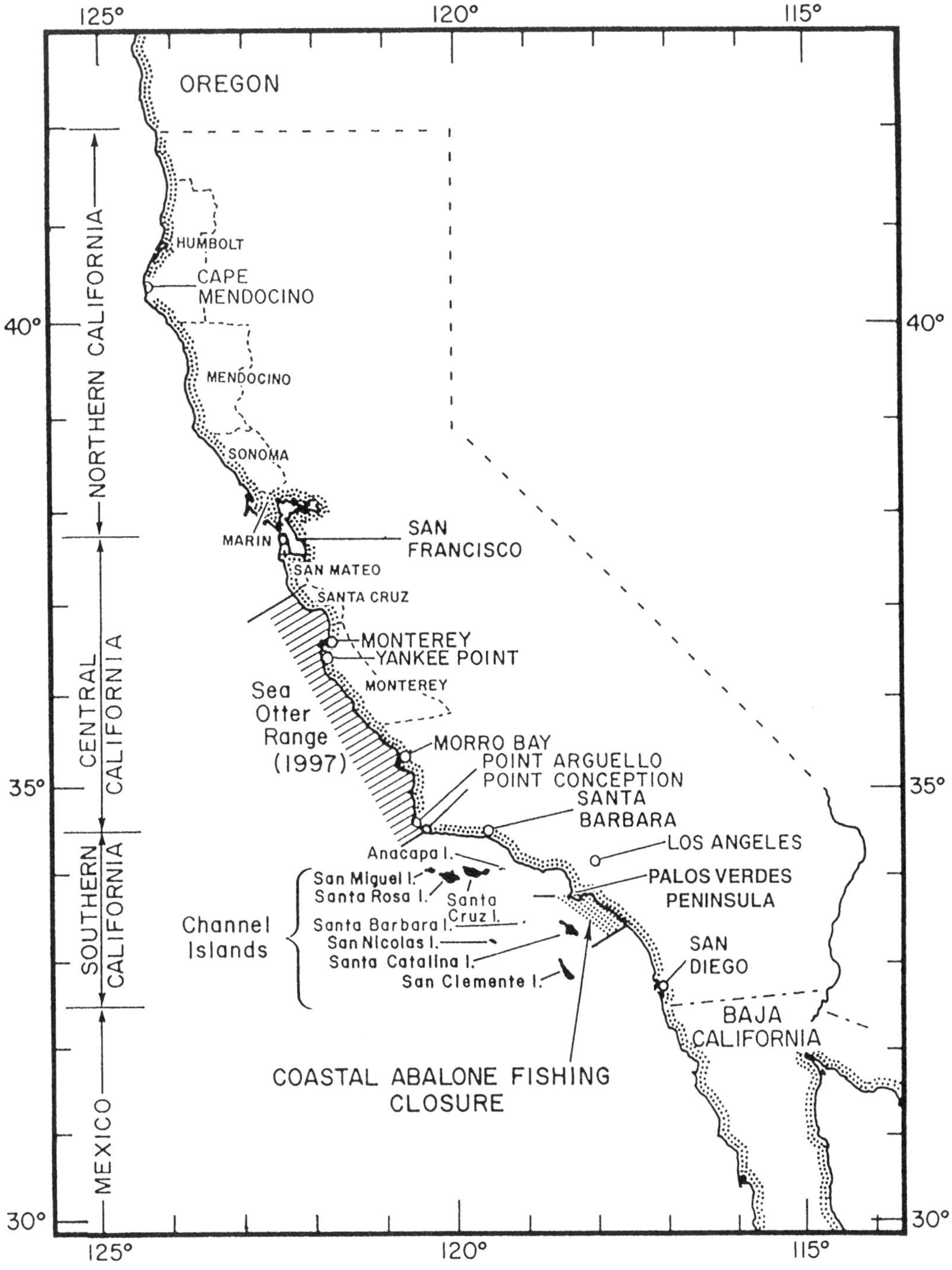

ent management issues in 1999 compared with two decades ago. In the case of the white and probably the black abalone, the primary management goal is not enhancement of fishable stocks but ensuring the survival of the species. For these species, outplanting of hatchery-reared animals may be critical, but for the other species the establishment of brood stock populations in well-enforced marine reserves needs to be examined seriously. Finally, investment in restoration needs to be balanced against the issue of sea otter range expansion.

Seeding of hatchery-reared juveniles

Munro and Bell (1997) reviewed international efforts to enhance stocks of demersal marine resources and some of the questions that these efforts have raised. In many ways abalones should be excellent candidates for enhancement: they are clearly limited by the supply of juveniles in many areas, stocks suffer from recruitment overfishing, and the sea urchin fishery has generally ensured that habitat resources have not been taken over by competitors. Thus, seeding of hatchery-reared juveniles was pioneered by Japan and has been actively pursued or considered in the United States, Mexico, New Zealand, Australia, South Africa, and Canada. While the spectacular example of the Japanese giant scallop fishery suggests that stock enhancement can enable some stocks to surmount the recruitment threshold above which the stock becomes self-sustaining at higher levels of recruitment for succeeding generations (Munro and Bell 1997), the results are not nearly so convincing for abalones. Funka Bay is an example in Japan where transplantation of wild juveniles was successful in establishing a self-sustaining population of *H. discus hannai* in an area that did not naturally support abalones (Saito 1979). Despite ever-increasing production of hatchery-reared seed, however, the total Japanese abalone harvest has decreased from more than 6000 tons in the 1960s (Ikenoue and Kafuku 1992) to 3000 tons (Seki and Sano 1998). Seki and Sano (1998) conclude that transplantation of cultured seed does not contribute to the enhancement of natural abalone resources. Most of the drop in harvest took place in northern Japan, where seeding efforts have been greatest (Uki 1989). While recaptures of outplanted seed have led to a "faintly profitable" turnover in the case when larger (>3 cm SL, shell length) seed were transplanted, the wild population did not increase (Seki and Sano 1998).

The California Department of Fish and Game (CDFG) and the California Sea Grant Program conducted a variety of seeding experiments with hatchery-reared juveniles in southern California (reviewed by Tegner and Butler 1989; David Parker, CDFG, 330 Golden Shore, Long Beach, CA 90802, personal communication). These experiments included three species (reds *Haliotis rufescens*, pinks *H. corrugata*, and greens *H. fulgens*), a range of juvenile seed sizes, planting densities, and techniques and were conducted in both island and mainland habitats. Species choice was based on depth, habitat type, and what was native to the planting area. Because Japanese work suggested that seed survival increased with outplanting size (references in Tegner and Butler 1989), a range of juvenile seed sizes (and thus shell thicknesses) was also tested. Seeding techniques included hand planting of individuals, the use of oyster shell as a seed substrate to reduce handling, cinder block habitats, and concrete modules (see below).

Many California experiments were conducted on the Palos Verdes Peninsula (PVP) in Los Angeles County (Fig. 2) because of easy access and because most of it was closed to sport and commercial abalone harvest in 1976 (Tegner and Butler 1989). Predators were not manipulated during our initial seeding experiment with 10 000 red abalone and 173 mortalities were collected within the first month. Removal of predators and scavengers before and during a second experiment reduced the one month shell total to 36. When preliminary data suggested that long-term survival of 20–35 mm seed was very poor, we conducted an experiment with larger (40–80 mm SL) and thus thicker-shelled seed and added monthly predator control to the collections at time of seeding (Tegner and Butler 1985*a*). Octopuses were taken in 11 of 12 visits during the year-long experiment, suggesting continual migration into the area. Survival after one year was about 1% for both size classes of seed, no better than other experiments without predator control. Shells accounted for 44% of the total number of animals seeded; more than half of these large-shelled animals were unaccounted for and may have moved out of the area. Presumably many more seed would have died if predators were not removed, but the results suggested that nearly continuous predator control would be necessary to improve survival (Tegner and Butler 1985*a*, 1989).

In six large-scale seeding experiments with red and green abalones conducted on the PVP and on Santa Cruz and Santa Catalina Islands, there was poor long-term survival of hatchery-reared animals in the study sites (Tegner and Butler 1989). In each case, the estimated 1- or 2-year survival rate based on the recovery of live seed was 2.8% or less. There was no evidence that seed survival increased with size, despite an outplant range from 10 to 80 mm SL. Survival rates were based on censuses of live animals in planting sites and immediate surroundings; they do not account for potential larger-scale migration because of the problems of sampling rare, cryptic animals. All sites had supported abalone populations in the past and some had significant native populations at the time of the experiments. All sites had adequate kelp populations and good habitat for juveniles. There were no apparent environmental fluctuations, such as extreme temperatures, that could have affected the abalones, with the exception of one site innundated by sand during the severe winter of 1983. Schmitt and Connell attributed the failure of a red abalone plant near Santa Barbara to storm-induced food shortage leading to starvation (see references in Tegner and Butler 1989).

As with other candidate taxa for enhancement (Munro and Bell 1997), there is evidence for reduced fitness of hatchery-reared seed. Seed for a large red abalone experiment at Tyler Bight on San Miguel Island came from a hatchery with recirculating seawater, which led to severe shell erosion. Many of the seeds were fragile and as shiny as fishing lures, and in severe cases, the respiratory pores had expanded and fused into one large hole in the shell (Tegner and Butler 1989). Many hatchery-reared seed have healthy shells but are colored differently than natives; how this affects their fitness is unknown. The intensity of predation pressure apparent in southern California puts a premium on seed behavior that minimizes susceptibility to predation. Animals raised in a hatchery are exposed to substrates, lighting, and food availability very different from what they will encounter in nature. Laboratory experiments with hatchery-reared and native abalones of the same size-classes show subtle but significant differences in behavior of red (Schiel and Welden 1987) and green (Tegner and Butler 1989) abalones and in both cases significantly more cultured animals were consumed when predators were added to the experiments. Tegner and Butler (1989) review other evidence for behavioral differences between wild and cultured seed. Encouragingly, Schiel and Welden (1987) were able to demonstrate reduced predation on acclimated seed relative to animals fresh from a hatchery.

A variety of seeding devices (e.g., Ebert and Ebert 1988; McCormick et al. 1994; Davis 1995; Sweijd et al. 1998) has been proposed to efficiently collect, transport, and seed relatively large numbers of abalone while also reducing handling stress and facilitating adaptation to the bottom and ultimately increasing survival. These often show short-term advantages, but have rarely been tested for longer than a few weeks against hand-planted controls. Parker et al. (1989; David Parker, CDFG, 330 Golden Shore, Long Beach, CA 90802, unpublished manuscript), using the modules of Ebert and Ebert (1988), which were designed to open after an 18-hour equilibration period, compared survival with hand-planted abalones of three species in experiments lasting six and fourteen months on the PVP. They found that long-term post-release mortality of module-seeded abalone was comparable to or higher than hand-seeded animals and that predators responded to the modules and preyed heavily on the released seed. Parker and colleagues concluded that the modules reduced the personnel time required for seeding and provided short-term protection from predation. Because of the high mortality in the natural habitat regardless of method of release, however, releasing modules did not improve the success of abalone seeding in these experiments (Parker et al. 1989; David Parker, CDFG, 330 Golden Shore, Long Beach, CA 90802, unpublished). In combination with the PVP experiment with the larger animals described above in which more than half of the recovered shells exhibited growth after planting (Tegner and Butler 1985*a*), these results suggest that abalone recovery off the PVP may be limited by predation, not the initial adaptation of the hatchery-reared seed.

Davis (1995) seeded 7200 juvenile red abalones into artificial habitats made up of concrete blocks inside a wire mesh frame, structures designed to sample abundance of juvenile abalones for fishery management purposes, at three sites on Santa Rosa and Santa Cruz islands in 1989. At the end of the first and second years, 3.9 and 0.8%, respectively, of the hatchery-reared abalones were still living in the artificial habitats; these figures represent minimum survival rates but other animals may have moved onto nearby reefs. Recovery rates of marked shells scattered around the habitats at each site were used to adjust the maximum mortality rates based on the assumption that all animals not accounted for were dead. Averaging minimum and maximum mortality rates led to estimates of 32% survival after the first year and 24% after the second year, figures considerably higher than survival rates based on recovery of live animals in other studies. Comparisons of size–frequency distributions of seeded animals with subsequent collections of 2977 empty shells yielded no evidence of a size differential in survival, despite a seed range of 15–91 mm SL at the time of outplanting (Davis 1995).

Biogenic structure is important for the survival of native juvenile abalones, which shelter under the spine canopies of adult sea urchins in California (Tegner and Dayton 1977), Japan (Kojima 1981), and South Africa (Tarr et al. 1996). The importance of this shelter was dramatically illustrated in a region of South Africa where a large increase in rock lobster abundance led to dramatic declines in sea urchins and simultaneously in juvenile abalones (Tarr et al. 1996). Day (1998) showed experimentally that urchin removal resulted in dramatic reductions in abalone juveniles and new recruits in the same area. She found that juvenile abalones benefit from their association with urchins through protection from predators, an enhanced food supply, and reduction in the amount of time they need to spend exposed while feeding. In Japan (Kojima 1981) and in northern California (Rogers-Bennett and Pearse 1998), the presence of sea urchins also enhanced the survival of hatchery-reared juvenile outplants. Rogers-Bennett and Pearse (1998) seeded 50 000 red abalones averaging 8 mm SL into six sites in central and northern California in October 1995. Abalones were planted by hand on red algal substrates and in the modules of Ebert and Ebert (1988). Recovery at six months, one and two years accounted for less than one percent of the total number seeded. At one and two years, significantly more seeds were recovered from sites with red sea urchins (*Strongylocentrotus franciscanus*), especially at sites designed as refugia from the commercial sea urchin harvest, compared with sites lacking urchins (Rogers-Bennett and Pearse 1998). Thus, sea urchins are an important component of the natural habitat and ecology of abalones and may be an essential element in abalone restoration in habitats of low rugosity.

Gaffney et al. (1996) presented genetic evidence that the 1979 outplanting of 42 431 red abalones at Tyler Bight on San Miguel Island was successful, based on allozyme analysis of animals fished commercially in 1992. They found a highly significant heterozygote excess at the GPI locus; the other three loci tested were not different from wild populations. Their suggestion that outplants dominated the 1992 sample, however, requires that natural recruitment of similar year-classes over more than 10 km of the island was minimal, an unlikely assumption for this region where red abalones remain relatively abundant (Haaker 1994). The peculiar nature of the broodstock that would be necessary to produce the observed results for the four loci tested, along with the possibility that sample degradation may have produced the GPI results (Ronald Burton, Scripps Institution of Oceanography, University of California, San Diego, La Jolla, California 92093–0202, personal communication), suggests that these results may be an artifact.

Larval seeding

The high cost, behavioral problems, and low survival associated with outplanted juvenile seed that have spent months or more in an artificial hatchery environment suggested the release of competent larvae as an alternate approach. This approach was based on the premise that the very high mortality rate characteristic of this early stage could be offset by the large numbers that could be produced at low cost. Preece et al. (1997) reviewed the history of larval seeding from the earliest efforts in Mexico. Tong et al. (1987) and Schiel (1992) presented short-term experiments in New Zealand and Preece et al. (1997) established that survival of post-larvae was density dependent in experiments lasting 11 months in South Australia. Altstatt and Raimondi (Jessie Altstatt, Marine Science Institute, University of California, Santa Barbara, CA 93106, personal communication) have been conducting larval seeding experiments with red abalone off the coast of Santa Barbara (Fig. 2). They report encouraging results on the time scale of days to months from releasing competent larvae into tented crates of cobble and observed that high sediments associated with storm waves caused abrasion and poor success; final analysis is pending. Efforts to undertake larval seeding by commercial interests in California have been hampered by the difficulty of distinguishing outplanted from wild stock.

Hatchery problems

Two serious abalone health problems, withering syndrome (WS) and sabellid polychaete infestations, in hatchery stocks led to a ban on outplanting in 1996. WS, a fatal disease, is characterized by an atrophied foot muscle, discoloration of the epipodium, reduced activity, and loss of ability to adhere tightly to the substratum (Altstatt et al. 1996). First observed on Santa Cruz Island in 1985, it spread throughout the Channel Islands with mortality approaching 95% relative to 1985 levels and then north of Point Conception into central California (Fig. 2). Lafferty and Kuris (1993) found that the geographic and temporal patterns of the spread of WS best supported an infectious disease process independent of environmental factors. WS is transmissible among black abalone cohabitating aquaria and studies found a correlation between WS and the presence of an intracellular bacterium, a rickettsial-like procaryote, in these experiments (Carolyn Friedman, CDFG Fish Health Laboratory, Bodega Marine Laboratory, P.O. Box 247, Bodega Bay, CA 94923, personal communication).

WS has been documented in wild red abalones on San Miguel Island and similar symptoms have been reported in pink and green abalones as well. During the recent El Niño, cultured red abalones exposed to elevated water temperatures suffered from WS (Carolyn Friedman, CDFG Fish Health Laboratory, Bodega Marine Laboratory, P.O. Box 247, Bodega Bay, CA 94923, personal communication) and the disease has now spread through most commercial abalone aquaculture facilities in California (Fred Wendell, Marine Aquaculture Coordinator, CDFG, 213 Beach Street, Morro Bay, CA 93442, personal communication). In August 1998, the Director of CDFG imposed restrictions on the transfer of stock among aquaculture facilities to keep the problem from getting worse. No abalone with rickettsial infections may be shipped from any farm to another farm outside the known WS range without histological inspection by the CDFG Fish Health Laboratory. Evidence to date suggests that the primary mode of infection of WS is fecal–oral contamination among larger animals, so larvae are not considered to be carriers (Fred Wendell, Marine Aquaculture Coordinator, CDFG, 213 Beach Street, Morro Bay, CA 93442, personal communication).

In the late 1980s, some aquaculturists began observing deformed shells with a domed teratology often lacking respiratory pores and slow growth rates in their abalone stocks (Oakes and Fields 1996; Culver et al. 1997; Kuris 1998). These resulted from infestation of a sabellid polychaete that settles on the growing margin of the aperture of the abalones, which stimulates the mantle to secrete new laminar shell to form a tube for the worm. Infestations do not affect the quality of the meat but deform the shell to the point where the abalone's growth slows or virtually ceases. When heavily parasitized, the respiratory pores close, leading to increased mortality. Native to South Africa, these undescribed worms were accidentally introduced when infested abalones intended for commercial research were imported. The problem was spread throughout most of the abalone mariculture industry by the mid-1990s as seed and large animals without obvious symptoms were exchanged among farms (Culver et al. 1997; Kuris 1998).

The sabellids spread from at least one central California culture facility to wild populations in the rocky intertidal community below its seawater discharge. Kuris (1998) detected infested cultured abalones as well as infections in three species of *Tegula* at this site in 1995. Laboratory experiments showed that most prosobranch gastropods exposed to the worms become infested, with archaeogastropods seemingly most susceptible. Kuris developed an eradication program for hatcheries, which focuses on screening to prevent susceptible hosts from being released from the facility (based on the premise that the sabellid's larva would not survive the cascade from the discharge site) and removal of infested and healthy hosts from the intertidal zone adjacent to the facility where the sabellid escaped. While the intertidal eradication program appears to have been successful, it has obvious implications for outplanting of hatchery-reared seed. The CDFG now requires that all abalone aquaculture facilities develop a sabellid eradication plan and that animals be certified as sabellid free before transfers are allowed among different facilities. All outplantings were banned in January 1996 to prevent further infestations in the wild (Kuris 1998; Fred Wendell, Marine Aquaculture Coordinator, CDFG, 213 Beach Street, Morro Bay, CA 93442, personal communication).

The CDFG is interested in promoting abalone enhancement efforts but also must protect native stocks from infection. For future outplantings to go forward, either a veliger larvae or a sabellid-free certification is required. The larval certification is based on one inspection by CDFG staff and is good for 12 months. The sabellid-free certification requires three inspections over a 24-month period. These certifications allow culture operations to sell or release veliger or seed abalones to holders of valid "Private Stocking Permits" for release in state marine waters. As of 21 January 1999, no abalone culture facilities had passed the initial sabellid-free inspection; thus outplantings of hatchery-reared seed are at least two years away. Similarly no facilities have yet to receive veliger permits, but it is believed that several facilities could pass this inspection soon (Fred Wendell, Marine Aquaculture Coordinator, CDFG, 213 Beach Street, Morro Bay, CA 93442, personal communication). Thus, larval seeding is the only type of outplanting likely to be feasible in the near future.

Native adult transplants

Emmett and Jamieson (1988) investigated the biological and economic feasibility of transplanting large numbers of sublegal northern abalone (*H. kamtschatkana*) from exposed sites which supported poor growth to once productive commercial grounds. Survival and growth were evaluated nine months later. Their data and economic model indicated that a reasonable rate of return on investment would result from instantaneous natural mortality values of 0.80 or less, as calculated from the recovery of live abalones. One of their two experimental sites was projected to show a reasonable rate of return of legal-sized animals two years after transplantation. This study did not consider the potential of transplanted abalones to enhance reproduction and recruitment in the transplant sites (Emmett and Jamieson 1988).

A severe environmental bottleneck eliminated giant kelp (*Macrocystis pyrifera)* and badly stressed populations of lower-standing kelps on the PVP from the mid-1950s until the mid-1970s; all indications are that abalone populations were similarly affected (Tegner 1993). Improvements in sewage treatment and kelp restoration led to the recovery of kelp forests, but the response of abalone populations was minimal. Green abalones, which had supported a fishery on the PVP before the bottleneck, were abundant on nearby islands. The short planktonic period of green abalone larvae suggested that long-distance larval dispersal between isolated populations is rare. A drift tube experiment supported this hypothesis and stressed the importance of local brood stock (Tegner and Butler 1985*b*). To test the hypothesis that brood stock is required for the recovery of green abalone populations off PVP, 4453 reproductively mature green abalones were transplanted in 1981–1982 into two sites within the closed area where the drift tube data indicated a high probability of larval retention. The abalones, tagged and clustered to facilitate fertilization success, suffered less than 10% natural mortality in the first year and showed visual evidence of reproduction. In 1985, a large number of juvenile green abalones within the full-size range expected from the timing of the brood stock transplants and growth rates were found in a scale of hundreds of metres to four kilometres along the PVP. There was no equivalent increase in green abalone recruitment at two sites not likely to have been affected by the brood stock transplant at PVP (Tegner 1992).

Despite the increase in green abalone recruitment, the long-term success of the brood stock transplant was dependent on survival of the adults. Abundance of these large, tagged animals dropped precipitously between 1983 and 1985 and no live tagged abalones were found on surveys in 1989 and 1990. The virtual lack of tagged shells indicated that poaching was the major source of mortality. Similarly, Henderson et al. (1988) transplanted 517 adult pink abalones to an isolated reef within an area closed to the take of invertebrates but near a small boat anchorage on Santa Catalina Island to create a concentration of brood stock. After one year they were able to locate less than 5% of the animals; natural mortalities based on shell recoveries accounted for only 18% of the transplants.

Fishery closures

Fishery closures are an old management tool that has been adopted for different reasons in different areas, with different degrees of success. One particularly successful closure established adjacent to the William's Head Penitentiary near Victoria, BC, to minimize opportunity for inmate escapes protects a dense population of large, relatively old northern abalones, *H. kamtschatkana*. Beyond the region of 24 hours per day patrolling, poaching reduces individual size and density in this closed fishery (Wallace 1997). The 68-km closure along the mainland of southern California that began in 1977 was renewed every five years until all abalone fishing was closed in 1997. Surveys of the PVP in the late 1970s found low to moderate numbers of red, pink, green, and black abalones, including concentrations of red and black abalone that seemed large enough to be self-sustaining. The green abalone brood stock transplant brought 4453 adults to the PVP in 1981–1982. By 1988–1989, subtidal abundances were less than one green, red, or pink abalone per diver hour (Tegner 1993). Miller and Lawrenz-Miller (1993) documented the decimation of intertidal black abalone populations on the PVP. While environmental problems could have contributed to these declines, poaching appeared to be severe and enforcement was minimal (Tegner et al. 1992; Miller and Lawrenz-Miller 1993). The PVP is adjacent to heavily populated Los Angeles, the area is popular with recreational divers, and the commercial sea urchin fishery was not affected by the abalone closure.

These southern California results contrast dramatically with the situation on the north coast of the state. Despite serious poaching pressure (Daniels and Floren 1998), the north coast fishery is the most productive abalone fishery in North America. Commercial fishing is prohibited and sport take is restricted to free diving or shore picking (Karpov et al. 1998). The ban on scuba or hookah establishes a *defacto* closed area or harvest refugia in habitat below the depths worked by recreational divers. Additional protections include the distance from human population centers and cold water, rough sea conditions. Recreational red abalone landings averaged 920 metric tons for 1985–1989, comparable to the average of the commercial fishery in southern California from 1931 to 1969, and were worth $10 000 000 to local economies. Red abalone abundances have actually increased in heavily harvested areas in recent years, coincident with the red sea urchin fishery, suggesting competitive release (Karpov et al. 1998).

Marine reserves or protected areas, in which no fishing is allowed, are an increasingly discussed approach to fisheries management. Tasmania established four marine reserves in 1991 to conserve representative marine habitats, to provide reference locations for understanding anthropogenic impacts, and to create fish propagation areas. Within a year, sea urchin density and mean abalone size were significantly greater than in matched controls (Edgar and Barrett 1997).

Sea otters

Any discussion of the future of abalone enhancement efforts in southern California would be incomplete without consideration of sea otters (*Enhydra lutris*). The short life of many abalone fisheries in the absence of otters has fueled debate over whether otters or overexploitation led to the status of abalone populations within the sea otter's range in central California (Estes and VanBlaricom 1985). Regardless of the outcome of that debate, there is agreement that abalone populations persist in deep crevices and other protected habitat within the otter range (Hines and Pearse 1982); however, abundance and size–frequency distributions are not adequate to support commercial fishing. In perhaps the best documented transition, sea otters reduced the abundance and size-distribution of red abalones at Point Estero on the central California coast between 1967 and 1971. Densities declined by a factor of ten and overall population size was estimated to have stabilized at 7% of pre-otter levels by 1993 (Wendell 1994). This pattern was repeated as otters moved further south towards the Point Conception region (Laur et al. 1988). To provide a second population of these threatened mammals, 139 otters were reintroduced to San

Nicolas Island between 1987 and 1990, but only a few animals remain; most apparently returned to central California. As part of the law that permitted the reintroduction plan, otters that were found south of Point Conception but not at San Nicolas Island were to be removed, but containment was suspended in 1993. Until 1998, the southern range boundary of the main California otter population remained in central California. In April 1998, 101 otters moved south of Point Conception foraging as far east as western Santa Barbara (U.S. Fish and Wildlife Service 1998; Harry Liquornik, 1421 Robbins St., Santa Barbara, CA 93101, personal communication). The Coho region near Point Conception was one of the last remaining healthy red abalone beds in southern California and the otters have apparently affected stocks there as previously described for central California (Harry Liquornik, 1421 Robbins St., Santa Barbara, CA 93101, personal communication).

The future of sea otter range expansion is unresolved. The California population, classified as threatened under the Endangered Species Act, has declined more than 10% over the last three years (U.S. Fish and Wildlife Service 1998). The Fish and Wildlife Service, the agency charged with otter recovery, views the recent range expansion into southern California as a possible replacement for the failed San Nicolas Island transplant, placing a portion of the population at less risk from a catastrophic oil spill. Thus, with the uncertain population status of this threatened member of the "charismatic megafauna," the Fish and Wildlife Service and the public are unlikely to support curtailing range expansion. Because the northern Channel Islands, which supported a major portion of the commercial abalone fishery, are easily visible at water level from the coastline east of Point Conception, it is likely that the otters will reoccupy these habitats soon unless measures are taken to reserve some areas for shellfish.

Discussion

Seeding of hatchery-reared animals was (and remains) an enormously appealing concept for several reasons: (*i*) the well-known successes with salmon and trout; (*ii*) the large-scale abalone seeding efforts in Japan; (*iii*) it is an alternative to restrictions on fishing effort; (*iv*) seeding was potentially much faster than natural recovery; and (*v*) it could prove self-sustaining (Tegner and Butler 1989). There are also important negative aspects of seeding hatchery- reared abalones to consider: (*i*) the prospects of enhancement may divert resources from management and needed assessment; (*ii*) seeding may deter harder decisions to reduce fishing effort to protect native stocks; (*iii*) potential spread of disease such as WS or introduction of exotic parasites such as the sabellid worm into wild populations; and (*iv*) potential loss of genetic diversity (McCall 1989; Tegner and Butler 1989). There is short-term evidence that seeding can prove effective in certain situations (Schiel 1993), but the only long-term results are from Japan. While the Japanese fishery has certainly not collapsed like those in southern and central California, the long-term decline in take in the face of ever-increasing rates of seeding (Uki 1989; Seki and Sano 1998) — coupled with poor results in California — strongly suggests that we should not view seeding as the main approach to restoring and enhancing depleted stocks.

The southern California seeding experiments reviewed above included three species in a range of sizes, island and mainland sites, and studies of devices to ease acclimation, but evidence for long-term survival was uniformly poor. Given that predation pressure generally increases from high to low latitudes, Tegner and Butler (1989) speculated that seeding might be more successful in higher latitudes, but the experiments of Rogers-Bennett and Pearse (1998) in central and northern California do not support that hypothesis. Two aspects of seeding seem to merit further research, acclimation of hatchery-reared seed before outplanting and larval seeding. Schiel and Welden (1987) found evidence that hatchery-reared abalones acclimated to reefs in laboratory aquaria after several days, but their only suggestion of how to do this in the field was to use the modules of Ebert and Ebert (1988). As described above, these modules did not provide long-term benefit in either southern (David Parker, CDFG, 330 Golden Shore, Long Beach, CA 90802, personal communication) or central and northern California (Rogers-Bennett and Pearse 1998). Munro and Bell (1997) reviewed how modifications of hatchery substrata and exercise conditioning were able to precondition outplants of other taxa; but whether this can be achieved for abalones is unknown. While the results of abalone larval releases to date are not dramatic, the very clear evidence for recruitment limitation in southern California abalone populations today supports further experimentation with larvae on the Channel Islands. With these two exceptions, I argue that it is time that California move beyond seeding and put its emphasis on other enhancement approaches.

I am especially concerned about critically depleted white abalones (Davis et al. 1998). Given the low numbers of animals known to still exist, the difficulty in finding and collecting them, and the apparent age of remaining animals, hatchery-propagation of juveniles and reseeding must be attempted. However, the problems keeping white abalones alive in hatcheries to date and the potential for problems with WS and sabellid infestations argue that, if at all possible, we should aggregate a group on a protected reef in case of disasters on land and to allow for natural reproduction and recruitment.

Richards and Davis (1993) observed that black abalone recruitment virtually ceased when adult populations dropped below 50% of their initial abundance. Shepherd and Brown (1993) estimated the minimum viable population size for an isolated population of *H. laevigata* to be about 1000 individuals. They showed that *H. laevigata* at low densities lose the ability to aggregate for spawning, so the effective population size declines much more rapidly than the true population size as density declines. Most local densities of pink and green abalone throughout southern California and of red abalone at locations other than San Miguel Island (Haaker 1994; Karpov et al. 1998) are probably too low for successful fertilization and thus many healthy animals are functionally sterile.

Future enhancement efforts should center on the interrelated lessons offered by the persistence of the north coast red abalone fishery (Karpov et al. 1998), the early successes of the green abalone brood stock transplant (Tegner 1992,

1993), and the minimum viable population estimates of Shepherd and Brown (1993). The north coast fishery appears to persist because an adequate density of brood stock is protected by the *de facto* depth refuge. The brood stock transplant at PVP established two sites with apparently minimum viable populations based on the widespread recruitment of green abalones observed three years later. However, this experiment ultimately failed due to poaching; the abalone fishing closure was not adequate to protect the incipient recovery (Tegner 1992, 1993). Brood stock aggregation can work if we develop the political will to establish and enforce marine protected areas (MPAs) closed to all fishing. Meaningful MPAs are necessary for the establishment of high density patches of large abalones, the requirement for the production of large numbers of eggs, and their successful fertilization. The important role of sea urchins in maintaining areas of encrusting coralline algae for abalone settlement and the role of the spine canopy in juvenile abalone survival (e.g., Tegner and Dayton 1977; Day 1998; Rogers-Bennett and Pearse 1998), along with the poaching problems on the north coast associated with urchin fishing (Tegner et al. 1992), indicate why all fishing should be prohibited in these areas.

California currently has a confusing network of 104 MPAs with 12 different designation categories related to the establishing authorities and allowed uses (McArdle 1998). In 93 of these some level of recreational fishing is allowed and in 94 some level of commercial fishing is permitted. Only nine of these MPAs, representing about 0.2% of state waters, restrict all species from being taken. Comparison of four intertidal MPAs with four unprotected areas in southern California found no difference between these categories, i.e., that the combination of heavy visitor use with the lack of awareness of and enforcement of MPA rules offered no protection to these ecosystems (Murray 1998). Given cutbacks in the CDFG enforcement staff, it is likely that this is the case for most MPAs, regardless of designation. A statewide task force is completing an evaluation of the existing system and developing recommendations for simplifying classifications. It is hoped that when this evaluation is complete, the California Fish and Game Commission and (or) the state legislature will support establishment of a scientific panel to recommend a network of MPAs for a study of their role in fisheries management.

The efficacy of MPAs is an active area of research in the 1990s as increasing numbers of scientists become concerned that existing management schemes are not working (e.g., Dugan and Davis 1993). There is widespread agreement that MPAs will protect biodiversity, lead to larger and more abundant stocks and more reproductive output of species that remain in the reserves, and importantly, provide reference sites, which scientists can use to separate anthropogenic impacts from natural changes. The degree to which reproduction inside MPAs can enhance stocks outside reserves remains to be demonstrated for most taxa, but the sustained success of the north coast fishery with its depth refugium strongly supports further use of this approach for abalones. Reserves cannot replace traditional forms of fishery management but do provide insurance against failures of that management. This is especially important for sedentary broadcast spawning species such as abalones for which annual stock assessment is prohibitively expensive and yet reproduction and recruitment of these molluscs appears to be sensitive to the climate shifts caused by the increased numbers of El Niño events since 1977 (Tegner and Dayton 1987; Tegner et al. 1996). Continued fishing of brood stock during periods of poor recruitment risks driving local abalone populations below the minimum viable population size. MPAs can protect abalone spawning aggregations, the dense patches of larger abalones necessary for successful reproduction — just what the fishery targets.

The California Abalone Association, an organization of commercial fishermen, is an active player in abalone management. They have planted seed abalones at several sites to augment common property resources (Pleschner 1984); however, quantitative evaluations have not been made. From a tax on landings, they sponsored the juvenile seeding research of Rogers-Bennett and Pearse (1998) and the larval work of Raimondi and Altstatt (Jessie Altstatt, Marine Science Institute, University of California, Santa Barbara, CA 93106, personal communication), described above. They have funds for additional enhancement efforts and remain interested in further studies of larval outplantings. The group is linking their support for MPA proposals to a resolution of sea otter range expansion that protects some areas for shellfisheries. (Harry Liquornik, 1421 Robbins St., Santa Barbara, CA 93101, personal communication).

Thus, the future of abalone enhancement and indeed, abalone fishing, in California is at a crossroads. We have learned a great deal about abalones since the 1970s, much of it the hard way. Perhaps the most significant lessons are the importance of brood stock aggregations for successful recruitment (e.g., Shepherd and Brown 1993) and that seeding will not be the hoped for panacea for depleted stocks (e.g., Tegner and Butler 1989). With abalone fisheries closed in the southern and central parts of the state, California faces hard political decisions. If we invest in MPAs with realistic levels of public education, enforcement, and penalties for poaching and then create brood stock aggregations, stock enhancement may be possible for species in which adequate levels of brood stock can be collected. For white and perhaps black abalones, hatchery seed production may be the required primary approach, followed by outplanting at very large sizes (Davis et al. 1998). The choice of sites for enhancement activities can be based on previous rates of sea otter range expansion and distance from the location of the herd. Identifying habitat of high rugosity, such as that off Hopkins Marine Station which supports a substantial population of red abalones within the sea otter range (Hines and Pearse 1982), may be very important for clustering rare brood stock as the otter range expands, as well as for establishing MPAs. The issue of sea otter range expansion and whether some areas should be set aside for fishing is also a political choice. One can only hope that somehow the decision will be made on the basis of objective data about the values of shellfisheries in some regions of southern California as well as the emotional appeal of the charismatic megafauna.

Acknowledgments

I thank A. Campbell for the invitation to participate in this workshop, J. Altstatt, R. Burton, C. Friedman, P. Haaker, K. Karpov, H. Liquornik, D. Parker, K. Riser, S. Shepherd, and F. Wendell for their help in gathering information, and the Pew Fellows Program in Marine Conservation for its financial support. A. Campbell, K. Karpov, L. Rogers-Bennett, P. Haaker, N. Sweijd, and S. Shepherd provided thoughtful suggestions for the manuscript and I am grateful to S. Shepherd and P. Haaker who filled in for me in Nanaimo when the flu struck.

References

Altstatt, J.M., Ambrose, R.F., Engle, J.M., Haaker, P.L., Lafferty, K.D., and Raimondi, P.T. 1996. Recent declines of black abalone, *Haliotis cracherodii*, on the mainland coast of central California. Mar. Ecol. Prog. Ser. **142**: 185–192.

Babcock, R., and Keesing, J. 1997. Fertilization ecology of *Haliotis laevigata*. Abstracts of the Third International Abalone Symposium, Monterey, California, 26–31 October 1997. California Sea Grant College System, La Jolla, Calif.

Culver, C.S., Kuris, A.M., and Beede, B. 1997. Identification and management of the exotic sabellid pest in California cultured abalone. University of California Sea Grant College System, La Jolla, California, Publication No. T-041. 29 p.

Daniels, R., and Floren, R. 1998. Poaching pressure on northern California's abalone fishery. J. Shellfish Res. **17**: 859–862.

Davis, G.E. 1995. Recruitment of juvenile abalone (*Haliotis* spp.) measured in artificial habitats. Mar. Freshw. Res. **46**: 549–554.

Davis, G.E., Haaker, P.L., and Richards, D.V. 1998. The perilous condition of white abalone, *Haliotis sorenseni* Bartsch 1940. J. Shellfish Res. **17**: 871–876.

Day, E.G. 1998. Ecological interactions between abalone (*Haliotis midae*) juveniles and sea urchins (*Parechinus angulosus*), off the south-west coast of South Africa. Ph.D. dissertation, University of Cape Town, Republic of South Africa.

Dugan, J.E., and Davis, G.E. 1993. Application of marine refugia to coastal fisheries management. Can. J. Fish. Aquat. Sci. **50**: 2029–2042.

Ebert, T.B., and Ebert, E.E. 1988. An innovative technique for seeding abalone and preliminary results of laboratory and field trials. Calif. Fish Game, **74**: 68–81.

Edgar, G.J., and Barrett, N.S. 1997. Short term monitoring of biotic changes in Tasmanian marine reserves. J. Exp. Mar. Biol. Ecol. **213**: 261–279.

Emmett, B., and Jamieson, G.S. 1988. An experimental transplant of northern abalone, *Haliotis kamtschatkana*, in Barkley Sound, British Columbia. Fish. Bull. (U.S.) **87**: 95–104.

Estes, J.A., and VanBlaricom, G.R. 1985. Sea otters and shellfisheries. *In* Marine mammals and fisheries. *Edited by* J.R. Beddington, R.J.H. Beverton, and D.M. Lavigne. George Allen and Unwin, London. pp. 187–235.

Gaffney, P.M., Rubin, V.P., Hedgecock, D., Powers, D.A., Morris, G., and Hereford, L. 1996. Genetic effects of artificial propagation: signals from wild and hatchery populations of red abalone in California. Aquaculture, **143**: 257–266.

Haaker, P.L. 1994. Assessment of abalone resources at the Channel Islands. *In* The Fourth Channel Island Symposium: Update on the Status of Resources. *Edited by* W.L. Halvorson and G.J. Maender. Santa Barbara Museum of Natural History, Santa Barbara, Calif. pp. 83–95.

Henderson, K.C., Parker, D.O., and Haaker, P.L. 1988. The survival and growth of transplanted adult pink abalone, *Haliotis corrugata*, at Santa Catalina Island. Calif. Fish Game, **74**: 82–86.

Hines, A.H., and Pearse, J.S. 1982. Abalones, shells and sea otters: dynamics of prey populations in Central California. Ecology, **63**: 1547–1560.

Ikenoue, H., and Kafuku, T. 1992. Modern methods of aquaculture in Japan. Kodansha Ltd., Tokyo, Japan.

Karpov, K.A., and Tegner, M. J. 1992. Abalones. *In* California's living marine resources and their utilization. *Edited by* W.S. Leet, C.M. Dewees, and C.W. Haugen. California Sea Grant Extension Publication UCSGEP-92-12. California Sea Grant College System, La Jolla, Calif. pp. 33–36.

Karpov, K.A., Haaker, P.L., Albin, D., Taniguichi, I.K., and Kushner, D. 1998. The red abalone, *Haliotis rufescens*, in California: importance of a depth refuge to abalone management. J. Shellfish Res. **17**: 863–870.

Kojima, H. 1981. Mortality of young Japanese black abalone, *Haliotis discus discus*, after transplantation. Bull. Jpn. Soc. Sci. Fish. **47**: 151–159.

Kuris, A. 1998. Management and risk assessment of sabellid polychaete infestations in abalone mariculture. California Sea Grant, Report of Completed Projects, 1994–97, California Sea Grant College System, La Jolla, California. Publication No. R-044. pp. 89–92.

Lafferty, K.D., and Kuris, A.M. 1993. Mass mortality of abalone, *Haliotis cracherodii*, on the California Channel Islands: tests of epidemiological hypotheses. Mar. Ecol. Prog. Ser. **96**: 239–248.

Laur, D.R., Ebeling, A.W., and Coon, D.A. 1988. Effects of sea otter foraging on subtidal reef communities off central California. *In* The community ecology of sea otters. *Edited by* G.R. VanBlaricom and J.A. Estes. Ecological Studies, Vol. 65. Springer Verlag, New York. pp. 151–168.

Levitan, D.R. 1995. The ecology of fertilization in free-spawning invertebrates. *In* Ecology of marine invertebrate larvae. *Edited by* L. McEdward. CRC Press, Boca Raton, Florida. pp. 123–156.

McArdle, D.A. 1998. The status of California marine protected areas. *In* California and the World Ocean '97. *Edited by* O.T. Magoon, H. Converse, B. Baird, and M. Miller-Henson. American Society of Civil Engineers, Reston, Va. pp. 74–85.

McCall, A.D. 1989. Against marine fish hatcheries: ironies of fishery politics in the technological era. Calif. Coop. Oceanic Fish. Invest. Rep. **30**: 46–48.

McCormick, T.B., Herbinson, K., Mill, T.S., and Altick, J. 1994. A review of abalone seeding, possible significance, and a new seeding device. Bull. Mar. Sci. **55**: 680–693.

Miller, A.C., and Lawrenz-Miller, S.E. 1993. Long-term trends in black abalone, *Haliotis cracherodii* Leach 1814, populations along the Palos Verdes Peninsula, California. J. Shellfish Res. **12**: 195–200.

Munro, J.L., and Bell, J.D. 1997. Enhancement of marine fisheries resources. Rev. Fish. Sci. **5**: 185–222.

Murray, S.N. 1998. Effectiveness of marine life refuges on southern California shores. *In* California and the World Ocean '97. *Edited by* O.T. Magoon, H. Converse, B. Baird, and M. Miller-Henson. American Society of Civil Engineers, Reston, Va. pp. 1453–1465.

Oakes, F.R., and Fields, R.C. 1996. Infestation of *Haliotis rufescens* shells by a sabellid polychaete. Aquaculture, **140**: 139–143.

Parker, D.O., Haaker, P.L., and Barsky, K.C. 1989. Evaluation of experimental abalone seeding modules. First International Symposium on Abalone Biology, Fisheries, and Culture, 21–25 November 1989, La Paz, B.C.S., Mexico. Abstract 58.

Pennington, J.T. 1985. The ecology of fertilization of echinoid eggs: the consequences of sperm dilution, adult aggregation, and synchronous spawning. Biol. Bull. (Woods Hole) **169**: 417–430.

Pleschner, D.B. 1984. The great abalone plant. Oceans, **17**: 27–31.

Preece, P.A., Shepherd, S.A., Clark, S.M., and Keesing, J.K. 1997. Abalone stock enhancement by larval seeding: effect of larval density on settlement and survival. Moll. Res. **18**: 265–273.

Richards, D.V., and Davis, G.E. 1993. Early warnings of modern population collapse in black abalone, *Haliotis cracherodii*, Leach 1814, at the California Channel Islands. J. Shellfish. Res. **12**: 189–194.

Rogers-Bennett, L., and Pearse, J.S. 1998. Experimental seeding of hatchery-reared juvenile red abalone in northern California. J. Shellfish Res. **17**: 877–880.

Saito, K. 1979. Studies on the propagation of the Ezo abalone, *Haliotis discus hannai* Ino. I. Analysis of the relationship between transplantation and catch in Funka Bay Coast. Bull. Jpn. Soc. Sci. Fish. **45**: 695–704.

Schiel, D.R. 1992. The enhancement of paua (*Haliotis iris* Martyn) populations in New Zealand. *In* Abalone of the world: biology, fisheries and culture. *Edited by* S.A. Shepherd, M.J. Tegner, and S.A. Guzmán del Próo. Fishing News Books, Blackwell Science Ltd., Oxford, U.K. pp. 474–484.

Schiel, D.R. 1993. Experimental evaluation of commercial-scale enhancement of abalone, *Haliotis iris*, populations in New Zealand. Mar. Ecol. Prog. Ser. **97**: 167–181.

Schiel, D.R., and Welden, B.C. 1987. Responses to predators of cultured and wild red abalone, *Haliotis rufescens*, in laboratory experiments. Aquaculture, **60**: 173–188.

Seki, T., and Sano, M. 1998. An ecological basis for the restoration of Japanese abalone populations. Bull. Tohoku Natl. Fish. Res. Inst. **60**: 23–40. [English abstract.]

Shepherd, S.A., and Brown, L.D. 1993. What is an abalone stock: implications for the role of refugia in conservation. Can. J. Fish. Aquat. Sci. **50**: 2001–2009.

Sweijd, N., Snethlage, Q., Harvey, D., and Cook, P. 1998. Experimental abalone seeding in South Africa. J. Shellfish Res. **17**: 897–904.

Tarr, R.J.Q., Williams, P.V.G., and McKenzie, A.J. 1996. Abalone, sea urchins, and rock lobster: a possible ecological shift that may affect traditional fisheries. S. Afr. J. Mar. Sci. **17**: 319–323.

Tegner, M.J. 1989. The California abalone fishery: production, ecological interactions, and prospects for the future. *In* Marine invertebrate fisheries: their assessment and management. *Edited by* J.F. Caddy. John Wiley & Sons, New York. pp. 401–420.

Tegner, M.J. 1992. Brood stock transplants as an approach to abalone stock enhancement. *In* Abalone of the world: biology, fisheries and culture. *Edited by* S.A. Shepherd, M.J. Tegner, and S.A. Guzmán del Próo. Fishing News Books, Blackwell Science Ltd., Oxford, U.K. pp. 461–473.

Tegner, M.J. 1993. Southern California abalones: can stocks be rebuilt using marine harvest refugia? Can. J. Fish. Aquat. Sci. **50**: 2010–2018.

Tegner, M.J., and Butler, R.A. 1985*a*. The survival and mortality of seeded and native red abalones, *Haliotis rufescens*, on the Palos Verdes Peninsula. Calif. Fish Game, **71**: 150–163.

Tegner, M.J., and Butler, R.A. 1985*b*. Drift tube study of the dispersal potential of green abalone (*Haliotis fulgens*) larvae in the Southern California Bight: implications for the recovery of depleted populations. Mar. Ecol. Prog. Ser. **26**: 73–84.

Tegner, M.J., and Butler, R.A. 1989. Abalone seeding. *In* Handbook of culture of abalone and other marine gastropods. *Edited by* K. Hahn. CRC Press, Boca Raton, Fla. pp. 157–182.

Tegner, M.J., and Dayton, P.K. 1977. Sea urchin recruitment patterns and implications of commercial fishing. Science (Washington, D.C.), **196:** 324–326.

Tegner, M.J., and Dayton, P.K. 1987. El Niño effects on southern California kelp forests. Adv. Ecol. Res. **17**: 243–279.

Tegner, M.J., DeMartini, J.D., and Karpov, K.A. 1992. The California red abalone fishery: a case study in complexity. *In* Abalone of the world: biology, fisheries and culture. *Edited by* S.A. Shepherd, M.J. Tegner, and S.A. Guzmán del Próo. Fishing News Books, Blackwell Science Ltd., Oxford, U.K. pp. 370–383.

Tegner, M.J., Dayton, P.K., Edwards, P.B., and Riser, K.L. 1996. Is there evidence for long-term climatic change in southern California kelp forest communities? Calif. Coop. Oceanic Fish. Invest. Rep. **37**: 111–125.

Tong, L.J., Moss, G.A., and Illingworth, J. 1987. Enhancement of a natural population of the abalone, *Haliotis iris*, using cultured larvae. Aquaculture, **62**: 67–72.

Uki, N. 1989. Abalone seeding production and its theory. Int. J. Aquacul. Fish. Technol. Part 1, pp. 3–15; Part 2, pp. 125–132; Part 3, pp. 224–231.

United States Fish and Wildlife Service. 1998. Sea otters in the management zone—an update. Report to the Marine Mammal Commission, December 1998. Washington, D.C. 7 p.

Wallace, S. 1997. The role of marine reserves in the management of non-migratory species in coastal British Columbia. *In* The design and monitoring of marine reserves. University of British Columbia, Vancouver, Fisheries Centre Res. Rep. **5**: 24–25.

Wendell, F. 1994. Relationship between sea otter range expansion and red abalone abundance and size distribution in central California. Calif. Fish Game Bull. **80**: 45–56.

Wyner, A.J., Moore, J.E., and Cicin-Sain, B. 1977. Politics and management of the California abalone fishery. Mar. Pollut. Bull. **1**: 326–339.

Rehabilitation of northern Japanese abalone, *Haliotis discus hannai*, populations by transplanting juveniles

Tetsuo Seki and Kazuya Taniguchi

Abstract: Transplant experiments with juveniles of the northern Japanese abalone, *Haliotis discus hannai* were undertaken in the test fishing ground of Ooya, Miyagi Prefecture, Japan, during 1996–1998. This fishing ground contains representatives of three typical algal communities on rocky reef, i.e., cobblestone habitat, the *Eisenia bicyclis* forest, and coralline flats. In two experiments, 100 000 seed abalone (mean 24.5 mm shell length, SL) in 1996 and 26 000 (mean 16.5 mm SL) in 1997 were introduced to a cobblestone nursery habitat. In addition, two identical groups of 20 000 seed abalone (mean 16.5 mm SL) were separately introduced in 1998 into cobblestone habitat and coralline flats. For three years after the first transplant, site selection, growth, and survival of abalone were monitored by scuba diving. At 853 days after the first transplant, 26.7% of transplanted abalone survived in the test fishing ground. The abalone grew to 76.6–87.3 mm SL and the mean daily growth rate of abalone was higher in the *Eisenia* forest than in the cobblestones. Most large abalone >40 mm SL were within the cobblestone habitat or the *Eisenia* forest, while transplants <40 mm SL remained within the cobblestone habitat. The *Eisenia bicyclis* forest played an important role as a growing habitat for abalone >30–40 mm SL.

Résumé : Des expériences de transplantation de juvéniles d'ormeau rouge du nord, *Haliotis discus hannai*, ont été menées en 1996–1998 dans la pêcherie expérimentale d'Ooya, préfecture de Miyagi, au Japon. Cette pêcherie présente des zones représentatives de trois communautés algales typiques des récifs rocheux : habitat de galets, herbier d'*Eisenia bicyclis* et fonds colonisés par les algues corallines. Lors de deux expériences, 100 000 naissains d'ormeau (longueur de coquille moyenne de 24,5 mm) en 1996, et 26 000 (LC moyenne de 16,5 mm) en 1997, ont été introduits dans un habitat de nourricerie à fond de galets. De plus, nous avons implanté séparément en 1998 deux groupes identiques de 20 000 naissains (LC moyenne de 16,5 mm) dans un habitat de galets et sur des fonds colonisés par des algues corallines. Pendant les trois ans qui ont suivi la première transplantation, des plongeurs ont surveillé le choix du site, la croissance et la survie des ormeaux. Au 853e jour après la première transplantation, 26,7 % des ormeaux transplantés survivaient dans la pêcherie expérimentale. Ils avaient atteint la taille de 76,6–87,3 mm LC, et le taux moyen de croissance quotidienne était plus élevé dans l'herbier d'*Eisenia* que sur les galets. La plupart des gros ormeaux >40 mm LC se retrouvaient dans l'habitat de galets ou dans l'herbier d'*Eisenia*, tandis que les ormeaux transplantés <40 mm LC restaient dans l'habitat de galets. L'herbier d'*Eisenia bicyclis* jouait un rôle important comme habitat de croissance pour les ormeaux >30–40 mm LC.

[Traduit par la Rédaction]

Introduction

Transplantation of juvenile abalone into fishing grounds has been used historically in Japan in an attempt to rehabilitate abalone resources. Of the four commercially important abalone species in Japan, the biology and culture of the northern Japanese abalone, *Haliotis discus hannai*, has been studied the most intensively in the last few decades. The development of seed production techniques for this species, based on biological studies (Uki et al. 1996; Seki 1997), has resulted in the establishment of large-scale seed production in hatcheries throughout Japan. Seed from the hatcheries has been transplanted into fishing grounds at a rate of about 1.5 × 10^8 individuals per year since the early 1980s. However, the commercial catch of this species has declined from 3000 to 1000 t per year. Although abalone seed recaptures were marginally profitable when transplanted at a size of 30 mm shell length (SL) or more, the wild abalone population did not increase. Hence further research was needed to find methods to replenish northern Japanese abalone populations.

The natural distribution of *H. discus hannai* overlaps with much of the *Laminaria* spp. distribution, but only with the northern limit of *Eisenia bicyclis* (Fig. 1). Since the importance of algal communities as a food source for released seed has been recognized only recently, the relationship between algal communities and abalone has not been well

T. Seki. National Research Institute of Aquaculture, 422-1 Nansei, Watarai, Mie, 516-0193, Japan.
K. Taniguchi. Tohoku University, Faculty of Agriculture, Amemiya, Tsutsumidori, Sendai, 981-8555, Japan.
Correct citation: Seki, T., and Taniguchi, K. 2000. Rehabilitation of northern Japanese abalone, *Haliotis discus hannai*, populations by transplanting juveniles. *In* Workshop on Rebuilding Abalone Stocks in British Columbia. *Edited by* A. Campbell. Can. Spec. Publ. Fish. Aquat. Sci. 130. pp. 72–83.

Fig. 1. The distribution of *Haliotis discus hannai* and dominant *Laminaria* species in northern Japan.

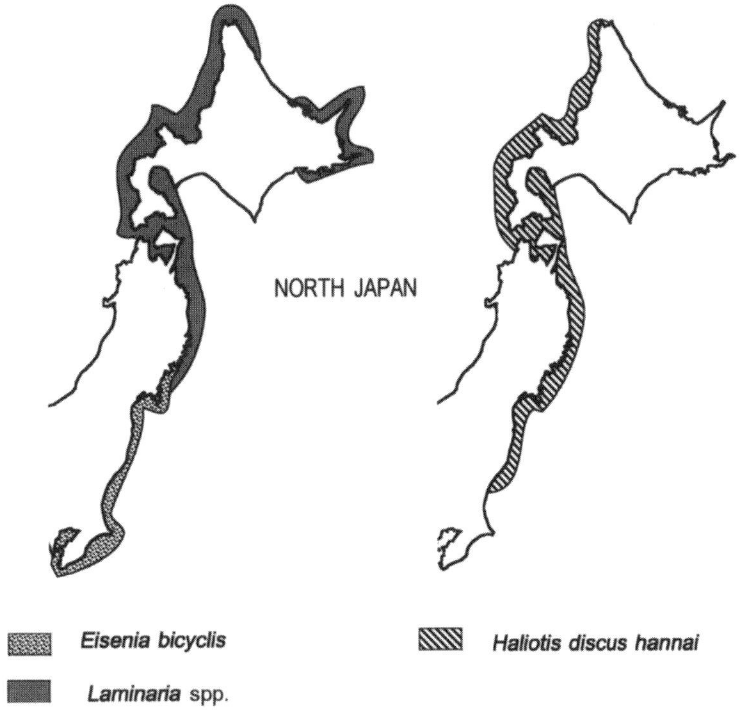

understood. Past abalone transplant practices in Japan have been designed to maximize recovery of the released seed rather than to enhance recruitment to the natural population (Seki and Sano 1998). However, recent studies on the chemical signaling role of marine algae have provided new insights into the interactions between algal communities and herbivores (Taniguchi et al. 1991, 1992, 1994; Seki and Taniguchi 1996). These studies show that most marine algae may survive strategically by controlling herbivore life with various chemical effects of metabolites (Geiselman and McConnell 1981; Steinberg 1984, 1985, 1988; Hay and Fenical 1988; Shiraishi et al. 1990, 1991; Taniguchi et al. 1992, 1994). Although studies on the recruiting process of *H. discus hannai* provide valuable information on the natural occurrence of pelagic larvae related to the coastal geomorphology (Sasaki and Shepherd 1995), the role of algal communities in larval settlement and subsequent juvenile growth is still unknown. The purpose of this study was to determine the site selection of 166 000 abalone juvenile seed transplanted into algal communities during 1996–1998. We describe the long-term change in distribution, growth, and survival of transplanted abalone in relation to the algal community in which they lived.

Materials and methods

Test fishing ground

The transplant experiment was undertaken at the test fishing ground of Ooya, Miyagi Prefecture (Fig. 2) where the wild abalone population was depleted due to recent overexploitation. This fishing ground is located near the northern limit of *Eisenia bicyclis* (Fig. 1) and contains representatives of three typical algal communities on rocky reef, i.e., cobblestone habitat, the *E. bicyclis* forest, and coralline flats.

Abalone juvenile production

The seed abalone were produced at Fukushima Prefectural Aquaculture Center (Okuma-cho, Futaba-gun, Fukushima 979-1301, Japan) with warm effluent seawater obtained from a nearby power plant. Juvenile abalone were reared to the required average size of 25 mm SL before early summer. In order to ensure genetic compatibility, 35 abalone adults were collected from the test fishing ground in the previous year (July 10, 1995) as parent brood stock. A total of 100 000 juvenile abalone (range 13–35 mm SL, mean 24.5 mm SL) were produced at the Fukushima Center in 1996 from these adults after they were conditioned in the hatchery. The abalone juvenile produced in 1997 and 1998 were also produced at the Fukushima Center with the parent abalone brood stock separately collected from the Ooya test fishing ground each year.

Transport, acclimation, and transplanting of seed

Units of about 2000 abalone were packed in refrigerated styrofoam boxes after seed were wrapped in cotton cloth and sealed in a vinyl bag with seawater moisture and oxygen gas. The seed were first transported to the intermediate abalone rearing facility at the Ooya Fisheries Cooperation on July 23, 1997 and acclimated by holding them in plastic cages (50 × 50 × 30 cm) suspended from a floating raft at the Ooya fishing port near the test fishing ground for 7 days. A layer of empty oyster shells was placed on coarse plastic netting (mesh size 3 × 3 cm) in the plastic cages. The cages were set in the raft 10 days before the arrival of juvenile abalone to allow diatoms to cover the oyster shells. When the seeds were introduced into the plastic cages, they crawled

Fig. 2. Location of the Ooya experimental site for abalone seed transplants and transect survey lines in Miyagi Prefecture.

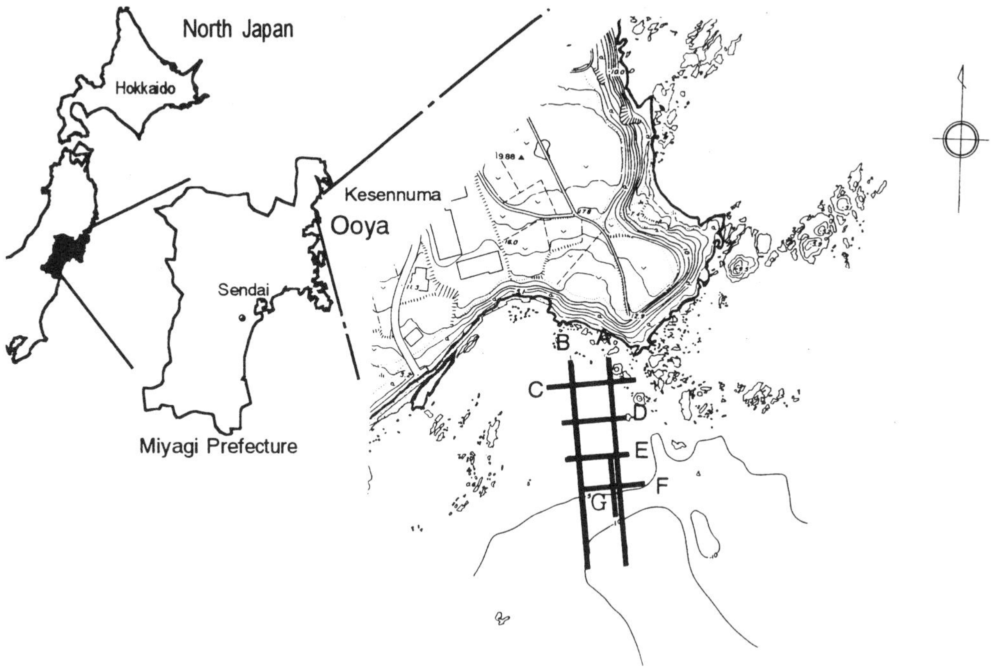

onto the oyster shells and remained there. A few fronds of *Eisenia* were placed in each cage every two days to promote acclimation of the seed to the natural environment.

The transplant experiments were designed to introduce abalone seed at the microalgae feeding stage into the cobblestone habitat, which was believed to be the main nursery area prior to the macroalgal feeding stage, in order to observe their choice of habitat and measure their growth. Seeds were transplanted annually during 1996–1998. In the first year, a total of 100 000 juvenile abalone (range 13–35 mm SL, mean 24.5 mm SL) were placed by hand, while still attached to oyster shells, into three 5-m^2 areas of the cobblestone zone by divers on July 31, 1996. The oyster shells were taken out of the study area on the next day with netting. Assuming that the survival rate of transplanted 3-year-old juveniles was <10%, 100 000 seeds would be required to achieve a final density of 1 m^{-2} in a 1-ha experimental area. This assumption was based on the highest survival of past transplants in Iwate Prefecture where the recapture rate of transplanted 30-mm-SL abalone was 25–30% (Takeichi 1988; Inoguchi 1993). The season of transplanting was adjusted to the first summer when wild 0+ juveniles started feeding on macroalgae. Because the size of wild abalone in early summer (June–July) in this area was estimated to be 14–17 mm SL (Sakai et al. 1975), the average size of the transplanted juveniles in 1997 was reduced to 16.5 mm SL. A total of 26 000 juvenile abalone was similarly introduced into the cobblestone habitat on June 30, 1997. At the final transplant on July 3, 1998, 20 000 juveniles (mean 16.5 mm SL) were transplanted into the cobblestone habitat and another 20 000 (of the same size) into the coralline flats to examine the roles of both habitats. Those placed in the coralline flats were distinctively tagged with a binary coded wire tag (Northwest Marine Technology, Inc., Washington, D.C.) fixed to the shell with an instant hardening epoxy resin (Erehon HR-123R, H-30, Erehon Synthesis Industry Co. Ltd., Oita, Japan). In independent rearing experiments over 5 months, tag shedding was <10% and the metal detector (for the coded wire tags) proved completely reliable.

Line transect survey

The distribution and growth of the transplanted and wild abalone population were derived from line transect data obtained by scuba divers at the experimental 1-ha site. The number, length, and positions of 1-m-wide line transects consisted of two parallel north–south lines (170 m, Fig. 2, A, B) and three parallel east–west lines (70 m and 50 m, Fig. 2, C–E) crossing the north–south lines at 30 m intervals. Most of the abalone seeds were placed near the intersection of lines A and C, located 20 m south from the north end in line A and 20 m west from the east end in line C. This intersection was set as the zero point of the XY axes to plot abalone distribution. In 1998, another set of transects (50 m each, Fig. 2, F, G) was added to follow the juveniles transplanted into the coralline flat. At the final survey, on December 1, 1998, the distribution of abalone in the other lines was also examined to determine abalone dispersal from the regular observation area. A new 100-m north–south line was set at 70 m eastward from zero point with three 20-m intersecting lines located at 30, 60, and 90 m from the north end. At the same time, a pair of 20-m intersecting lines was also placed in the eastern neighboring cove.The number of abalone individuals was counted by scuba divers along the line transect without disturbing the abalone. In order to determine the start and end points of each line transect for each diver survey, iron bolts were fixed at both end points of line transect A and B, where screw drilled holes were prepared in the rock substrate using a compressed air driving underwater drill. The transplanted juveniles were distinguished by their green colored shells from wild abalone until the shells were covered by coralline algae at around one year after the transplant. After this period, abalone were recorded in 4 size

Fig. 3. Density of *Eisenia bicyclis* along line transect A and the depth of transect.

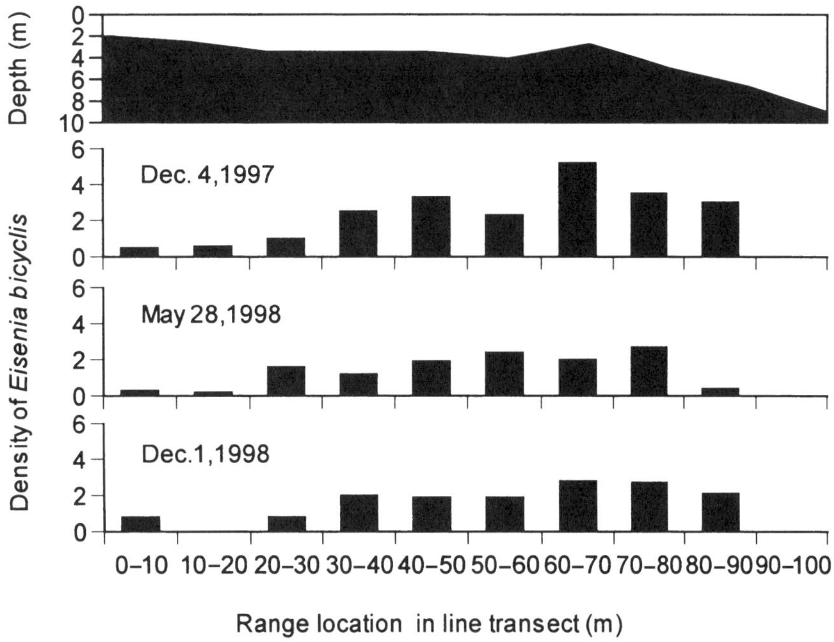

groups (0–30, 31–60, 61–90, and >90 mm SL). Diving observations of abalone distribution were implemented every two months unless inclement weather or rough seas interrupted the survey. Besides the regular surveys, five censuses were undertaken in which abalone were collected from the line transects, brought to the boat, and shell length recorded. Fouling coverage was scratched off to check for shell color and the origin of each juvenile was confirmed. All the juveniles were returned to their sampled location. The origin of the transplanted area was confirmed by detecting the existence of coded-wire tags with the 1998 juveniles on December 1, 1998. The area of *Eisenia* forest as well as other algal community areas within the test fishing ground were verified based on the density measurement of the *Eisenia* individuals in line A and the aerial photograph of the test field obtained by Kesennuma Miyagi Prefectural Fisheries Experimental Station in 1994. The surviving number of abalone at each algal community was calculated based on the average abalone density and the image analyzed area of each algal community. The sea surface area of each of the component habitat types in the aerial photograph of the site, including all line transects in the cove and neighboring coves, was estimated by computer image analysis. The estimated areas of cobblestone habitat, *Eisenia* forest, and coralline flats were 7 836, 16 691, and 5 047 m^2, respectively, for a total area of 29 574 m^2.

Results

Range of algal communities

Although the area and the density of the *E. bicyclis* community varied slightly during 1996–1998, higher densities always occurred at the 20–85 m points along line transect A (Fig. 3). Despite a marked fluctuation of the living *Eisenia* density at both edges of the forest belt, the borders of *Eisenia* forest belt at both sides between the cobblestone habitat and the coralline flat remained stable.

Transplanted abalone behavior and dive survey accuracy

The transplanted abalone started moving actively when the oyster cultch was placed on the bottom. Within two hours of release, most abalone had moved off the oyster shells on to the substratum and aggregated at the points of release either staying on the upper surface of boulders or moving into narrow crevices between them. Few seeds remained on the oyster shells one day after release. To estimate the dive survey accuracy, three crossing line transects of 20 m were set covering the center of each released circle one day after release. The average density in the 1200-m^2 area within three transects placed in the cobblestone area was 41.69 m^{-2}. No released abalone was observed from the *Eisenia* forest or coralline flats. Despite their small size, we estimated that 50.0% of the seed were detectable by divers due to the distinctive green color of the shells.

Long-term change in distribution of transplanted abalone

The distribution and density of seed along the line transect over time are shown in Figs. 4–6 and percentage survival in Fig. 7. Survival estimates were based on the mean abalone densities in the three habitats in transects A–G. The data points for the final observations on December 1, 1998, (Fig. 7) include the density data obtained from added lines. In the 1996 experiment, seed were densely aggregated at placement sites for about one month. From about November 1996, some of the seed then dispersed from the cobblestones into the *Eisenia* forest in a southerly and westerly direction. Although many seed remained in the cobblestone habitat during the study, abalone densities in the *Eisenia* forest gradually increased. On December 1, 1998 (853 days after

Fig. 4*a*. Long-term change of abalone distribution and density at the experimental site with the juveniles transplanted in 1996 (August 5, 1996 – May 7, 1997). Numbers in brackets are days elapsed after transplanting seed.

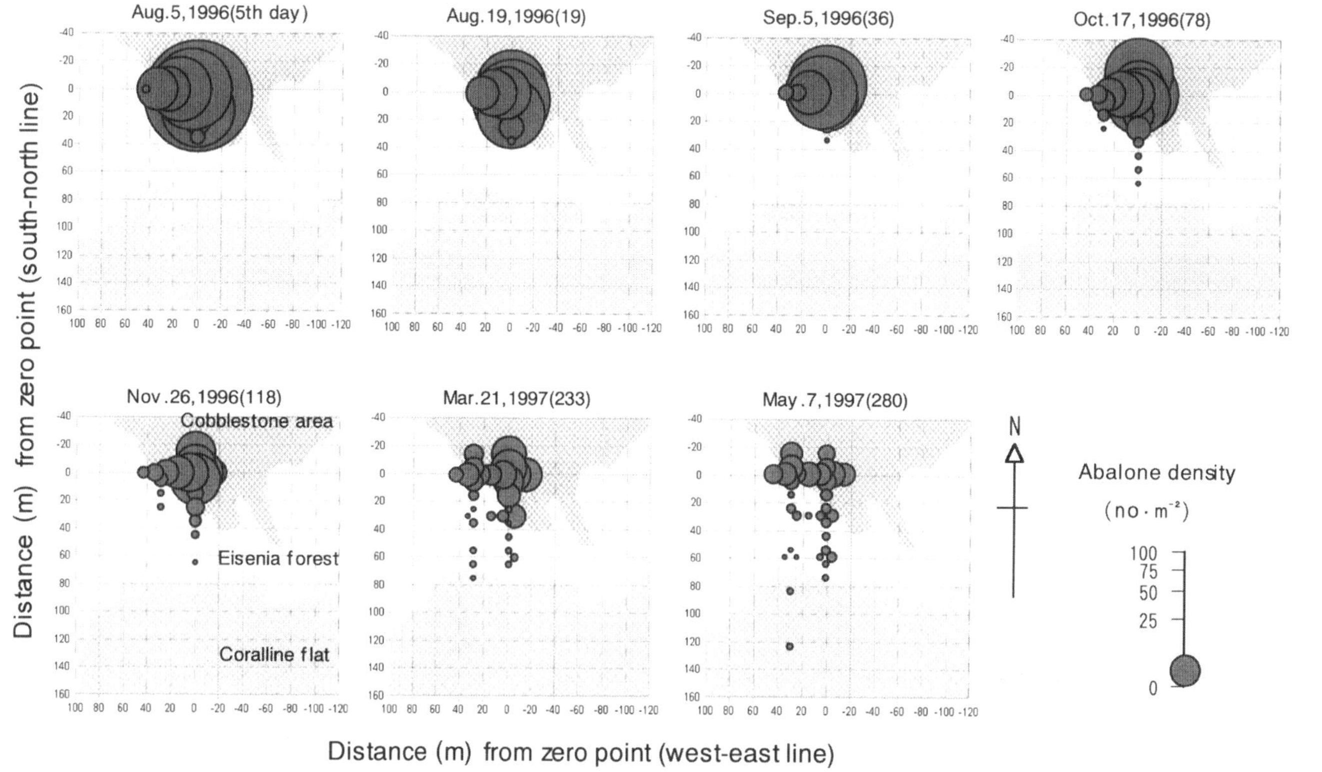

Fig. 4*b*. Long-term change of abalone distribution at the experimental site with the juveniles transplanted in 1996 (July 11, 1997 – May 28, 1998). Numbers in brackets are days elapsed after transplanting seed.

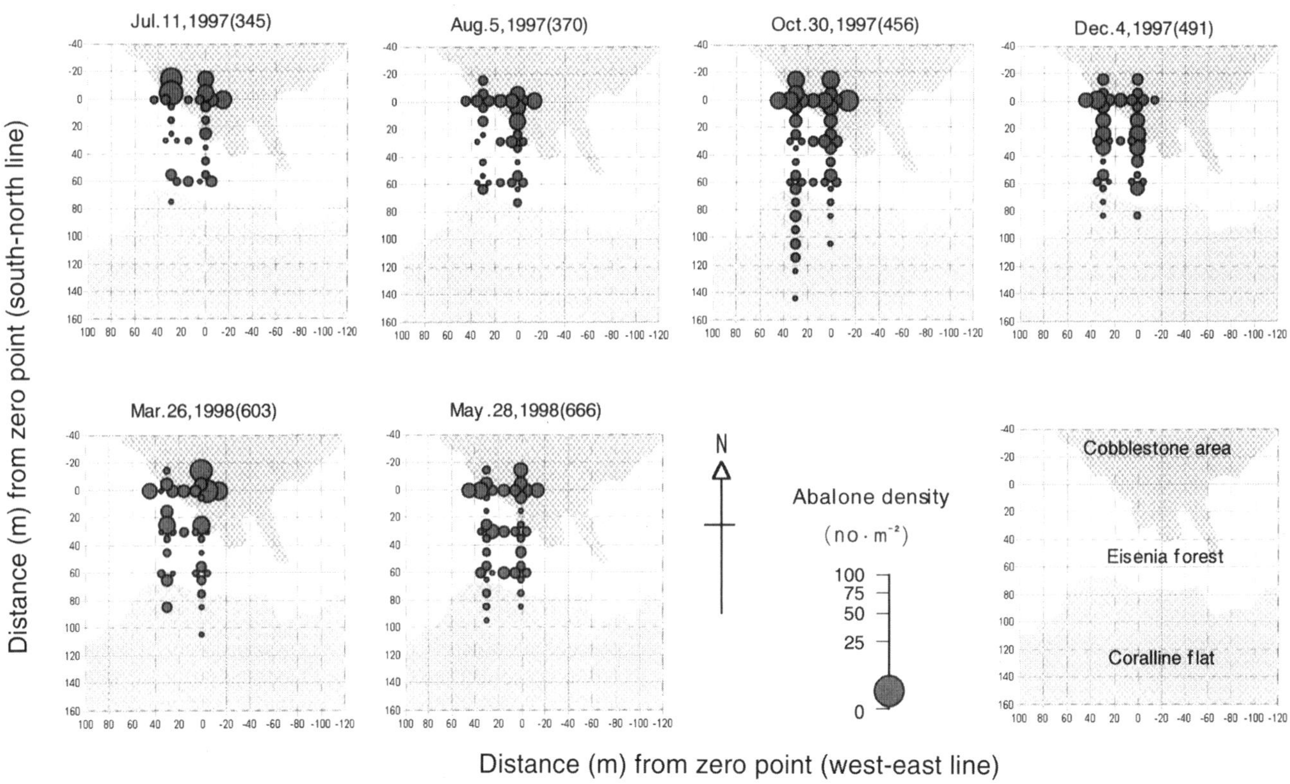

Fig. 4*c*. Long-term change of abalone distribution at the experimental site with the juveniles transplanted in 1996 (August 7, 1998 – December 1, 1998). Numbers in brackets are days elapsed after transplanting seed.

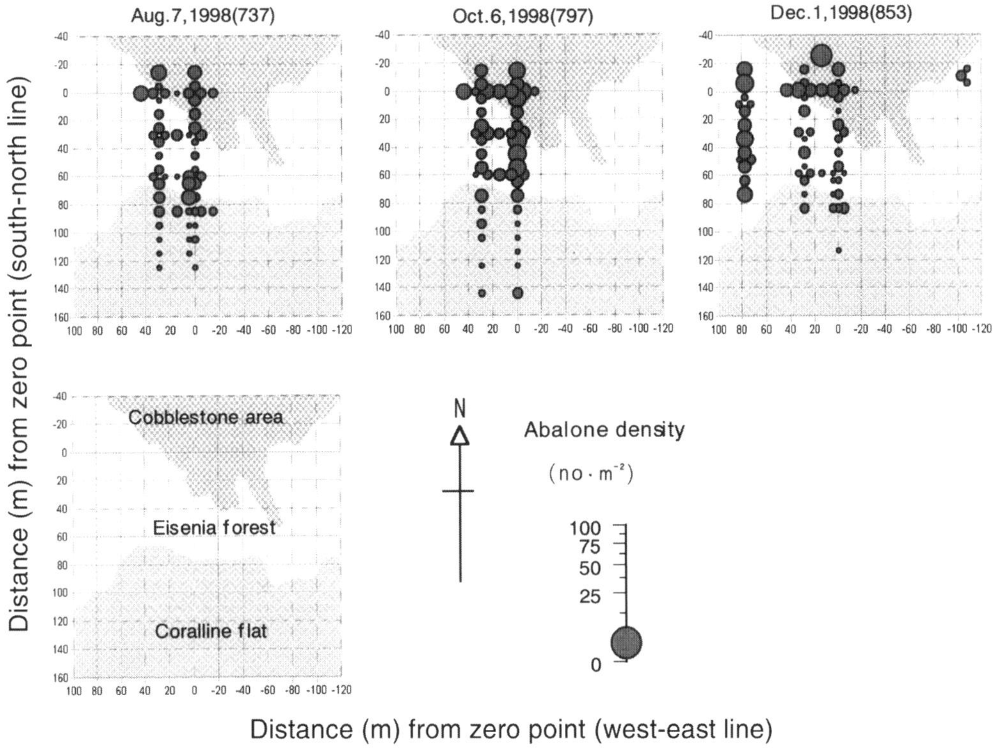

Fig. 5. Change of abalone distribution at the experimental site with the juveniles transplanted in 1997. Numbers in brackets are days elapsed after transplanting seed.

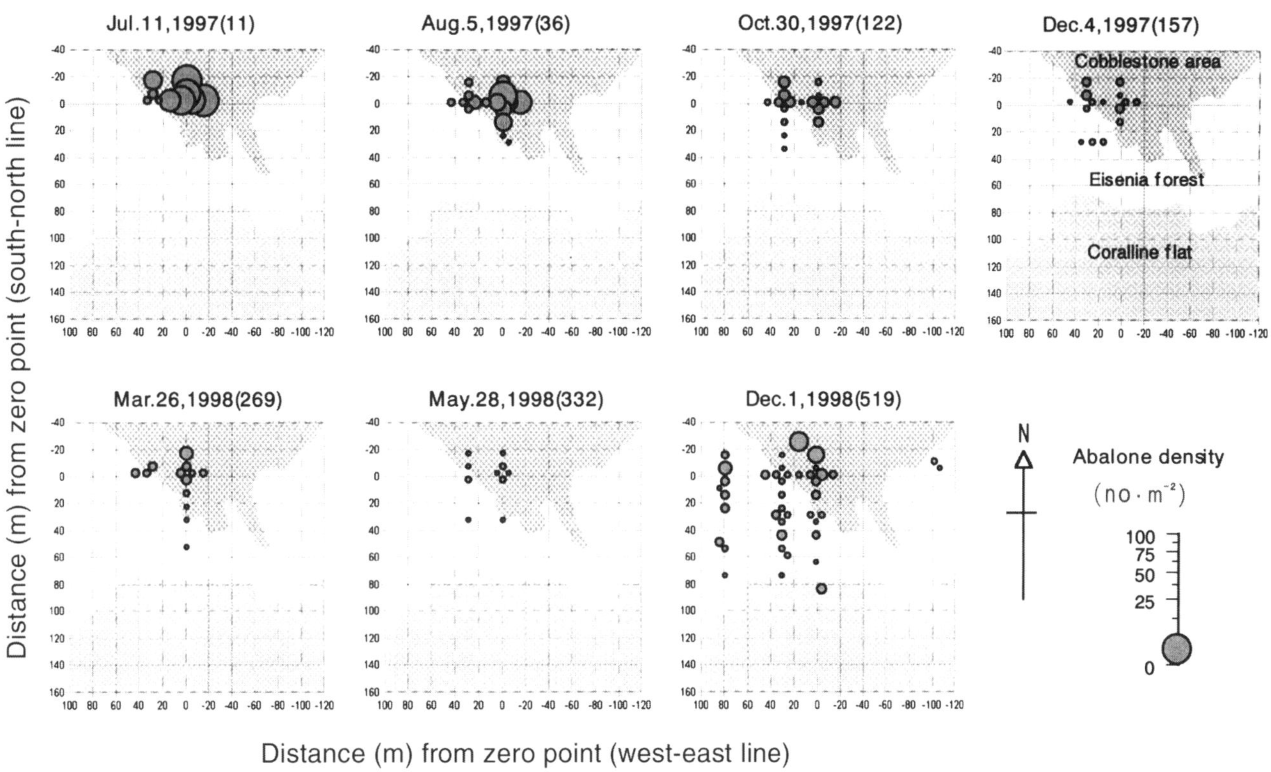

Fig. 6. Change of abalone distribution transplanted to two different algal communities in 1998. A–C: transplanted to cobblestone area, D–E: transplanted to coralline flat area. Numbers in brackets are days elapsed after transplanting seed.

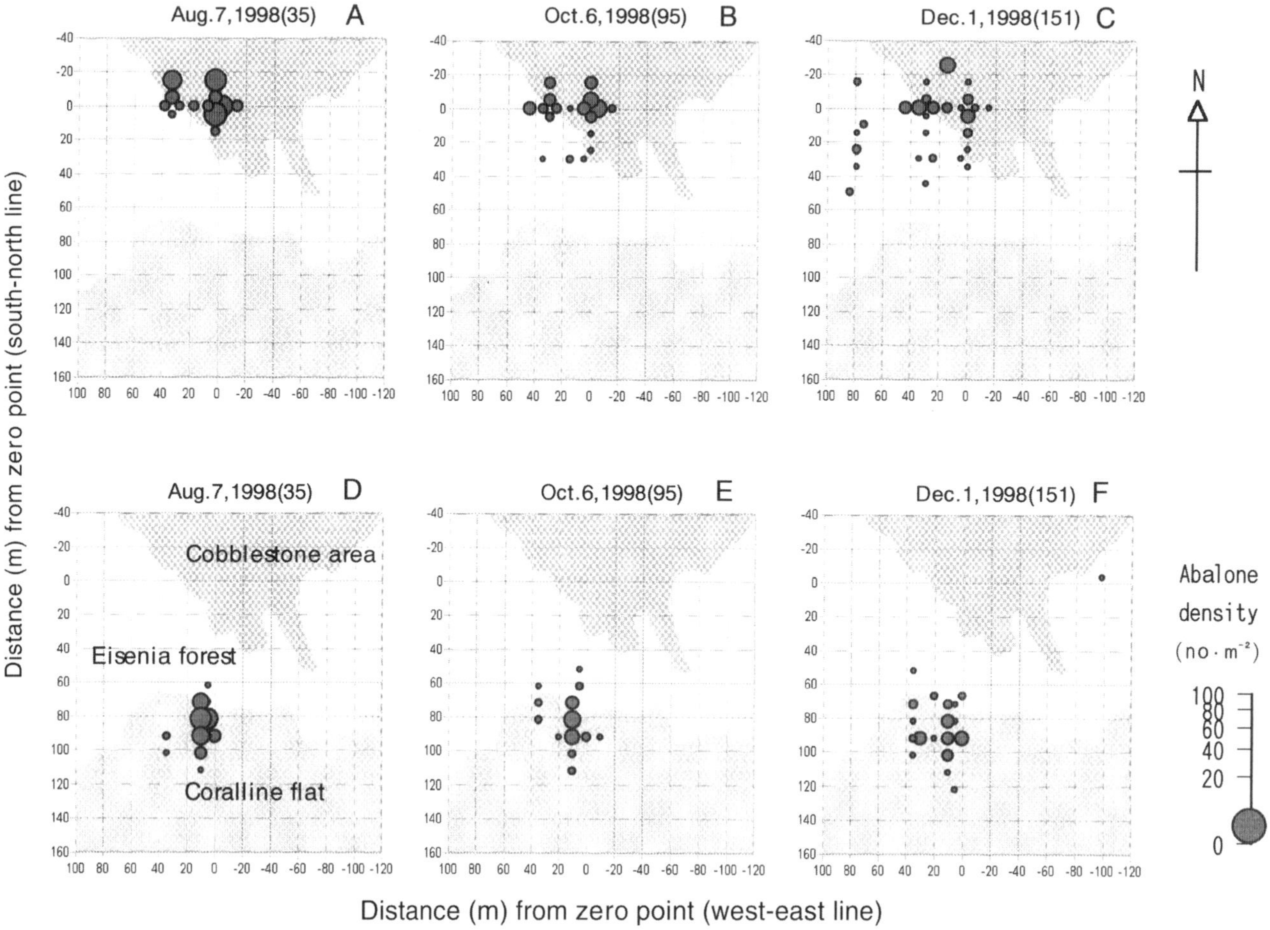

Fig. 7. Estimated percent survival rate of transplanted abalone at the experimental site after transplanting seed. The same symbols, with no line connection, show the survival rate for transplanted abalone collected from neighboring coves during the final observation on December 1, 1998.

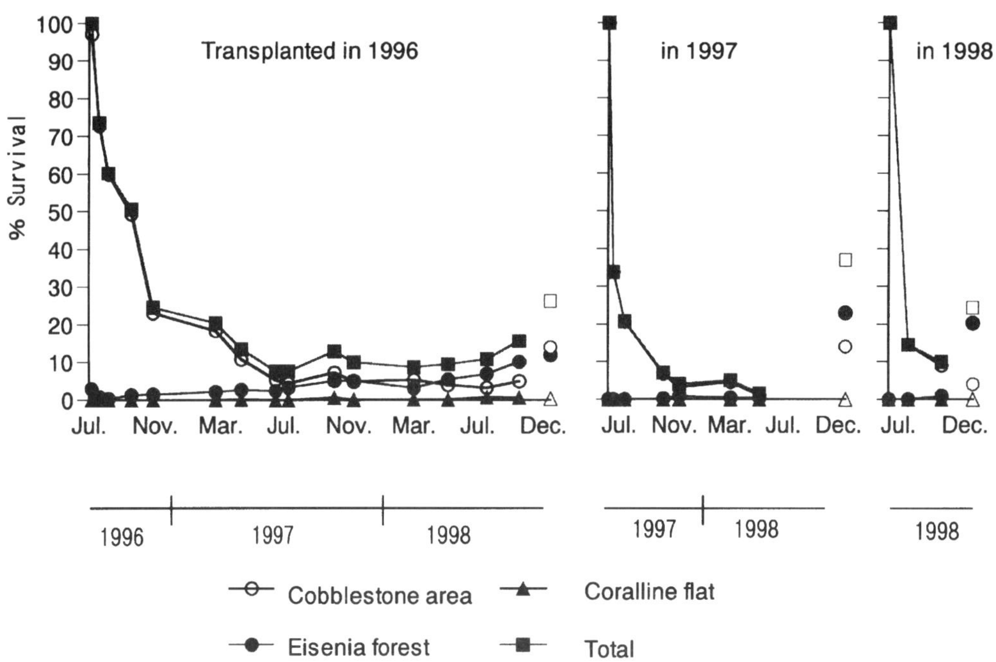

Fig. 8. Growth and size frequency of the transplanted abalone.

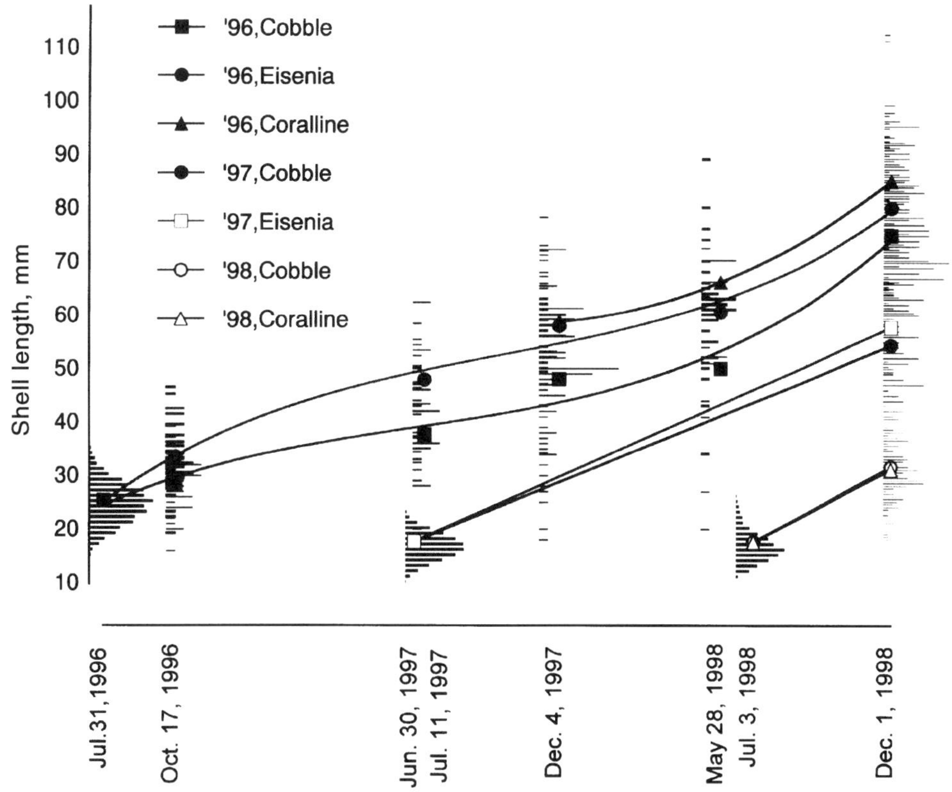

transplantation), 26.7% of the total abalone released were estimated to be living in the experimental site. Of these 14.0% were in the *Eisenia* forest, 11.9% were in the cobblestone habitat and 0.4% were in the coralline flats.

In the 1997 experiment, the seed dispersed out of the cobblestones in the same manner as the seed transplanted in the previous year. Although the 1997 seed were in low densities on May 28, 1998 (332 days after transplantation), a survival of 37.0% was estimated at day 518 in December 1998. At the final date, 23% were recorded in the cobblestone habitat and 14% in the *Eisenia* forest.

In the 1998 experiment, a movement pattern by seed transplants similar to earlier experiments was observed. On the 151st day after transplanting, overall survival of seed was estimated to be 24.4% of which 20.3% were in the cobblestone habitat and 4.0% in the *Eisenia* forest. Only 10.8% of the abalone transplanted to the coralline flats survived, of which 7.6% had moved into the *Eisenia* forest leaving only 3.2% in the deeper coralline flats.

Growth of seed

The size frequency, mean SL, and the growth curves of the seed transplants for each year (Fig. 8) show that growth rates of seed transplanted in 1996 varied according to season and habitat. The abalone in the *Eisenia* forest consistently grew faster than those in the cobblestone habitat. The daily mean growth rate of seed in the *Eisenia* forest was 75 μm during winter and spring and 11–181 μm during summer. Although abalone were scarce in the coralline flats, the daily growth rate there was always higher than in the other habitats. The daily mean growth rate of the abalone after the third year of transplantation was 197 μm and mean SL was 87.3 mm on December 1, 1998. A similar growth rate was obtained with abalone transplanted in 1997, although their growth in the *Eisenia* forest was less. However, the daily growth rate of abalone transplanted in 1998 was high in both the cobblestone habitat and the coralline flats (97 and 94 $\mu m \cdot day^{-1}$, respectively) and did not differ significantly.

Interrelation between algal communities and abalone distribution

The distribution of abalone in the three habitats of both transplanted and wild individuals varied with size. Wild abalone >90 mm SL occurred mainly in the coralline flats near the border of the *Eisenia* forest on December 1, 1998. Wild abalone at 30–90 mm SL, however, occurred widely in all three habitats. Abundance of this size class was highest in the *Eisenia* forest and next highest in the coralline flats (Fig. 9). The distribution of transplanted and wild abalone was analyzed in detail along Transect A in relation to the corresponding density of *Eisenia* sporophytes (Fig. 10). Most of the 1996 and 1997 seed transplants, as well as wild abalone, were found in the *Eisenia* forest (30–90 m south from zero point in Transect A). However, only a few larger abalone (>40 mm SL) of the 1998 seed dispersed into the *Eisenia* forest. Most of the smaller individuals of this cohort (>40 mm SL) remained within the cobblestone habitat (0–30 m south in Transect A). In the coralline flats, however, there were a few scattered abalone. The cobblestone habitat was inhabited by abalone of all sizes, but the *Eisenia* forest

Fig. 9. Distribution of different size groups of wild abalone (A. >90 mm SL; B. 30–60 mm SL; C. 60–90 mm SL; D. 0–30 mm SL) at the test fishing ground on December 1, 1998.

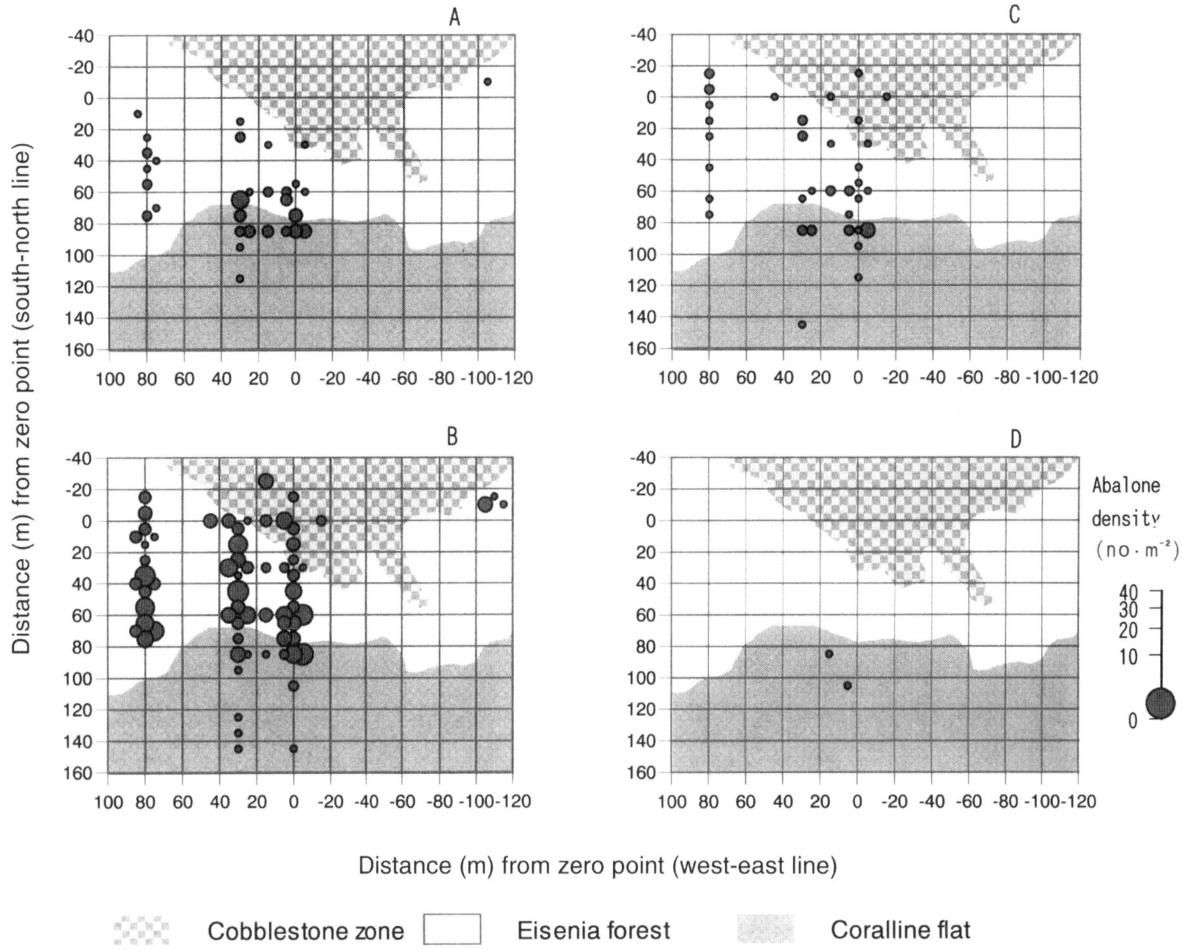

was particularly preferred by large abalone (>30–40 mm SL), whereas most of the coralline flat area was avoided by abalone except at sites near the edge of the *Eisenia* forest.

Discussion

A marked decline in the annual commercial catches of abalone has occurred since 1980 when the intensive transplantation of seed abalone was increasingly undertaken throughout Japan. Transplanting of seed, however, was not designed to rehabilitate wild abalone populations but rather to use the fishing grounds to raise a generation of released seed (Seki and Sano 1998). Most ecological studies of Japanese abalone during the period prior to 1980 considered the role of the substratum type in abalone growth and survival but ignored the role of the algal communities (Inoue 1973*a*, *b*). When larger abalone aggregations were observed in sea bottom areas where drift algae were abundant (Inoue 1973*c*), a new understanding on the importance of algae as a food supply emerged. Kojima (1974) observed higher juvenile densities in the shallow cobblestone habitat than elsewhere and concluded that juvenile and adult *H. discus* had different habitat requirements.

Kishigami (1894), Ino (1952), and Uno (1972) reported that the southern Japanese abalone changed their habitat seasonally. Abalone in Kominato, Chiba Prefecture, moved toward the deeper *Sargassum* spp. habitat in October and stayed there until April. Similar seasonal movements were observed in the northern Japanese abalone, *H. discus hannai,* in Shiriya, Aomori Prefecture, where the abalone moved from deeper water to the intertidal zone in May to June (Asuke et al. 1974). The reason for these seasonal movements is unknown. Although the habitat and behavior of other abalone in the world have been reported (e.g., Sakai 1962; Shepherd 1973, 1975; Breen and Adkins 1979; Sloan and Breen 1988; Steinberg 1989), the relation between algal communities and the distribution of particular size groups of abalone has not been examined. Following the demonstration by Taniguchi et al. (1989) that several herbivores on rocky reefs were dependent on specific algal communities, this study specifically examined their role for abalone and is the first to show the importance of algal communities as a habitat for young and adult abalone.

Two aspects of our results demonstrate the importance of the *Eisenia* forest as a habitat. First, the movement of seed at a size of 30–40 mm SL from the cobblestone habitat, dominated by small perennials and *Sargassum* spp., to the *Eisenia* forest, where wild abalone of about the same size occurred (Fig. 10), suggests a size-dependent habitat preference. Second, the higher growth rates of seed in the *Eisenia* forest

Fig. 10. Interrelated distribution of *Eisenia bicyclis* and *Haliotis discus hannai* in the test fishing ground.

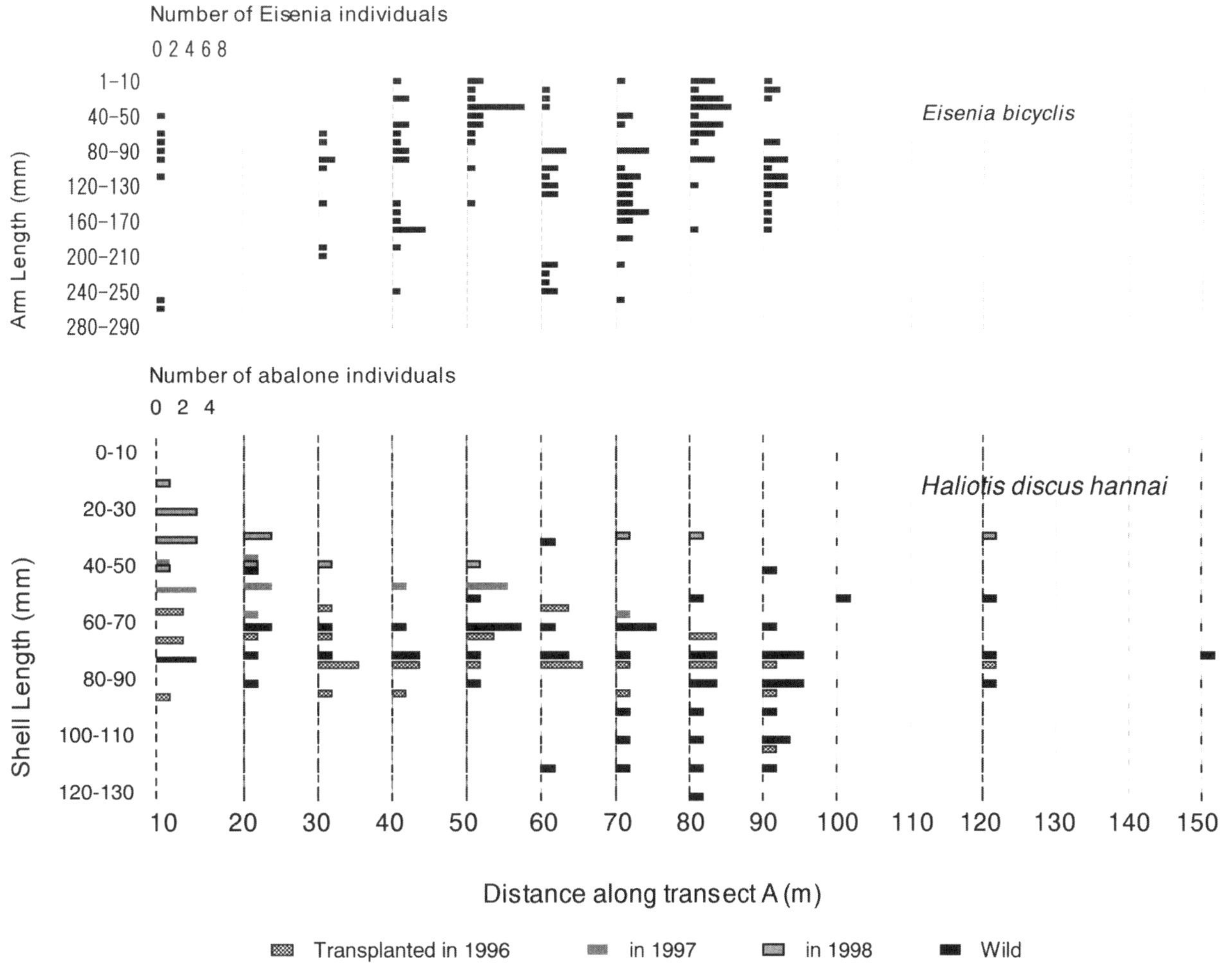

compared with the cobblestone habitat (Fig. 8) suggest that the latter habitat is suboptimal for abalone >30–40 mm SL.

Seki (1997) and Takami et al. (1997*a*, *b*) showed a transition in diet for *H. discus hannai* during the late juvenile stage, so this may be the size at which a shift to a macroalgal diet occurs. In previous attempts to transplant abalone in Japan, abalone have typically been reared to a size of 30 mm SL before seeding with the aim of increasing seed survival. This requirement was met if abalone were transplanted into the *Eisenia* forest. The present study suggests that abalone change habitat as they seek a macroalgal diet, according to their growth requirements. Introducing abalone to the deeper coralline flats caused lower abalone survival unless they encountered the *Eisenia* forest. The high growth of abalone in the coralline flats (Fig. 8) may have been due to a good supply of drift algae, since coralline surfaces are unlikely to provide enough nutrients. The patchy distribution of abalone in the coralline flats may be explained by the patchy abundance of algal drift. The coralline flats are unlikely to be a preferred habitat of *H. discus hannai* except in proximity to the *Eisenia* forest.

In this study, the estimated abalone survival rate of 26.7%, in the test fishing ground at 853 days after transplantation, is higher than most previous studies of transplanted *H. discus hannai* (Tsuchida et al. 1971; Miyamoto et al. 1982; Sasaki et al. 1983; Uchida 1994). These studies recorded maximum survival rates of 25–30% at a transplant size of >30 mm SL (Takeichi 1988; Inoguchi 1993). Our method of seeding a specific habitat, combined with careful transplant and acclimation procedures offers a valuable means of achieving high survival at a smaller size of seed. Moreover, our experiments demonstrate the value of planned transplanting of abalone at the premacroalgal feeder stage to the cobblestone habitat in early summer. Since abalone transplanted into deeper coralline flats suffered higher mortality, the crevice-rich shallower cobblestone habitat may have a key role as a nursery in protecting juveniles from predators. This is supported by studies that have reported a high abundance of juveniles among cobblestones (Kojima 1974; Saito 1981). The closely related species, *H. kamtschatkana*, settles in deeper coralline areas in British Columbia (Breen and Adkins 1982; Sloan and Breen 1988). However, the differences in coastal geomorphology and high variability in salinity in shallow waters of British Columbia make direct comparisons of habitats difficult. The northern Japanese abalone, *H. discus hannai,* has a different, although overlapping, distribution than that of *Eisenia*, e.g., in Iwate Prefecture and the eastern coast of Hokkaido. In the *Eisenia* growing area in Miyagi Prefecture, the sea urchin *Strongylocentrotus nudus* uses deeper coralline flats as a substratum for settlement and as a nursery and the *Eisenia* forest as a habitat during the later growing stage (Sano et al. 1998). It is possible that *H. discus hannai* uses the cobblestone habitat as a nursery area to avoid niche competition during post-larval and juvenile stages with other herbivores. Our study demonstrates the benefits of transplanting seed into optimal habitat in overfished areas, but more research is necessary to achieve the goal of rehabilitating abalone populations. Currently the rehabilitation of abalone stocks in Japan is wholly dependent on a national transplant program with hatchery seed. The huge cost involved in producing 30 mm SL seed and transplanting them to each fishing ground could be substantially reduced by using smaller abalone seed <30 mm SL. A theoretical cost–size analysis of seed transplantation (Seki and Sano 1998) strongly supports this idea. Erroneous ideas have been pursued for too long in commercial seed transplantation, while habitat requirements of abalone seed and environmental effects on them have been largely ignored. We urge studies of the factors which are important for increasing and maintaining recruitment in exploited populations.

Acknowledgments

We thank Minji Fukuda for his technical assistance in setting the underwater survey and tracking observation by scuba diving. We also thank the Ooya Fisheries Cooperation and the Kesennuma Fisheries Experimental Station of Miyagi Prefecture for their kind assistance in the field survey. *H. discus hannai* juveniles were kindly supplied by the Fukushima Prefectural Aquaculture Fishery Center. This research was supported by the Ministry of Agriculture, Fsorestry and Fisheries (MAFF), Japan. We thank Alan Campbell and Scoresby Shepherd for suggesting editorial improvements.

References

Asuke, M., Sawada, M., Miki, F., Ueki, T., Sudo, K., and Shikauchi, M. 1974. Investigation on the abalone resources in Shiriya frontage, Aomori. (Shiriya chisaki no awabi shigenryo chosa.) Aomori Suizo Jigyo Gaiyo, **3**: 126–130. (In Japanese.)

Breen, P.A., and Adkins, B.E. 1979. A survey of abalone populations on the east coast of the Queen Charlotte Island, August 1978. Fish. Mar. Serv. Manuscr. Rep. 1490. 125 p.

Breen, P. A., and Adkins, B.E. 1982. Observations of abalone populations on the north coast of British Columbia, July 1980. Can. Manuscr. Rep. Fish. Aquat. Sci. 1633. 55 p.

Geiselman, J.A., and McConnell, O.J. 1981. Polyphenols in the brown algae *Fucus vesiculosus* and *Ascophyllum nodosum*: chemical defenses against the herbivorous snail *Littorina littorea*. J. Chem. Ecol. **7**: 1115–1133.

Hay, M.E., and Fenical, W. 1988. Marine plant–herbivores interactions, the ecology of chemical defense. Ann. Rev. Ecol. Syst. **19**: 111–145.

Ino, T. 1952. Biological studies on the propagation of Japanese abalone (genus *Haliotis*). Bull. Tokai Reg. Fish. Res. Lab. **5**: 29–102.

Inoguchi, N. 1993. Abalone seedling and fishing ground utilization. (Awabi no shubyo horyu to gyojo riyo.) Bull. Tohoku Branch Jpn. Soc. Sci. Fish. **43**: 22–24. (In Japanese.)

Inoue, M. 1973*a*. Abalone inhabitation and sea bottom geomorphology. (Awabi on sumitsuki to kaiteichikei.) Aquaculture, **20**: 147–160. (In Japanese.)

Inoue, M. 1973*b*. Site selectivity of abalone inhabitation. (Awabi no sumitsuki to ba no senntaku ni tsuite.) Aquaculture, **20**: 173–180. (In Japanese.)

Inoue, M. 1973*c*. Difference in the abalone inhabitation by size. (Awabi no ookisa ni yoru sumiba no sa.) Aquaculture, **20**: 181–189. (In Japanese.)

Kishigami, K. 1894. The first report of abalone study. (Awabi kennkyu dai ippou.) Suisan chosahokoku, **3**: 1–25. (In Japanese.)

Kojima, H. 1974. The habitat of young black abalone, *Haliotis discus discus* Reeve in Kaifu-gun, Tokushima Prefecture. Michurin Seibutugaku Kenkyu, **10**: 155–160.

Miyamoto, T., Saito, K., Motoya, S., Nishikawa, N., and Monma, H. 1982. Experimental studies on the release of the cultured seeds of abalone, *Haliotis discus hannai* Ino in Oshoro Bay, Hokkaido. Sci. Rep. Hokkaido Fish. Exp. Stn. **24**: 59–89.

Saito, K. 1981. The appearance and growth of 0-year-old abalone. Bull. Jpn. Soc. Sci. Fish. **47**: 1393–1401.

Sakai, S. 1962. Ecological studies on the abalone *Haliotis discus hannai* Ino-III. Study on the mechanism of production of the abalone in the region of Onagawa Bay. Bull. Jpn. Soc. Sci. Fish. **28**: 891–898.

Sakai, S., Obata, K., Kikuta, A., and Suda, Z. 1975. Research on the abalone resource enhancement by means of seedling-II. Bull. Kesennuma Miyagi Prefect. Fish. Exp. Stn. **1**: 1–8. (In Japanese.)

Sano, M., Omori, M., Taniguchi, K., Seki, T., and Sasaki, R. 1998. Distribution of the sea urchin *Strongylocentrotus nudus* in relation to marine algal zonation in the rocky coastal area of the Oshika Peninsula, northern Japan. Benthos Res. **53**: 79–87.

Sasaki, R., Oikawa, S., Iizuka, T., and Kawamura, T. 1983. A case study on the effect of abalone seed release at Karakuwa, Miyagi Prefecture. (Miyagi ken Karakuwa-cho ni okeru awabi shubyo houryu kouka jirei.) Report of Tohoku Mariculture Research Conference. (Showa 57 Nendo Tohoku Block Zoyoshoku Kenkyu Renraku Kyogikai Houkokusho.) Tohoku National Fisheries Research Institute, Resource Enhancement Division, Shiogama, Japan. pp. 45–50. (In Japanese.)

Sasaki, R., and Shepherd, S.A. 1995. Larval dispersal and recruitment of *Haliotis discus hannai* and *Tegula* spp. on Miyagi coasts, Japan. Mar. Freshw. Res. **46**: 519–529.

Seki, T. 1997. Biological studies on the seed production of the northern Japanese abalone, *Haliotis discus hannai* Ino. Bull. Tohoku Natl. Fish. Res. Inst. **59**: 1–71.

Seki, T., and Sano, M. 1998. An ecological basis for the restoration of abalone populations. Bull. Tohoku Natl. Fish. Res. Inst. **60**: 23–40.

Seki, T., and Taniguchi, K. 1996. Factors critical to the survival of herbivorous animals during settlement and metamorphosis. *In* Survival strategies in early life stages of marine resources. *Edited by* Y. Watanabe, Y. Yamashita, and Y. Oozeki. A.A. Balkema, Rotterdam, Netherlands. pp. 341–354.

Shepherd, S.A. 1973. Studies on southern Australian abalone (genus *Haliotis*) I. Ecology of five sympatric species. Aust. J. Mar. Freshwater Res. **24**: 217–257.

Shepherd, S.A. 1975. Distribution, habitat and feeding habits of abalone. Aust. Fish. **34**: 1–4.

Shiraishi, K., Taniguchi, K., Kurata, K., and Suzuki, M. 1990. Effects on the methanol extracts from the brown alga *Dilophus okamurai* on avoidance behavior of young abalone (*Haliotis discus hannai*). Bull. Tohoku Natl. Fish. Res. Inst. **52**: 13–15.

Shiraishi, K., Taniguchi, K., Kurata, K., and Suzuki, M. 1991. Effects on the methanol extracts from the brown alga *Dictyopteris divaricata* on feeding by the sea urchin *Strongylocentrotus nudus* and the abalone *Haliotis discus hannai*. Nippon Suisan Gakkaishi, **57**: 1945–1948.

Sloan, N.A., and Breen, P.A. 1988. Northern abalone, *Haliotis kamtschatkana*, in British Columbia: fisheries and synopsis of life history information. Can. Spec. Publ. Fish. Aquat. Sci. 103. 46 p.

Steinberg, P.D. 1984. Algal chemical defense against herbivores: allocation of phenolic compounds in the kelp *Alaria marginata*. Science (Washington, D.C.), **223**: 696–698.

Steinberg, P.D. 1985. Feeding preference of *Tegula funebralis* and chemical defenses in marine brown algae. Ecol. Monogr. **55**: 333–349.

Steinberg, P.D. 1988. The effects of quantitative and qualitative variation in phenolic compounds on feeding in three species of marine invertebrate herbivores. J. Exp. Mar. Biol. Ecol. **120**: 221–237.

Steinberg, P.D. 1989. Biogeographical variation in brown algal polyphenols and other secondary metabolites, comparison between temperate Australasia and North America. Oecologia, **78**: 373–382.

Takami, H., Kawamura, T., and Yamashita, Y. 1997*a*. Survival and growth rates of post-larval abalone *Haliotis discus hannai* fed conspecific trail mucus and/or benthic diatom *Cocconeis scutellum var. parva*. Aquaculture, **152**: 129–138.

Takami, H., Kawamura, T., and Yamashita, Y. 1997*b*. Contribution of diatoms as food sources for post-larval abalone *Haliotis discus hannai* on a crustose coralline alga. Moll. Res. **18**: 143–151.

Takeichi, M. 1988. Recapture and survival rate of mass released seed of abalone, *Haliotis discus hannai*. (Tairyo horyu sareta ezoawabi jinnkoshubyo no kaishuuritu to seizannritu.) Saibaigiken, **17**: 27–36. (In Japanese.)

Taniguchi, K., Kojima, H., Kurata, K., Shiraishi, K., Uki, N., Omori, M., Asano, M., and Hagino, S. 1989. Marine afforestation of *Eisenia* and *Ecklonia* and its utilization. Marine Ranching Project. Agriculture, Forestry and Fisheries Research Council, Kouseisha-Kouseikaku Tokyo. pp. 275–358. (In Japanese.)

Taniguchi, K., Kurata, K., and Suzuki, M. 1991. Feeding-deterrent activity of some Laminariaceous brown algae against the Ezo-abalone. Nippon Suisan Gakkaishi, **58**: 577–581.

Taniguchi, K., Kurata, K., Suzuki, M., and Shiraishi, K. 1992. Feeding-deterrent activity of diterpens from the brown alga *Dilophus okamurai* against the abalone *Haliotis discus hannai*. Nippon Suisan Gakkaishi, **58**: 1931–1936.

Taniguchi, K., Kurata, K., Maruzoi, T., and Suzuki, M. 1994. Dibromomethane, a chemical inducer of larval settlement and metamorphosis of the sea urchin *Strongylocentrotus nudus*. Fish. Sci. (Toyko), **60**: 795–796.

Taniguchi, K., Kurata, K., and Suzuki, M. 1991. Feeding-deterrent activity of some Laminariaceous brown algae against the Ezo-abalone. Nippon Suisan Gakkaishi, **58**: 577–581.

Tsuchida, K., Takeichi, M., Hirose, T., Kanazawa, T., Omura, R., and Sakashita, T. 1971. On the growth, recapture ratio and distribution of released abalone, *Haliotis discus hannai* Ino. Bull. Iwate Prefect. Fish. Exp. Stn. **1**: 121–130.

Uchida, A. 1994. Stock enhancement of Ezo abalone *Haliotis discus hannai* in Yamada area, Iwate Prefecture. Saibai Giken, **23**: 11–17.

Uki, N., Omori, M., Kawahara, I., Ishida, R., and Yanagisawa, T. 1996. Seed production technology of abalone. Aquaculture technology series No. 2. "Awabi rui no shubyoseisan gijutu." Japan Sea-Farming Association, Tokyo. (In Japanese.)

Uno, K., Koike, Y., and Monma, H. 1972. Ecological studies on the propagation of the Japanese abalone (Genus *Haliotis*)-I. Distribution and movement of *Haliotis sieboldii* in the natural environment. Umi (La mer), **10**: 43–49. (In Japanese.)

Tired nature's sweet restorer? Ecology of abalone (*Haliotis* spp.) stock enhancement in Australia

S.A. Shepherd, P.A. Preece, and R.W.G. White

Abstract: We describe studies in which abalone (*Haliotis* spp.) larvae and juveniles were variously released experimentally at sites in Australia. Larvae released at different densities in South Australia showed variable survival of 0.02–7.8% 6–9 days after settlement. In the next 4–6 weeks, the mean natural mortality rate, *M*, was 1.8–2.8·month^{-1} and in the next 11 months mean *M* was 2.9·year^{-1}. Mortality in the first year was density-dependent. In the second year, *M* was 1.4·year^{-1}. Larval seeding is not recommended because the larval stage is a strongly density-dependent phase of the abalone life history. Seeding of 7-month-old juveniles in southern Tasmania showed ~20% mortality from handling stress in the first two days, elevated *M* in the next 12 days, and a lower and likely stable *M* thereafter. Over one year *M* was ~1.8 year^{-1}, which was substantially higher than *M* in wild populations. Problems of juvenile seeding were high *M* due to handling stress and behavioral deficits of seed in the presence of predators. In contrast, commercial experiments in which juveniles were seeded onto artificial modules in a seagrass habitat with few predators at Flinders Island, Tasmania, showed promise, with low *M*.

Consideration of the carrying capacity of abalone habitat and the ecosystem consequences of seeding suggest a limited role for stock enhancement, i.e., seeding juvenile abalone in places where the wild stocks have collapsed or where stock enhancement and habitat improvement can be combined with a marine-farming operation.

Résumé : Des études sont décrites pour lesquelles larves et juvéniles d'ormeaux (*Haliotis* spp.) ont été relâchés expérimentalemente à différents endroits d'Australie Merridionale et de Tasmanie. Les semis de larves à des densités variables près de Port Lincoln en Australie Merridionale ont montré que le taux de survie était densité-dependant lors d'échantillonages de substrat six à neuf jours plus tard. Dans les quatre a six semaines suivantes, le taux de mortalité naturells, *M*, etait de l'ordre de 1.8 à 2.8·mois^{-1}. Par conséquent, le taux de survie des post-larves la prémière année était fortement densité-dependant. Sauf dans le cas où le stock s'est effondré, il n'est pas recommandé de relâcher des larves car cette opération doit être effectuée pendant la période de vie des ormeaux où la mortalité est fortement dépendante de la densité. Le semis de juvéniles au sud de la Tasmanie ont montré que le stress lié à la manipulation entrainait une forte mortalité dans les deux premiers jours, une augmentation de la mortalité naturelle (*M*) dans les 12 jours suivants, et enfin une diminution et une relative stabilisation de *M* lors de la 3ème phase. *M*, lors de cette dernière phase, était cependant plus élevé que ce qu'on pourrait attendre pour des ormeaux sauvages. Les principaux problèmes liés au semis de juvéniles ont été identifiés comme étant les suivants : forte mortalité due au stress des manipulations, et augmentation de *M* liée au manque de réponses comportmentale en présence de prédateurs. En comparaison avec les semis de juvéniles à Flinders Island, l'expérience réalisée en Tasmanie a prouvé que l'operation réalisée dans un habitat presque dépourvu des prédateurs était réussite, avec un *M* très faible.

Considérant les capacités de l'habitat à soutenir des ormeaux et les conséquences des semis sur l'ecosystème, il apparaît qu'étant donnée la densité-dependance de *M* des post larves et des juvéniles ainsi que les autres facteurs de l'augmentation de *M*, l'intérêt du faire meilleur des stocks reste restreint. La amelioration des stocks apparaît seulement intéressant dans les zones où les stocks sauvages se sont effondrés ou quand l'amelioration des stocks et de l'habitat, par l'immersion de blocks de beton par exemple, peuvent être combinés lors d'une opération de marriculture.

Introduction

Most abalone fisheries of the world have suffered declines of all or some of their component stocks. In Australia, the declines have been near large human population centers (as in New South Wales) or in small or isolated stocks. The greenlip (*Haliotis laevigata* Donovan) fishery has proved to be less robust than the blacklip (*Haliotis rubra* Leach) fishery. In the southern states of Australia, many greenlip stocks have variously suffered declines of 25–90% but fewer

S.A. Shepherd and P.A. Preece. South Australian Research and Development Institute, P.O. Box 120, Henley Beach, S.A. 5022, Australia. e-mail address: shepherd.scoresby@saugov.sa.gov.au
R.W.G. White. School of Zoology, University of Tasmania, G.P.O. Box 252-05, Hobart, Tasmania, Australia 7001.
Correct citation: Shepherd, S.A., Preece, P.A., and White, R.W.G. 2000. Tired nature's sweet restorer? Ecology of abalone (*Haliotis* spp.) stock enhancement in Australia. *In* Workshop on Rebuilding Abalone Stocks in British Columbia. *Edited by* A. Campbell. Can. Spec. Publ. Fish. Aquat. Sci. 130. pp. 84–97.

Fig. 1. Map of southern Australia showing six seeding sites, numbered 1–6, for larvae in South Australia and seeding sites for juveniles at Blubber Head and Flinders Island, Tasmania. Bars are 10 km in each inset.

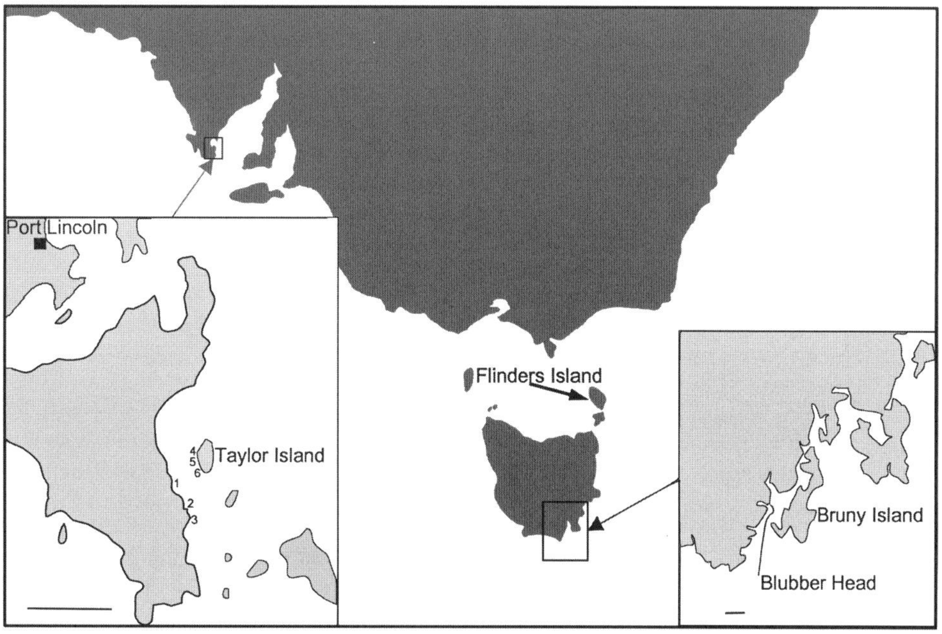

blacklip stocks have declined (Prince and Shepherd 1992; Shepherd and Baker 1998; S.A. Shepherd, unpublished data).

Release of larvae or juveniles has often been proposed as a panacea for depletion of overexploited abalone stocks and seeding juveniles has been practised in Japan for many decades with some success (Kojima 1995; Masuda and Tsukamoto 1998). Larval release has also been practised sporadically in Mexico since the 1960s (Ortiz-Quintanilla 1980), but its success has never been evaluated. In Australia, only a few seeding experiments have been attempted and these are described in this paper. The two methods described above were used in independent sets of experiments in South Australia and Tasmania, respectively. In the South Australian experiments, larvae were released onto the rocky bottom as described by Preece et al. (1997, 2000) and we summarize the results of those experiments, full details of which may be found in Preece et al. (2000). In Tasmania, juveniles were released onto the rocky bottom or artificial substratum in two experiments in southern Tasmania and during long-term experiments at Flinders Island, Bass Strait. First, we evaluate these experiments and then consider more generally the ecological requirements of juvenile abalone and their major predators within the conceptual framework of the carrying capacity of the habitat and the ecosystem consequences of artificially increasing the abundance of a herbivore. Finally, we propose a protocol for seeding abalone in Australia in light of the theory and the results presented.

Materials and methods

Larval release experiments

Seeding sites satisfying the following criteria were chosen, i.e., within 2–3 h travelling time from the hatchery, on long-shore boulder slopes of 500–1000 m^2 with boulders of a size amenable to sampling, sheltered from oceanic swell, and in abalone habitat with at least some wild abalone present. Boulders 20–30 cm in diameter suitable for sampling typically had 20–30% cover of crustose coralline algae and smaller boulders had up to 100% coralline cover. Post-larvae remain in this microhabitat up to three months (Shepherd and Daume 1996). Six seeding sites, named sites 1–3 for *H. rubra* and sites 4–6 for *H. laevigata*, at depths of 7–12 m were selected near Port Lincoln, South Australia (Fig. 1). Tidally upstream control sites were chosen and marked in similar habitat adjacent to and within 100 m of each of the six sites. The limited extent of suitable boulder habitat permitted one experiment per site; however, we seeded three similar sites per day. Sites 1–3 were seeded on 20 August 1994 and 29 September 1995 and sites 4–6 on 8 November 1994 and sites 4–5 again on 29 November 1995. All sites were subject to commercial fishing although adult greenlip densities were in decline (see Shepherd and Baker 1998 for catch data for northern Thorny Passage).

Larvae competent to settle were transported to the sites on damp 100-µm mesh screens in insulated containers at 14°C, resuspended in water at the desired densities, and released onto the experimental sites over an area of 500 m^2 by a plastic hose controlled by a diver (Preece et al. 1997).

In the 1994 experiments, early (6–9 days after release) and late (27–49 days after release) post-larval surveys were made by taking 5–7 boulders (average surface area 516 cm^2 per boulder) from sampled subareas at each site and control, sealing the boulders in plastic bags *in situ*, and then transporting them to the laboratory for washing and counting post-larvae. The surface area of the boulders was measured and densities estimated in numbers·m^{-2}. The mean surface area sampled per site and per control was 11 094 cm^2 (SE 261 cm^2). These values were later converted to densities·m^{-2} planar area (i.e., per m^2 quadrat) using a conversion factor of 1.7 for greenlip sites (see Preece et al. 1997) and 2.15 for blacklip sites (Preece et al. 2000). Subsequent nondestructive surveys were carried out at the sites by carefully searching boulders within 24 (for blacklip) and 80–120 (mean 100) (for greenlip) 0.25-m^2 quadrats placed haphazardly throughout the release and control areas.

For the 1995 seeding experiments, all done at release densities of 4000·m^{-2}, design 2 artificial larval collectors (Rodda et al. 1997) were also set out, four per site, at all sites about 7 days before seeding to ensure conditioning of collectors with diatoms and bacteria. The collectors were made of black polycarbonate material with a complex surface area of ~1 m^2, forming channels that provided a variety of dark microhabitats for settling larvae. A subset of collectors was removed for examination in the same manner as boulders within a week of seeding and the rest were removed at 39 days. We also conducted four experiments to determine the relationship between the number of larvae that settled in collectors and the number that settled on natural substrata nearby during seeding trials. The results were variable but showed that a mean of 1.6 (SE 0.8) times the number of larvae settled on collectors compared with natural substrata. For the 1995 experiments, using collectors rather than monitoring natural substrata to estimate densities of settlers, we divided the mean number of post-larvae per collector by the conversion factor of 1.6 to crudely estimate settlement densities on natural substrata for the purpose of estimating the percentage of settlement and survival to day 6.

During 1994 and 1995, the 11 larval release experiments, 5 with *H. laevigata* and 6 with *H. rubra*, were monitored for 1 year, and 9 of them for 2 years. We also obtained data for each site on the densities of conspecifics (*H. rubra* at sites 1–3 and *H. laevigata* at sites 4–6), potentially competing herbivorous gastropods (see Clarkson and Shepherd (1984) for main taxa), and sedentary predators in the underboulder fauna by counting their numbers in 0.25-m^2 quadrats while estimating juvenile abalone abundance. We estimated the density of the wrasses *Notolabrus tetricus* and *Pictilabrus laticlavius* by visual counts along 10 × 50 m transects through the seeded sites. These wrasses have a defined home range (Barrett 1995; Shepherd and Clarkson, unpublished data) and census data give a good estimate of their abundance at any site.

Juvenile seeding experiments

Experiment 1

Two seeding sites at Blubber Head, Port Esperance, Tasmania, were chosen for seeding with juvenile *H. rubra*. Site A was exposed to south-easterly swell with high algal cover and site B was more sheltered from the swell within Port Esperance. The sites were boulder slopes going to a depth of 12 m with at least three vertical layers of boulders (termed upper, middle, and lower) with abundant crustose coralline cover. Both sites were fished by abalone divers but site A had higher natural recruitment of *H. rubra* than site B, which had more epiflora and sediment. The experiments were conducted during January–March of 1984 when sea surface temperatures were 15–18°C.

At each site the experiments were done in 4 gridded areas, each 12 × 14 m subdivided into 42 squares of 2 × 2 m. Each area was assigned 1 of 4 recovery periods: 2, 6, 14, or 28 days after release when the respective areas were sampled destructively. There were 4 density treatments, each with 10, 20, 30, or 40 seeds per square, and one density treatment allotted to a square. There were 3 replicates of each density treatment randomly assigned to each area. The range of densities chosen reflected the range of abalone densities commonly found in boulder habitats. Abalone in each density treatment were distinctively marked with a color-coded plastic tag fixed to the shell with cyanoacrylate adhesive (super glue). The seed abalone were 7 months old and had a mean shell length (SL) of 12.1 mm (range 7–22 mm SL) and a mean soft body weight of 104 mg. Seed abalone were taken to the bottom in trays and released by one method per replicate of each density treatment. The release method was (*i*) on abalone shells, (*ii*) inside oval cavities 25 × 75 mm in concrete bricks, or (*iii*) by hand into the mid or lower level of boulders. In the first two cases, the shells and bricks were then concealed under boulders to protect them against searching vertebrate predators like monacanthid and labrid fishes. Handling time was about the same for each method of release. To measure handling stress, a control group of 100 abalone, handled in the same way as the seed, was placed at site A in barrels filled with rocks and covered with mesh to exclude predators.

At each census date, seeded squares were carefully searched by divers who recorded the following for each marked abalone: (*i*) the substratum of attachment, whether gravel or rock, and in the latter case whether bare or encrusted with pink or faded coralline or with an epifloral cover; (*ii*) location, whether on an upper, middle, or lower boulder; and (*iii*) orientation, whether on the upper or lower side, on the side of a boulder, or within a crevice. Microhabitat selection caused by the juxtaposition of boulders could not be assessed because searching required their removal. Incremental growth of seeded abalone was recognizable by the distinct color difference in the shell due to the switch from a hatchery diet to a natural one.

The three methods of release did not differ significantly in survival after 2 days (one way ANOVA, $F = 0.6$; $P = 0.9$), an unsurprising result given that the animals rapidly left the shells and blocks on which they were seeded. Therefore, method of release was excluded as a treatment leaving a 2 × 4 × 4 array with 3 replicates for analysis of effect of site, time, (treated as random factors) and density (a fixed factor). A three-way hierarchical analysis of variance (ANOVA) was performed after arcsine transformation of proportions surviving (Zar 1974) to test for significant effects of site, time, and density of seeding on seed survival. We assumed that searching efficiency was 100% and that there was no emigration from the squares. Searching efficiency, although not 100%, was probably high because the colored tags were conspicuously placed dorsally on the small abalone. Emigration out of the release squares was negligible (see Results).

Experiment 2

A second experiment, similar in design to experiment 1, was carried out to estimate longer-term mortality. The experiment was done at the same sites and began at the same time. Squares were sampled at 4, 8, and 12 months after seeding.

Experiment 3

Seeding with small abalone has been practised at Flinders Island, Tasmania, by Furneaux Aquaculture since the 1980s.

Fig. 2. Concrete module used for seeding juvenile *H. laevigata* at Flinders Island. The fiber glass rods at the apex of the module trap drift algae. The module has a base of 1 × 1 m and a height of 1 m.

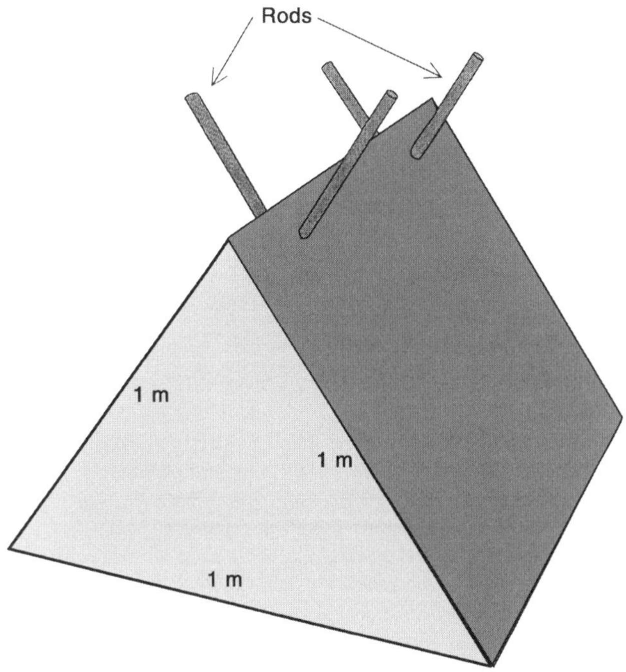

Here we describe one of many experiments for which survival data are available. Fifty seeds of *H. laevigata*, acquired from Tasmanian Univalve, were tagged at an age of 3 years and placed on two artificial concrete substrata (Fig. 2), among *Posidonia* beds at a depth of about 5 m in a mariculture lease site in Franklin Sound in December 1988. In February 1998, the survivors, then aged 9+ years, were recovered by one of the authors (SAS) on the same concrete modules. Observations were made on the fauna and flora of the release site at the time of recapture.

We have used the following notations: S is survival, calculated as a percentage of the initial number or density for the given time period; $\%M$ is finite mortality = $100 \cdot (1 - S)$; M is the instantaneous coefficient of mortality, expressed as a rate per month or per annum and related to S by the equation $S = e^{-M}$.

Results

Larval release experiments

The six 1994 seeding experiments are summarized in Tables 1 and 2. The five additional 1995 experiments, all seeded at larval densities of $4000 \cdot m^{-2}$, together with density and survival data up to two years after seeding, are summarized in Table 3. In the blacklip experiments (Tables 1 and 3), 0.02–1.2% (mean 0.6%) of the released larvae settled in the experimental areas, but only one estimate of minimum post-larval natural mortality (M) was possible (see Table 1) because some natural settlement occurred between day 6 and day 27 surveys and confounded the results. In the greenlip experiments (Tables 2 and 3), there was also high variability of larval settlement (0.1–7.8%, mean 2.1%) and, because negligible natural settlement occurred between surveys, plausible estimates of M of $1.8–2.8 \cdot month^{-1}$ resulted for the first 30–40 days after settlement (see Table 2). Only a minimum M (> 0.8) is known for site 1, but a significant correlation between M and the logarithm of post-larval densities exists if that datum is within the range 0.8–1.8 ($r > 0.88$; $P < 0.05$).

Abalone post-larval abundance in the samples was variable. Variance-to-mean ratios (Sokal and Rohlf 1995) for post-larval numbers, calculated from numbers per boulder, for both abalone species ranged from 18 to 765. In the 1995 experiments, post-larval densities were estimated from collector densities (Table 3). M was variable and in a few cases, mostly controls, densities appeared to increase over time giving negative M values (see Discussion). In four cases, three treatments and one control, we recorded zero densities at two years; here we assumed a final density of $0.1 \cdot m^{-2}$ for the purpose of estimating a minimum M. Given the difficulties in accurately estimating densities of cryptic animals, estimates of M must be considered crude. The mean M per annum for all sites for the first year was 2.9 (SE 0.3) and for the second year was 1.4 (SE 0.3). Variance-to-mean ratios of densities calculated on quadrat data for all sites combined were 1.2 for both species for one-year-old abalone and 1.2 for blacklip and 0.8 for greenlip for two-year-old abalone. This indicated that, at the scale sampled ($0.25\text{-}m^2$ quadrat), one-year-olds were randomly distributed and two-year-olds were either randomly distributed or tended to a regular distribution within the boulder habitat.

Annual M of the 0+ age-class, M_0, (estimated from the time of the second post-larval survey to one year) was positively correlated with the logarithm of post-larval density, D_0 (Fig. 3), and gave the regression

$$M_0 = 1.24 + 1.47 \log D_0 \ (R^2 = 0.59;\ P < 0.01)$$

On the other hand, no correlation was evident between the mortality rate of the 1+ age-class during its second year and initial density of that year-class ($R^2 = 0.16$; n.s.).

Densities of competing herbivores and predators (wrasses and sea stars) are given in Table 4. Among the competing gastropods, the most common species were *Scutus antipodes*, *Granata imbricata*, and *H. scalaris*, in that order. Predatory gastropods present were *Cominella* sp. and *Pleuroploca australasia* but both were in low abundance (each $<0.15\ m^{-2}$ at all sites). We saw only a few crabs under boulders at site 2 and found no empty chipped shells indicative of crab predation at any site despite extensive searching; therefore, we concluded that crab predation was low (see Shepherd 1998). Also, we did not find any whole empty shells suggestive of predation by the sea-star *Coscinasterias muricata* (against which small abalone have a well-developed escape response; see Day et al. 1995) and concluded that sea star predation was also low. Octopuses are not common in this habitat and predation by them has never been observed or suspected on small abalone. The only common predators at the seeding sites were wrasses, which fed preferentially on small abalone (Shepherd and Clarkson, unpublished data); therefore, these wrasses were probably responsible for most mortality of seeded abalone that were ≥3 months old. However, there was no relation between density of wrasses and M of small

Table 1. Larval *H. rubra* 1994 release experiments.

Site	Release density	Day 6 density	Day 19 density	Day 27 density	Percentage settlement	M·month^{-1}
1 T	20 000	25 (13)	—	14 (8)	0.13	>0.8
C	0	2 (2)	—	14 (7)		
2 T	1 600	—	19 (10)	—	1.2	—
T	16 000	—	61 (21)	—	0.4	—
T	80 000	—	16 (9)	—	0.02	—
C	0	9 (4)	10 (7)	—	—	—
3 T	20 000	57 (14)	—	76 (20)	0.4	—
C	0	16 (7)	—	28 (12)	—	—

Note: Larval release densities and densities of post-larval survivors in surveys variously done 6, 19, or 27 days after seeding at experimental and control sites, with estimates of percentage of larvae that settled and survived until the first post-larval sampling, and monthly mortality rates from day 6 to day 27. Densities are given in numbers·m^{-2}. Values in brackets are ±1 SE. A dash indicates data not obtained. Data for site 2 are extracted from Preece et al. (1997). T = seeding treatment, C = control.

Table 2. Larval *H. laevigata* 1994 release experiments.

Site	Release density	Day 6 (or 9)	Day 39	Day 49	Percentage settlement	M·month^{-1}
4 T	2 000	124 (45)	—	11 (4)	6.5	1.8
T	120 000	422 (90)	—	8 (3)	0.4	2.8
C	0	0	—	—	—	—
5 T	20 000	1567 (415)	99 (47)	—	7.8	2.8
C	0	5 (5)	5(3)	—	—	—
Col	20 000	380 (210)	24 (12)	—	1.9	2.5
C	0	3 (3)	3 (2)	—	—	—
T	20 000	21 (8)	29 (10)	—	0.1	—
C	0	1 (1)	2 (1)	—	—	—
Col	20 000	56 (22)	101 (19)	—	0.1	—
C	0	0.5 (0.3)	2.5 (1.6)	—	—	—

Note: Release larval densities and densities of survivors in surveys done 9 (6 for site 4), 39, or 49 days after seeding at experimental sites and on collectors, with estimates of percentage larval settlement, and monthly mortality rates between post-larval surveys. Values in brackets are ±1 SE. Dash indicates data not obtained, or in case of mortality estimates, no estimate was possible. Data for sites 4 and 5 are extracted from Preece et al. (1997). Densities of released larvae and post-larvae on natural substrata or in collectors are in numbers·m^{-2}. T = seeding treatment, C = control, Col = collector.

abalone, as found by Shepherd and Clarkson (unpublished data) at West Island.

Juvenile seeding experiments

Experiment 1

Almost all seed abalone vacated the shells and bricks in which they were initially placed and found refuge in the surrounding boulders. Of the 1666 recaptures, all except 5 were found within their 2 × 2 m release square indicating negligible horizontal movement during the 28-day period. At 2 days after release, the seeds were dispersed almost equally among the three levels of boulders but during the ensuing 26 days the majority of the survivors migrated to the upper level of boulders (Fig. 4). Overall, 48% of the seed were recaptured on the undersides of boulders, 30% on the sides, 16% on upper surfaces, and 6% within cracks. In terms of microhabitat choice, 55% were found on crustose corallines, 36% on bare rock, 6% on pebbles, and 2% among algae on rock.

The ANOVA showed no significant effect of density (F = 1.4; n.s.) or site (F = 1.0; n.s.), but found a significant effect of time (F = 4.3; P < 0.001), and a significant interaction between density and time (F = 1.8; P < 0.05). After pooling the site data (Fig. 5), a plot of presumed %M over the 28 days showed three phases of mortality. In the first phase, the mean %M was 21% compared with 19% for the control group in barrels. In the second phase (day 2–14), the mean %M was 12% and occurred during the migration from lower to upper boulders in the boulder field (Fig. 5). In the third phase (day 14–28), mean %M was 4.6% (equivalent to

Table 3. Combined densities of *H. rubra* (sites 1–3) and *H. laevigata* (sites 4–6) 1 and 2 years after seeding, with estimates of the annual instantaneous natural mortality rate, *M*, for the period from the last postlarval survey (19–49 days after seeding) until 1 year and from 1 to 2 years.

				After year 1		After year 2	
Site	Year	Type		Density	*M*	Density	*M*
	1994	T	14	1.75	2.1	0	>2.8
		C	14	0.75	2.9	0.32	0.9
1	1995	T	30	0.62	3.9	1.50	—
		C	1	0.67	0.4	1.50	—
2	1994	T	32[a]	1.75	2.9	1.00	0.6
		C	10	5.50	2.6	0.17	3.5
2	1995	T	0.8[b]	2.0	—	1.50	0.3
		C	0	1.6	—	1.67	—
3	1994	T	76	2.75	3.3	1.67	0.5
		C	28	2.50	2.4	1.33	0.6
3	1995	T	41	1.00	3.7	0.67	0.4
		C	0.5	0.80	—	0	>2.1
4	1994	T	10	0.82	2.5	1.0	—
		C	0	0.40	—	0.17	0.9
4	1995	T	5.3	0.33	2.8	—	—
		C	0.2	0.67	—	—	—
5	1994	T	99	0.9	4.7	0	>2.2
		C	5	0.2	3.2	0.5	—
5	1995	T	5	0.17	3.4	—	—
		C	0	0.33	—	—	—
6	1994	T	29	0.47	4.1	0	>1.5
		C	2	0.33	1.4	0.5	—

Note: 1994 post-larval densities from Tables 1 and 2 and 1995 values inferred from collector data (see text). A dash indicates that there were no data, or in the case of estimates of *M*, that values were not possible or meaningless. Minimum estimates of *M* are indicated with '>'. Densities are in numbers·m^{-2}. T = seeding treatment, C = control. Larval release densities of all 1995 experiments were 4000·m^{-2}.

[a] Mean of 3 larval release density treatments.

[b] Due to malfunction (air bubble) within hose, few seeds were released around collector, although remainder of seeding site was seeded satisfactorily.

an annual instantaneous rate of 1.2) and possibly representing the longer-term mortality rate. The density–time interaction was due to the different survival trends with different densities and not to differential size-dependent mortality between densities because mean initial lengths, calculated for the highest density group, did not differ significantly at the beginning and at the end of the experiment (*t* test; $P > 0.25$). The mean shell increment over the 28 days was 1.9 mm and there was no significant difference in growth rate between the low and high density treatments (Mann–Whitney *U* test; $P > 0.05$).

Experiment 2

In the long-term experiment %*M* was 84–92% after 4 months. At 8 and 12 months, recoveries did not differ significantly from the number found at 4 months, giving an annual *M* of 1.8. Maximum movement was 8.9 m of an individual after 12 months but the vast majority of tagged abalone remained in their square of release. The mean size of seeded abalone increased from 12.6 mm (SE 1.0) SL at 13 months to 42.9 mm (SE 3.4) SL at 25 months.

Fig. 3. Plot of the annual instantaneous natural mortality rate, *M*, of the 0- to 1-year age-class, versus the logarithm of post-larval density in numbers·m^{-2} (data for *H. laevigata* and *H. rubra* combined on the same graph are from Table 3). See text for the equation describing the regression.

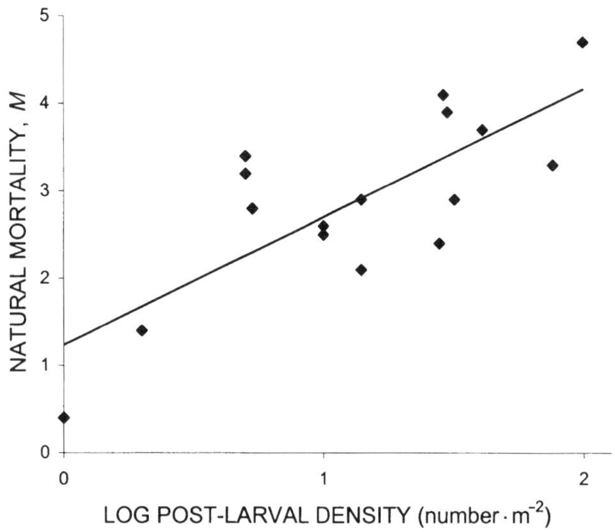

Experiment 3

In the single Flinders Island experiment with marked abalone, 17 tagged abalone were recaptured on the same modules after 6.2 years at a mean size of 149 mm SL and a mean total weight of 427 g. The mean annual mortality rate was 0.18.

Discussion

Larval seeding experiments

Larval settlement

The larval seeding experiments provided valuable information on poorly understood post-settlement processes in abalone. Notable features were the intense patchiness of post-larvae (previously reported only by Tutschulte 1976) and the highly variable settlement and survival to day 6. Many factors may contribute to these results. Before settlement, larvae may be advected from the site by water currents and eddies characteristic of topographically complex substrata such as boulders (Guichard and Bouchet 1998). Preece et al. (1997) used tents to retain larvae and increase settlement; they inferred that in the experiments without tents many larvae were advected by water currents elsewhere. Schiel (1992) also used tents but thought that the larvae that failed to settle had died. Larval behavior may play a role because Daume et al. (1999) showed experimentally that *H. laevigata* preferentially settled on crustose coralline algae where there were no recently settled conspecifics; however, where there were older conspecifics, larvae settled gregari-

Table 4. Estimated mean relative densities of wrasses >15 cm length (mean numbers per 50 × 10 m transect census), grazing gastropods (see text), juvenile abalone (<60 mm SL), chitons, and *Coscinasterias*.

Site	Wrasses	Grazing gastropods	Abalone	Chitons	*Coscinasterias*
1	22	5.6 (1.1)	2.1 (0.8)	22 (7)	1.2 (0.8)
2	35	3.4 (0.9)	0.9 (0.6)	14 (5)	0.6 (0.4)
3	32	3.7 (1.1)	2.0 (0.7)	10 (3)	0.4 (0.7)
4	20	0.6 (1.4)	0.8 (0.3)	12 (3)	0
5	27	0.5 (1.0)	0.1 (0.1)	1.6 (1.6)	0
6	22	0.8 (0.8)	0.1 (0.05)	2.5 (1.4)	0

Note: Densities of invertebrate fauna are in numbers·m^{-2} (±1 SE in brackets). Abalone are *H. rubra* (sites 1–3) or *H. laevigata* (sites 4–6).

Fig. 4. Schematic diagram of three levels of boulders and movement of small *H. rubra* within the 3 levels. Recaptures is the percentage of the total number of tagged abalone released.

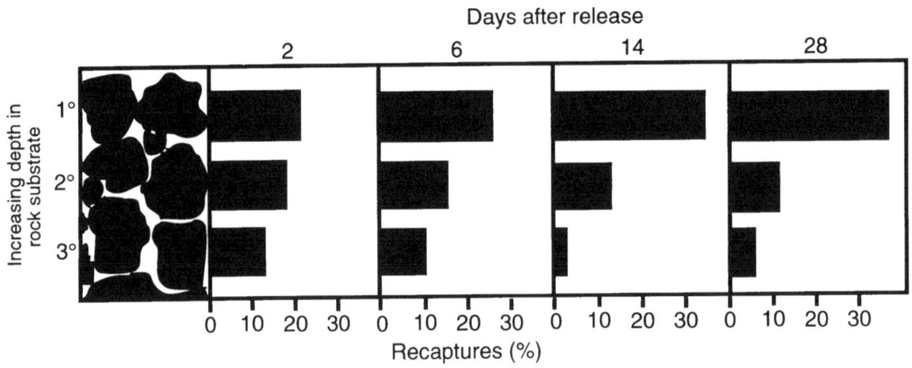

ously. Therefore, larvae may avoid settling in high densities, except in specific circumstances, and remain in the water column until advected elsewhere.

Post-settlement factors that affect distribution are the patchy occurrence of suitable crustose corallines (Shepherd and Daume 1996; Daume et al. 1999) and the patchy predation by post-larval predators, e.g., terebellids, nemerteans, and polyclads, as shown in tropical habitats by Keesing and Halford (1992) and Stoner et al. (1998). These three factors, larval advection and patchy settlement and survival, are sufficient to cause the observed variability. We suggest that future larval seeding may be more effective at lower densities than we tested, at least in the boulder habitats we used. However, a relatively high settlement success rate of 10%, after seeding at 20 000·m^{-2}, was achieved by Schiel (1992) for *H. iris* in New Zealand. This emphasizes the danger of extrapolating results from one place to another.

Post-larval survival

Shepherd (1998) reviewed six studies of *M* on abalone from settlement to 6 months of age. In the first 3 months the range of *M* was 1–3·month^{-1} and in the second 3 months, *M* was as low as 0.2–0.4·month^{-1}. Post-larval *M* values (day 6–49) in this study (Tables1 and 2) were within the range reported in the above review. The correlation between the logarithm of density and monthly *M* suggested density-dependence, but the suggestion was faint given our crude estimates. The estimates of *M* from ~1 month to 2 years of age also must be considered crude for two reasons. First, detectability of small abalone increased with size and varied with swell conditions. Shepherd (1998) found an 8% increase in detectability of *H. scalaris* from 20 mm SL to 50 mm SL. Second, dispersal out of the seeded site may have occurred, although this was unlikely given the size of the seeded sites and the fact that they were bounded on at least two sides by unsuitable habitat. The two effects tended to bias estimates of *M* in opposite directions. Nevertheless, post-larval survival rates from ~1 month to 1 year of age suggested strong density-dependence. What was the likely mechanism for density-dependence?

Shepherd (1998), in a 13-year study of juvenile abalone mortality in a similar underboulder habitat, found that wrasses were responsible for 87% (SE 2.3%) of the natural mortality of small *H. laevigata* (aged from 6 months to two years). Crabs at densities of 12–13 m^{-2} accounted for the remainder of *M* but were relatively more important predators of abalone <6 months old than older ones. Abalone mortality due to wrasses was correlated with wrasse density (Shepherd 1998; S.A. Shepherd, unpublished data). Shepherd and Clarkson (unpublished data) showed that wrasses preferred small abalone to most other underboulder prey. The possibil-

Fig. 5. Three-dimensional plot of mean percentage mortality of small *H. rubra* for four density treatments (numbers per 2 × 2 m^2) over 28 days.

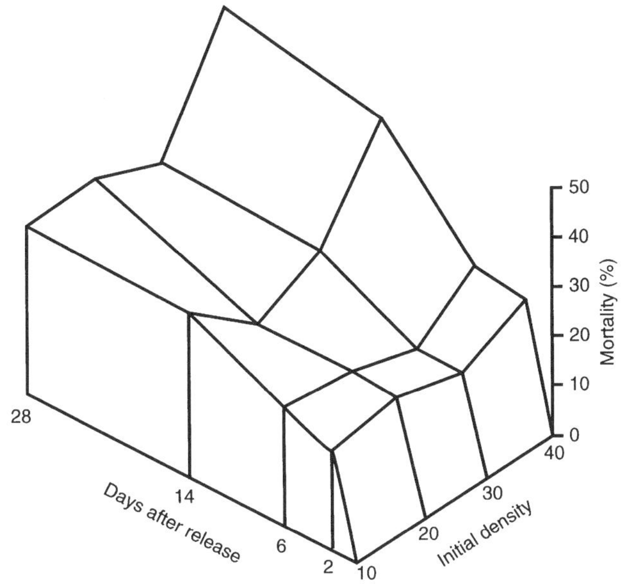

ity that wrasses have a functional response to small abalone as their density increases is considered below (see Ecosystem considerations).

The shift in the dispersion pattern, from a strongly aggregated pattern as post-larvae to a random or regular pattern at 2 years of age was a predictable outcome of random movement and mortality. Shepherd and Godoy (1989) found that *H. laevigata* and *H. scalaris* at this age moved randomly in the underboulder habitat at a rate of about 0.5 m·month^{-1}. If *M* was density-dependent as suggested by this study, then random dispersal tended to produce a less aggregated dispersion pattern that was clearly adaptive. This behavior contrasted sharply with that described by Ray and Stoner (1994) and Stoner and Glazer (1998) for the gastropod *Strombus gigas* where *M* was strongly inversely density-dependent, i.e., *M* was lower within dense aggregations.

Juvenile seeding experiments

The boulder field seeding experiment showed that handling stress could cause substantial losses in such trials but that the method of release into the habitat in bricks, on shell, or directly onto boulders, was of no significance. This suggested that mortality from handling stress, ~20% in our experiments, possibly augmented by the tagging process, resulted from stress during transport combined with failure to adapt to the habitat, rather than placement in a microhabitat where seeds were vulnerable to predators. Hence, the use of seeding modules offered no advantage in terms of reducing *M*, as Parker, cited in Tegner (2000), also found, although the use of modules for seeding did reduce diving time. Schiel (1993) recorded an average %*M* of 1.5% during transport of small abalone in chilled styrofoam containers but a maximum of 47% when abalone were packed at excessive density; clearly there is much scope in developing methods for minimizing handling stress during transport. However, Schiel (1993) did not estimate *M* in the first few days after outplanting. Tegner and Butler (1989) and McCormick et al. (1994) stressed the importance of maintaining abalone in good physical condition. In most cases, McCormick et al. (1994) found that survival was variously 92–99% up to 5 weeks after outplanting, although in the worst case only 61% survived a 48-h voyage from the hatchery to the field and only 30% of the original number were alive after 10 days. Schiel (1992) found in small-scale experiments that survival of hatchery juveniles was ~25% lower than that of wild juveniles after 5 weeks; the difference could have resulted from handling stress, failure of juveniles to adapt, or both. The variability in the stress factor and its magnitude emphasize the value of controls in order to separate this source of *M* from other causes.

In our study during the second phase (day 2–14), *M* was still elevated compared to the third phase. This may have been due to the mortality of survivors already weakened by stress or their failure to find suitable microhabitats. In the third phase, however, *M* was only 1.2·year^{-1} (= 4.6% over 14 days) and not much higher than levels recorded in other studies (see below). The high *M* for the highest density treatment was unlikely to have been due to food limitation because there was no difference in growth rates between treatments. Possible explanations included a functional response by predators to high densities of abalone or high *M* of weaker abalone with poor growth rates.

The overall *M* of ~1.8 over one year in the long-term experiment was similar to that of other seeding experiments (see below). However, in the commercial-scale experiments of Schiel (1993), *H. iris* was seeded at 5–20 mm SL (aged ~6 months) for periods at liberty of 1–2 years; except for two sites with very high *M* (4.4 and 2.3·year^{-1} respectively), the mean *M* was 1.2·year^{-1} (SE 0.2; $N = 6$), which was much lower than *M* estimated in our Tasmanian experiments, but not much higher than *M* of wild populations (see below).

Transplanting juveniles onto concrete modules has been practised at Flinders Island for many years. If commercial results matched those of the reported experiment, we have an intriguing example of the commercial potential of seeding in a novel habitat in seagrass where there is an abundant food supply of drift epiphytic algae and seagrass. The seeding site was on a sandy bottom among scattered *Posidonia* beds in a place of moderate current, with few predators of small abalone (see below). The rods on the modules trapped algal and seagrass drift; therefore, abalone movement off the modules in search of food was unnecessary. The site was similar to the habitat (reef of low relief outcropping among sand and seagrass) of some very productive populations of greenlip abalone in southern Australia (e.g., Franklin Sound, Tasmania, and Thorny Passage and Tiparra Reef, South Australia (see Shepherd 1973); Shepherd and Baker 1998) where detached seagrass epiphytes provide food for abalone. This habitat appears promising for enhancement of greenlip abalone populations, although the predatory spider crab *Leptomithrax gaimardii* may be a source of mortality in warmer waters.

Estimates of juvenile abalone mortality on reefs

There are only four studies of *M* for the period from ~6 months to 3 years in natural abalone populations. For *H. rubra*, Prince et al. (1988) and Day and Leorke (1986) recorded annual *M* of 0.7–0.9 and, for *H. tuberculata*, Clavier and Richard (1985) found a value of 1.7·year^{-1} (reviewed in Shepherd and Breen 1992). In a long-term study over 13 years, Shepherd (1998) found a mean *M* of 1.0·year^{-1} (SE 0.2) for *H. laevigata* and a mean of 0.4·year^{-1} (SE 0.1) for *H. scalaris*; the considerable interannual variation of *M* in both species was correlated with wrasse density. Shepherd (1998, Table 8) also reviewed 12 seeding studies of at least a year including the same age range as the studies above and found *M* values ranging from 0.2 to 11.9·year^{-1}. Excluding the latter extreme value and the study of Davis (1995) whose estimate of *M* (0.7·year^{-1}) was ambiguous, mean *M* for all studies of seeded abalone was 1.8·year^{-1} (SE 0.5; $N = 21$), indicating that mortality of seed juvenile transplants has, thus far, been higher than that for wild abalone of the equivalent age group.

Other studies

The only other molluscan species for which attempts were made to enhance populations by juvenile seeding in Australia has been the trochid, *Trochus niloticus*. Crowe et al. (1997) reviewed eight studies involving 12 trials of which four were reported to have positive outcomes. Annual *M* in those with positive outcomes ranged from 1.2 to 5.8 after a year; these results span a range similar to those of seeded juvenile abalone. Lowest mortalities were achieved when *Trochus* juveniles were placed in coral rubble or were protected by mesh. According to reviewers, the data were inadequate to draw any conclusions about the feasibility of seeding or the optimal size of release. However, Castell (1997), who conducted many of the reviewed trials was more positive. She considered that the optimal seeding density was ~5 individuals·m^{-2} and advocated release of seed at ~5 mm SL in coral rubble (Castell 1993). In contrast to juvenile seeding, translocation of adults has been remarkably productive, as exemplified by the long history of introductions in the Pacific islands (reviewed by Gillett 1993; Clarke and Ianelli 1995).

Future abalone seeding in southern Australia

The two studies we have described, together with the commercial experience at Flinders Island, illustrate the problems and potential for enhancement of abalone. A useful framework in which to consider enhancement is the concept of carrying capacity (reviewed by Kashiwai 1995), defined as the general productivity of the ecosystem.

Kashiwai (1995) considers that the carrying capacity of a given reef is determined by all the factors influencing recruitment and mortality. The three most important are the availability of food and shelter and the extent of predation. All three factors are likely to act in a strongly density-dependent way. For abalone we distinguish two critical microhabitats of juveniles: (*i*) the crustose coralline algal (CCA) crusts on which larvae settle and (*ii*) the cryptic habitat of juveniles before they reach the emergent phase at 2–3 years in the case of *H. laevigata* and *H. rubra*.

Little is known about the limiting density of post-larvae on CCA crusts or their predators, except in a very general way (Shepherd and Daume 1996; review of Kawamura et al. 1998). In laboratory studies, T. Kawamura (3-27-5 Shinhama, Shigama, Miyagi, Japan, personal communication) estimated that the maximum densities of small abalone that could be sustained by diatoms under optimal conditions for diatom growth was ~3000·m^{-2} for abalone of 2 mm SL and ~14·m^{-2} for abalone of 11 mm SL. In wild populations, the highest natural densities of post-larvae recorded were in the range of 50 to 200·m^{-2} (Sasaki and Shepherd 1995; Shepherd 1998), although McShane (1991) once recorded densities of *H. rubra* an order of magnitude higher on sheet vertical surfaces with 100% cover of CCA. However in most boulder habitats, CCA cover is much lower and typically 20 to 35%, rarely to 60%, of boulders (Shepherd and Turner 1985; Shepherd and Daume 1996; Daume 1998). Small abalone are bound to this habitat for ~3 months (Shepherd and Cannon 1988; Shepherd and Daume 1996) where they feed on surface polysaccharides, bacteria, diatoms, and later the deeper CCA crusts. Small abalone compete with chitons and gastropods in the same microhabitat. Roberts et al. (2000) considered that the very presence of CCA implied that grazing pressure was intense and considered that competition for food among abalone and other grazers may be a major source of mortality.

Sasaki and Shepherd (1995) and McShane (1991) found, as we have, that post-larval survival was strongly density-dependent. Overall, in five out of nine of our seeded experimental sites (Table 3), there was no enhancement after two years, although some enhancement still persisted at one year in some of the sites. On this empirical evidence, we consider that the carrying capacity of CCA crusts was probably <100 post-larvae·m^{-2} in our boulder habitats so that optimal seeding densities were likely to be about the same order of magnitude.

The cryptic habitat of abalone has been studied mostly in boulder fields where the complex topography provided a range of crevices for shelter for small abalone as well as guilds of other grazers and predators and also promoted a rich epiflora (Guichard and Bourget 1998). Hence, availability of shelter is unlikely to be a limiting factor. At West Island, where one of the authors (SAS) has worked for many years, the density of abalone among boulders was strongly influenced by the amount of drift algae transported into the habitat. At sites with little algal drift there were no abalone and very few grazers, but where drift was abundant the upper limit of density of small abalone <80 mm SL was ~25·m^{-2} (Shepherd 1998). At the sites seeded in this study, densities of juvenile abalone were quite low, as were the guilds of grazing chitons and archaeogastropods, compared with West Island. The sites chosen may have been close to their carrying capacity in terms of food availability for small abalone. Overall, our experience, mostly with greenlip abalone, suggests that for juveniles aged 6 months to 3 years, the carrying capacity of boulder habitats supplied with drift algae is ~5–20·m^{-2}.

In a seminal paper, Hilborn (1998) argued that species with a strong density-dependent larval or juvenile stage are unsuitable for enhancement. Increasing the density of juveniles will not increase the net production because they will be lost to predation via density-dependent predation or competition. Hillborn (1998) offered two exceptions, one being that natural stocks have been so reduced that density-dependence plays no role in the survival of young. Tegner (1989), in reference to sea-urchin enhancement, argued similarly and predicted that enhancement efforts that ignored the study of the limiting factors in advance were doomed to failure. Schiel (1993) seeded at carefully chosen sites with abundant boulders providing good juvenile habitat, but even so, found successful enhancement at only three of eight sites. He found no evidence of density-dependent *M* suggesting that *H. iris* may have been recruitment-limited at some of these sites; at other sites inundation by sand caused high *M*. The second exception offered by Hilborn (1998) is to release seed at a size above which density-dependent processes operate. This is the practice in Japan (Masuda and Tsukamoto 1998) following numerous studies showing increased survival by rearing seed abalone in culture to a size of 20–40 mm SL (reviewed by McCormick et al. 1994; Shepherd 1998).

Seeding with larvae or juveniles?

The fragility of abalone larvae imposes a serious liability on the widespread use of larval seeding for enhancement. Once the larvae are competent to settle at 5–7 days after fertilization, then seeding cannot be postponed for more than a few hours because of adverse weather or other factors. We have not tested the viability of larvae for different periods of exposure, but our experience in this study, when larvae were transported at optimal temperatures of ~14°C, showed no larval deterioration for up to 5 h of emersion; however, after 20 h of emersion only ~10% remained viable. In practice, larval seeding may be limited to sites within a day's travelling time of the hatchery.

The other main weakness of larval seeding is the increased unpredictability compared with seeding juveniles. Larvae can be advected by the vagaries of oceanographic currents into unsuitable habitat, where the presence and suitability of CCA microhabitats for post-larval survival are uncertain. The combination of these uncertainties reduces the probability of enhancement success with larvae compared with seeding juveniles in habitats whose ecology may be better understood.

Yet enhancement with juvenile abalone also poses formidable problems. Handling stress is potentially serious but, as shown by McCormick et al. (1994) and Schiel (1993), can be reduced to a minimal mortality of a few percent by transporting juveniles in insulated containers at low densities in air enriched with oxygen. Whang and Chang (1977), cited in Schiel (1993), noted a direct relation between survival and transport time.

A less tractable problem is the persistent elevated mortality suffered by hatchery-reared juvenile seed for many months after seeding as exemplified in our study in southern Tasmania. The behavioral deficits of hatchery-reared animals that are naive to predators are better known for fish (reviewed by Olla et al. 1998) than for molluscs. Schiel and Weldon (1987) and Tegner and Butler (1989) found that hatchery-born juvenile abalone moved about more, rested in the open more, and were more susceptible to predation than wild abalone. Osumi (1998), cited in Roberts and Nicholson (1998), compared the behavior of hatchery versus wild juveniles for two different sizes of abalone on boulders within sea cages. Whereas larger ones (~35 mm SL) were slow or failed to find shelter and had poor survival (16% after 33 days) compared with wild ones with 66% survival, smaller seed (~20 mm SL) showed no differences in behavior but still had lower survival (47% compared with 71% for wild ones).

In our southern Tasmanian experiment, the presence of 46% of the hatchery seed on the sides or tops of boulders suggested naivity to predators and no doubt contributed to the high initial mortality. Stoner and Glazer (1998) similarly found that hatchery-reared queen conch were less cryptic and suffered higher mortality than their wild counterparts. Olla et al. (1998) suggested that the behaviour of hatchery stock could be improved by exposure to predators or their odor before release and cited examples of substantial improvement in the case of small salmon. This has not been tested yet for abalone.

Hilborn (1998) considered that the only justifiable purpose of stock enhancement is to rebuild depleted stocks and that attempts to augment existing stocks to increase productivity are almost certainly a waste of time and money. Seki and Sano (1998) suggested that transplantation of juveniles was less costly than larval seeding, but even that was only faintly profitable. They concluded that nowhere in Japan has seed transplantation enhanced wild stocks. Similarly, Tegner (2000), after reviewing Californian experiments in the last decade, was pessimistic about juvenile seeding. While we concur with these authors, we consider that the seeding of artificial substrata, which effectively extends abalone habitat into adjoining areas limited by absence only of suitable substratum, may be cost effective because it achieves the dual purpose of both habitat and stock improvement and may have corresponding dual benefits. The problems of enhanced mortality of hatchery-produced seed may be surmounted only by placing juveniles in a novel or virtually predator-free environment, as at Flinders Island, or in the experiments of Sweijd et al. (1998) who seeded *H. midae* outside its natural range, or in the few reported, successful Japanese commercial seeding projects (e.g., Kojima 1995). Enhanced populations may also increase egg production of the wild populations in the long term where these have declined through overfishing.

Ecosystem considerations

A consideration of the role of the main biotic components in the habitat of juvenile abalone is important both to assess the suitability of potential sites for enhancement and to predict the effects of enhancement within the system (Wahl et al. 1995). Here we consider specifically the underboulder habitat with which we have the most experience, but other cryptic habitats, which are intractable to study, largely share the same fauna. The major components of the food web are shown in Fig. 6. Small abalone share the habitat with a rich fauna of chitons, gastropods, and crabs (Clarkson and Shepherd 1984; Kangas and Shepherd 1984; Mower and Shepherd 1988), all of which are preyed upon by wrasses (Family Labridae), mainly *Notolabrus tetricus* in south and south-eastern Australia and the blue groper *Achoeroedus gouldii* in the eastern Great Australian Bight. Wrasses are preyed upon by sea lions and fur seals (S.A. Shepherd, unpublished data). The chiton and gastropod grazing guilds are important because: (*i*) they maintain the CCA microhabitat (Clarkson and

Fig. 6. Simplified trophic relations relating animals in the underboulder habitat to higher-order predators in South Australia. Arrows show major predator–prey interactions and indicate direction of flow of food and energy. See text for references to species' guilds in boxes.

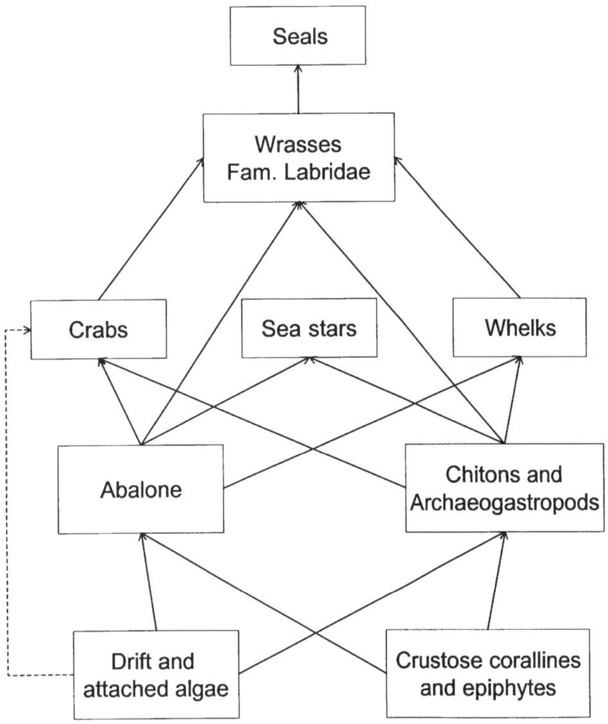

Shepherd 1984) and (*ii*) they provide alternative prey for crabs and wrasses and may to that extent reduce predation on small abalone (Murdoch and Oaten 1975).

The addition of cultured juvenile abalone to reefs could increase predation on abalone and other underboulder fauna via an aggregative numerical response as reported by Shepherd (1990) for larger abalone where more predatory myliobatid rays were seen in abalone habitat at high abalone densities (also see Schmitt 1987). This may be the reason for the failure of many clam-seeding projects (Mottet 1980).

Another likelihood is that predators will have a functional response to increased abundance of prey abalone. Highly mobile predators can respond to differences in prey density over much greater distances than sedentary predators. Hence the former are expected to have a type 3 response producing density-dependent *M* and the latter a type 1 response which is never density-dependent (Murdoch and Oaten 1975; Peterson 1982; Boulding and Hay 1984; Schiel and Weldon 1987; Kashiwai 1995). The wrasse *Notolabrus tetricus* has a home range of ~2000–3000 m^2 which is shared with all other members of the same haremic group (Barrett 1995; Shepherd and Clarkson, unpublished data). Therefore, this wrasse and its congener, *Pictilabrus laticlavius,* with similar behavior, could inflict a density-dependent *M* on preferred prey populations, whereas slow-moving or sedentary predators, such as most crabs, whelks, and sea stars, would have only a local impact. Whether or not this is the case will depend on other factors as well, e.g., their own density and the abundance and availability of alternative preferred prey.

Higher-order predators may also have a cascade effect lower down the food web. Our long-term data set at West Island showed that when seals arrived and established a seasonal haul-out site, wrasse abundance declined and abalone survival increased (Shepherd 1998; S.A. Shepherd, unpublished data). However, attempts to control voracious predators such as wrasses and blue groper could be counterproductive because their reduction may allow intermediate predators such as crabs, also a preferred prey of labrids, to increase in abundance. Termed meso-predator release, this effect could increase seed mortality (Kerfoot and Sih 1987; Soulé et al. 1988). In a review of studies of predator–prey systems, Sih et al. (1985) found that 40% showed positive rather than the expected negative effects of predation on their prey. Similarly, Schoener (1994) found that indirect effects were larger than direct effects in 25% of the experiments surveyed, although direct effects were more important overall.

Our comments touch upon only those few complexities in multispecies food webs that we have observed or suspected in abalone habitat. According to Holt and Lawton (1994), a humbling message from studies of food webs is that it is difficult to predict the effect of one species on another with so many potential routes for indirect interactions between species.

On the basis of our experience with small abalone in southern Australia, we make the following recommendations (inspired by Stoner and Glazer 1998) for stock enhancement.

- Sites should be selected after careful prerelease surveys of potential sites and on the basis of the history of past abalone catches rather than for reasons of convenience or suitability of substratum. However, we do not exclude the novel extension of abalone habitat. The surveys should estimate abundance of the major molluscan fauna, other herbivores, and crustacean, echinoderm, and vertebrate predators. Possibly the guild of gastropod herbivores may serve as surrogates from which to predict the carrying capacity of the habitat for abalone. Preliminary experiments with hatchery seed on the trophic suitability of sites should be undertaken with adequate control sites. The many sites where abalone populations once existed before they were overfished in the 1970s in southern Australia (see Shepherd and Baker 1998) are all candidates for restocking or enhancement.
- Abalone should be released at the optimal size, also determined by experiment, to minimize density-dependent effects and behavioral deficits. The release of very small seeds may be more cost effective depending on the habitat and the predators present (Osumi 1998; Sweijd et al. 1998). After seeding, the sites should be monitored frequently and collections made of empty shells to provide clues about the agents of mortality.
- Optimal release densities are probably quite low and rapid release techniques with devices, such as those used by McCormick et al. (1994) to minimize postrelease *M* and to disperse seed over wide areas by boat, have promise.
- Releases should be timed to minimize predation and handling stress. This is likely to be in the late afternoon when fish predators are less active (Shepherd and Clarkson, unpublished data) and in cool weather.

The quality of seed in a genetic sense and the genetic compatibility of seed with wild abalone populations in the vicinity are fundamentally important but are outside the scope of this paper (see Blankenship and Leber 1995 for an excellent review). Also, we do not consider the economic viability of enhancement, which should be assessed according to strict criteria requiring evaluation by mark–recapture experiments and determining effects on the wild stock. As emphasised by Hilborn (1998), restocking programs have rarely been economically successful and are never a panacea for restoring overfished stocks. Improved fishery management and transplants of adults, as advocated by Tegner (2000), are certain to be more cost effective for most declining stocks than enhancement with juveniles.

Nevertheless, despite our pessimism, we concede that there is some scope for enhancement. This is possible in places where existing stocks have disappeared and in places where dual goals can be achieved, i.e., enhancement of habitat and of stock in essentially a marine farming operation.

Acknowledgments

Jim Morrison from S.A. Mariculture and Rodney Grove-Jones from S.A. Abalone Developments supplied larvae for the experiments and professional divers Graham Ford, Billy Ford, Tom McNab, George Dixon, Doug Graske, John Mortlock, and John Yates assisted in various ways. Kate Rodda, Brian Foureur, and Tim Wolfe also helped with the diving. We acknowledge the contribution of Peter Whyte who set up the seeding experiments in southern Tasmania and supervised data collection for the first 28 days until he left the project. Stefan Delebarre and Dimitri Onof helped in the laboratory and Bruce Brown gave valued statistical advice. Jamie and Lynn Mason of Furneaux Aquaculture kindly gave permission to cite their data and generously provided accommodation for SAS on Flinders Island. Rodney Roberts provided information on seeding in New Zealand and Tomo Kawamura allowed us to cite his unpublished data. We thank Sabine Daume and Peter Clarkson for their helpful suggestions. Sylvain Huchette translated the abstract into French. Rob Day advised on the design of the Tasmanian seeding experiments and he, Mia Tegner, Dave Schiel, and an anonymous referee helpfully reviewed the manuscript. The seeding studies were funded by grants from Fisheries Research and Development Corporation in 1983–1985 and 1994–1996.

References

Barrett, N.S. 1995. Short- and long-term movement patterns of six temperate reef fishes (Families Labridae and Monacanthidae). Mar. Freshw. Res. **46**: 853–860.

Blankenship, H.L., and Leber, K.M. 1995. A responsible approach to marine stock enhancement. Am. Fish. Soc. Symp. **15**: 167–175.

Boulding, E.G., and Hay, T.K. 1984. Crab response to prey density can result in density-dependent mortality of clams. Can. J. Fish. Aquat. Sci. **41**: 521–525.

Castell, L.L. 1993. Some aspects of the ecology of juvenile *Trochus niloticus* relevant to population enhancement. *Trochus* Information Bulletin No. 2. South Pacific Commission, New Caledonia. pp. 2–4.

Castell, L.L. 1997. Population studies of juvenile *Trochus niloticus* on a reef flat on the north-eastern Queensland coast, Australia. Mar. Freshw. Res. **48**: 211–217.

Clarke, R.P., and Ianelli, J.N. 1995. Current paradigms in *Trochus* management and opportunities to broaden perspectives. *Trochus* Information Bulletin No. 4. South Pacific Commission, New Caledonia. pp. 3–29.

Clarkson, P.S., and Shepherd, S.A. 1984. Distribution and feeding of archaeogastropods in a boulder habitat at West Island, South Australia. J. Malacol. Soc. Aust. **7**: 35–44.

Clavier, J., and Richard, O. 1985. Études sur les ormeaux dans la région de Saint-Malo. Association pour la mise en valeur du littoral de la côte D'Emeraude, Dinard, France. 185 p.

Crowe, T.P., Amos, M.J., and Lee, C.L. 1997. The potential of reseeding with juveniles as a tool for the management of *Trochus* fisheries. *In* Trochus: status, hatchery practice and nutrition. *Edited by* C.L. Lee and P.W. Lynch. Proceedings of a workshop held at Northern Territory University, Darwin, Australia, 6–7 June 1996, Australian Council for International Agricultural Research Proceedings No. 79. pp. 170–177.

Daume, S.M. 1998. The ecological role of non-geniculate coralline red algae in abalone larval settlement. Ph.D. thesis, La Trobe University, Melbourne.

Daume, S., Woelklerling, W.J., and Brand-Gardner, S. 1999. Settlement of abalone larvae (*Haliotis laevigata* Donovan) in response to non-geniculate coralline red algae (Corallinales, Rhodophyta). J. Exp. Mar. Biol. Ecol. **234**: 125–143.

Davis, G.E. 1995. Recruitment of juvenile abalone (*Haliotis* spp.) measured in artificial habitats. Mar. Freshw. Res. **46**: 549–554.

Day, R.W., and Leorke, A.E. 1986. Abalone catches — what factors affect them. Aust. Fish. **45**: 32–36.

Day, R.W., Dowell, A., Saint, G., Klemke, J., and Shaw, C. 1995. Patchy predation: foraging behavior of *Coscinasterias calamaria* and escape responses of *Haliotis rubra*. Mar. Behav. Physiol. **26**: 11–33.

Gillett, R. 1993. Pacific Islands *Trochus* introductions. *Trochus* Information Bulletin No. 2. South Pacific Commission, New Caledonia. pp. 13–16.

Guichard, F., and Bourget, E. 1998. Topographic heterogeneity, hydrodynamics, and benthic community structure: a scale-dependent cascade. Mar. Ecol. Prog. Ser. **171**: 59–70.

Hilborn, R. 1998. The economic performance of marine stock enhancement projects. Bull. Mar. Sci. **62**: 661–674.

Holt, R.D., and Lawton, J.H. 1994. The ecological consequences of shared natural enemies. Annu. Rev. Ecol. Syst. **25**: 495–520.

Kangas, M., and Shepherd, S.A. 1984. Distribution and feeding of chitons in a boulder habitat at West Island, South Australia. J. Malacol. Soc. Aust. **6**: 101–111.

Kashiwai, M. 1995. History of carrying capacity concept as an index of ecosystem productivity (review). Bull. Hokkaido Natl. Fish. Res. Inst. **59**: 81–100.

Kawamura, T., Roberts, R.D., and Takami, H. 1998. A review of the feeding and growth of postlarval abalone. J. Shellfish Res. **17**: 615–625.

Keesing, J.K., and Halford, A.R. 1992. The importance of post-settlement processes for the population dynamics of *Acanthaster planci*. Aust. J. Mar. Freshwater Res. **43**: 635–651.

Kerfoot, K.C., and Sih, A. (*Editors*). 1987. Predation: direct and indirect impacts on aquatic communities. University Press of New England, Hanover, N.H.

Kojima, H. 1995. Evaluation of abalone stock enhancement through the release of hatchery-reared seed. Mar. Freshw. Res. **46**: 689–695.

Masuda, R., and Tsukamoto, K. 1998. Stock enhancement in Japan: review and perspective. Bull. Mar. Sci. **62**: 337–358.

McCormick, T.B., Herbison, K., Mill, T.S., and Altick, J. 1994. A review of abalone seeding, possible significance and a new seeding device. Bull. Mar. Sci. **55**: 680–693.

McShane, P.E. 1991. Density-dependent mortality of recruits of the abalone *Haliotis rubra* (Mollusca: Gastropoda). Mar. Biol. (N.Y.), **110**: 385–389.

Mottet, M.G. 1980. Research problems concerning the culture of clam spat and seed. Wash. Dep. Fish. Tech. Rep. 63.

Mower, A.G.J., and Shepherd, S.A. 1988. The crab fauna of West Island, South Australia: their abundance, diet, and role as predators of abalone. Trans. R. Soc. S. Aust. **112**: 83–86.

Murdoch, W.W., and Oaten, A. 1975. Predation and population stability. Adv. Ecol. Res. **9**: 1–125.

Olla, B.L., Davis, M.W., and Ryer, C.H. 1998. Understanding how the hatchery environment represses or promotes the development of behavioral survival skills. Bull. Mar. Sci. **62**: 531–550.

Ortiz-Quintanilla, M. 1980. Un sistema para inducir el desove de abulón en los campos pesqueros de Baja California, México. Memorias del segundo simposio latinoamericano de acuacultura, **1**: 871–881.

Osumi, K. 1998. Behaviour and survival of juvenile paua (*Haliotis iris*) in the field. N.Z. Mar. Sci. Soc. p. 87. [Abstract].

Peterson, C.H. 1982. Clam predation by whelks (*Busycon* spp.): experimental tests of the importance of prey size, prey density, and seagrass cover. Mar. Biol. **66**: 159–170.

Preece, P.A., Shepherd, S.A., Clarke, S.M., and Keesing, J.K. 1997. Abalone stock enhancement by larval seeding: effect of larval density on settlement and survival. Moll. Res. **18**: 265–273.

Preece, P.A., Shepherd, S.A., Clarke, S.M., and Keesing, J.K. 2000. Feasibility of enhancing abalone stocks by larval seeding. Fisheries Research and Development Corporation, Canberra, Australia, Report 94/005.

Prince, J.D., and Shepherd, S.A. 1992. Australian abalone fisheries and their management. *In* Abalone of the world: biology, fisheries and culture. *Edited by* S.A. Shepherd, M.J. Tegner, and S.A. Guzmán del Próo. Fishing News Books, Blackwell Science Ltd., Oxford, U.K. pp. 407–426.

Prince, J.D., Sellers, T.L., Ford, W.B., and Talbot, S.R. 1988. Recruitment, growth, mortality and population structure in a southern Australian population of *Haliotis rubra* (Mollusca: Gastropoda). Mar. Biol. (N.Y.), **100**: 75–82.

Ray, M., and Stoner, A.W. 1994. Experimental analysis of growth and survivorship in a marine gastropod aggregation: balancing growth with safety in numbers. Mar. Ecol. Prog. Ser. **105**: 47–59.

Rodda, K.R., Keesing, J.K., and Foureur, B.L. 1997. Variability in larval settlement of abalone on artificial collectors. Moll. Res. **18**: 253–264.

Roberts, R., and Nicholson, C. 1998. Inventory of written information on paua biology and culture. Report No. 434. Cawthron Institute, Nelson, New Zealand. 14 p.

Roberts, R., Kawamura, T., and Takami, H. 2000. Abalone recruitment — an overview of some recent research in New Zealand, Australia and Japan. Bull. Tohoku Natl. Fish. Res. Inst. **62**: 95–107.

Sasaki, R., and Shepherd, S.A. 1995. Larval dispersal and recruitment of *Haliotis discus hannai* and *Tegula* spp on Miyagi coasts, Japan. Mar. Freshw. Res. **46**: 519–529.

Schiel, D.R. 1992. The enhancement of paua (*Haliotis iris* Martyn) populations in New Zealand. *In* Abalone of the world: biology, fisheries and culture. *Edited by* S.A. Shepherd, M.J. Tegner, and S.A. Guzmán del Próo. Fishing News Books, Blackwell Science Ltd., Oxford, U.K. pp. 474–484.

Schiel, D.R. 1993. Experimental evaluation of commercial-scale enhancement of abalone *Haliotis iris* populations in New Zealand. Mar. Ecol. Prog. Ser. **97**: 167–181.

Schiel, D.R., and Weldon, B.C. 1987. Responses to predators of cultured and wild red abalone, *Haliotis rufescens*, in laboratory experiments. Aquaculture, **60**: 173–188.

Schmidt, R.J. 1987. Indirect interaction between prey: apparent competition, predator aggregation, and habitat segregation. Ecology, **68**: 1887–1897.

Schoener, T.W. 1994. On the relative importance of direct and indirect effects in ecological communities. *In* Mutualism and community organization: behavioral, theoretical and food web approaches. *Edited by* H. Kawanabe, J.E. Cohen, and K. Iwasaki. Oxford University Press, Oxford, U.K. pp. 365–411.

Seki, T., and Sano, M. 1998. An ecological basis for the restoration of Japanese abalone populations. Bull. Tohoku Natl. Fish. Res. Inst. **60**: 23–40.

Shepherd, S.A. 1973. Studies on southern Australian abalone (genus *Haliotis*) I. Ecology of five sympatric species. Aust. J. Mar. Freshwater Res. **24**: 217–257.

Shepherd, S.A. 1990. Studies on southern Australian abalone (genus *Haliotis*) XII. Long-term recruitment and mortality dynamics Aust. J. Mar. Freshwater Res. **41**: 475–492.

Shepherd, S.A. 1998. Studies on southern Australian abalone (genus *Haliotis*) XIX. Long-term juvenile mortality dynamics. J. Shellfish Res. **17**: 813–825.

Shepherd, S.A., and Baker, J. 1998. Biological reference points in an abalone (*Haliotis laevigata*) fishery. *In* Proceedings of the North Pacific Symposium on Invertebrate Stock Assessment and Management. *Edited by* G.S. Jamieson and A. Campbell. Can. Spec. Publ. Fish. Aquat. Sci. 125. pp. 235–245.

Shepherd, S.A., and Breen, P.A. 1992. Mortality in abalone: its estimation, causes and variability. *In* Abalone of the world: biology, fisheries and culture. *Edited by* S.A. Shepherd, M.J. Tegner, and S.A. Guzmán del Próo. Fishing News Books, Blackwell Science Ltd., Oxford, U.K. pp. 276–303.

Shepherd, S.A., and Cannon, J. 1988. Studies on southern Australian abalone (genus *Haliotis*) X. Food and feeding of juveniles. J. Malacol. Soc. Aust. **9**: 21–26.

Shepherd, S.A., and Daume, S. 1996. Ecology and survival of juvenile abalone in a crustose coralline habitat in South Australia. *In* Survival strategies in early life history stages of marine resources. *Edited by* Y. Watanabi, Y. Yamashika, and Y. Oozeki. Balkema, Rotterdam. pp. 297–313.

Shepherd, S.A., and Godoy, C. 1989. Studies on southern Australian abalone (genus *Haliotis*) XI. Movement and natural mortality of juveniles. J. Malacol. Soc. Aust. **10**: 87–95.

Shepherd, S.A., and Turner, J.A. 1985. Studies on southern Australian abalone (genus *Haliotis*) VI. Habitat preference, abundance and predators of juveniles. J. Exp. Mar. Biol. Ecol. **93**: 285–298.

Sih, A., Crowley, P., McPeek, P.J., Petranka, J., and Strohmeier, K. 1985. Predation, competition and prey communities: a review of field experiments. Annu. Rev. Ecol. Syst. **16**: 269–311.

Sokal, R.R., and Rohlf, F.J. 1995. Biometry: the principles and practice of statistics in biological research. 3rd edition. W.H. Freeman and Co., New York.

Soulé, M.E., Bolger, D.T., Roberts, A.C., Wright, J., and Sorice, M. 1988. Reconstructed dynamics of rapid extinctions of chaparral-requiring birds in urban habitat islands. Conserv. Biol. **2**: 75–92.

Stoner, A.W., and Glazer, R.A. 1998. Variation in natural mortality: implications for stock enhancement. Bull. Mar. Sci. **62**: 427–442.

Stoner, A.W., Ray-Culp, M., and O'Connell, S.M. 1998. Settlement and recruitment of queen conch, *Strombus gigas*, in seagrass meadows: associations with habitat and micro-predators. Fish. Bull. **96**: 885–899.

Sweijd, N., Snethlage, Q., Harvey, D., and Cook, P. 1998. Experimental abalone (*Haliotis midae*) seeding in South Africa. J. Shellfish Res. **17**: 897–904.

Tegner, M.J. 1989. The feasibility of enhancing red sea urchin, *Strongylocentrotus franciscanus*, stocks in California: an analysis of the options. Mar. Fish. Rev. **51**: 1–22.

Tegner, M.J. 2000. Abalone (*Haliotis* spp.) enhancement in California: what we've learned and where we go from here. *In* Workshop on Rebuilding Abalone Stocks in British Columbia. *Edited by* A. Campbell. Can. Spec. Publ. Fish. Aquat. Sci. 130. pp. 61–71.

Tegner, M.J., and Butler, R.A. 1989. Abalone seeding. *In* Handbook of culture of abalone and other marine gastropods. *Edited by* K.O. Hahn. CRC Press, Boca Raton, Fla. pp. 157–182.

Tutschulte, T.C. 1976. The comparative ecology of three sympatric abalones. Ph.D. thesis, University of California, San Diego, Calif.

Wahl, D.H., Stein, R.A., and Devries, D.R. 1995. An ecological framework for evaluating the success and effects of stocked fishes. Am. Fish. Soc. Symp. **15**: 176–189.

Whang, H.C., and Chang, K.O. 1977. Study on relationship between exposure time and mortality of young abalone, *Haliotis discus* (Reeve). Bull. Fish. Res. Dev. Agency (Busan), **18**: 123–129.

Zar, J.H. 1974. Biostatistical analysis. Prentice Hall, Englewood Cliffs, N.J.

Additional Pieces of the Rebuilding Puzzle

Genetic tools for identification and conservation of exploited abalone (*Haliotis* spp.) species

Ruth E. Withler

Abstract: Analyses of genetic diversity in the speciose abalone genus, *Haliotis*, are beginning to provide molecular genetic tools that will be useful in the conservation efforts underway for exploited and protected abalone species worldwide. For species for which no legal exploitation exists, mitochondrial DNA or nuclear loci encoding gamete recognition proteins can provide DNA sequences that differ among species to enable species identification of confiscated tissue samples, fresh or processed, lacking informative morphological features. For exploited species with areas closed to harvest, surveys of polymorphic nuclear DNA markers may enable identification of confiscated abalone samples illegally harvested from closed areas. Limited available genetic data indicate that abalone species are comprised of weakly to moderately differentiated local populations with F_{ST} values < 0.10, in spite of potentially extensive gene flow mediated by planktonic larval dispersal. Genetic surveys, combined with studies on larval dispersal, will be required to delineate phylogeographic subdivisions (evolutionarily significant units) and the geographic scale of demographically independent populations (management units) within species. This knowledge of abalone population structure will be important in defining marine reserves, rebuilding depleted populations, and developing brood stocks for aquaculture.

Résumé : Les analyses de la diversité génétique chez les espèces d'ormeau (Haliotis) commencent à fournir des outils de génétique moléculaire qui seront utilisés dans les efforts de conservation visant les espèces d'ormeau exploitées et protégées à l'échelle mondiale. Dans le cas des espèces qui ne sont pas exploitées légalement, l'ADN mitochondrial ou les loci nucléaires encodant les protéines permettant la reconnaissance des gamètes peuvent fournir des séquences d'ADN différentes d'une espèce à l'autre, ce qui permet de déterminer à quelle espèce appartiennent les échantillons de tissus saisis, frais ou traités, en l'absence de traits morphologiques informatifs. Dans le cas des espèces exploitées dans des zones interdites, les études des marqueurs d'ADN nucléaire polymorphes peuvent permettre l'identification des échantillons saisis d'ormeaux qui ont été illégalement prélevés dans des zones fermées. D'après les quelques données génétiques que nous possédons, les espèces d'ormeau constituent des populations locales faiblement à moyennement différenciées avec des valeurs FST <0,10, malgré un flux génétique qui peut être important en raison de la dispersion des larves planctoniques. Il faudra effectuer des études génétiques, ainsi que des études sur la dispersion des larves, pour préciser les subdivisions phylogéographiques (unités évolutives importantes) et l'échelle géographique de populations démographiquement indépendantes (unités de gestion) au sein des espèces. Ces connaissances de la structure des populations d'ormeau seront importantes pour la délimitation des réserves marines, le rétablissement des populations affaiblies, et le développement d'un cheptel reproducteur pour l'aquaculture.

[Traduit par la Rédaction]

Introduction

Declines in abundance of exploited abalone species (genus *Haliotis*) pose challenges to fishery managers on a worldwide basis (Davis et al. 1992; Prince and Shepherd 1992; Tarr et al. 1996; Campbell 1997; Shepherd et al. 1998). Although clearly an important factor, overfishing is not the sole culprit implicated in the population declines, which are of a nature sufficiently severe to raise concerns for species extinction in some cases (Davis et al. 1998). Because reproduction in abalone can fail completely once adult density declines below a critical level (Shepherd and Partington 1995), even cessation of fishing may not ensure recovery of a depleted population. Moreover, the high individual value of abalone makes illegal harvest an economically viable enterprise even when abundances are very low.

Critical to the successful conservation of abalone is an understanding of both the genetic population structure of each species (i.e., the partitioning of the neutral and adaptive genetic diversity across the geographic range of the species) and of the population dynamics involved with recruitment, survival, and reproduction in the species. Research on these interrelated aspects of abalone biology is underway for a number of the more important commercial species. Studies on larval dispersal and recruitment have not provided entirely consistent results (Prince et al. 1987, 1988; McShane 1992; Shepherd et al. 1992; Tegner 1993) and studies on the genetic structure of populations are notable for their scarcity (Brown 1991; Brown and Murray 1992; Shepherd and Brown

R.E. Withler. Department of Fisheries and Oceans, Pacific Biological Station, Nanaimo, BC V9R 5K6, Canada.
e-mail: withlerr@pac.dfo-mpo.gc.ca

Correct citation: Withler, R.E. 2000. Genetic tools for identification and conservation of exploited abalone (*Haliotis* spp.) species. *In* Workshop on Rebuilding Abalone Stocks in British Columbia. *Edited by* A. Campbell. Can. Spec. Publ. Fish. Aquat. Sci. 130. pp. 101–110.

1993; Hanna et al. 2000; Huang et al. 2000), but preliminary evidence indicates the existence of metapopulation structure in studied abalone species. Typically, recruitment may be primarily local, whereas genetic diversity may be partitioned only weakly among local populations separated by unsuitable habitat but connected by gene flow (Shepherd and Brown 1993).

An understanding of the genetic structure of abalone species is required to identify both evolutionary significant units (ESUs) and management units (MUs) (Moritz 1994). ESUs are important phylogeographic subdivisions within species (those based on historical separations or fluctuations in abundance that are still evident in the gene pool) whose maintenance might be recognized as critical for long-term conservation of biodiversity. MUs are individual metapopulations and (or) local populations, whose preservation is critical for short-term maintenance of the species range and abundance. Delineation of the genetic structure of a species is dependent on the development of appropriate genetic markers to quantify the partitioning of genetic diversity among the local populations, metapopulations, and ESUs.

The development of genetic markers for the study of evolutionary and ecological relationships in abalone (Tables 1 and 2) have fortuitous application in the enforcement of harvesting restrictions imposed for management of abalone populations. Mitochondrial and nuclear DNA sequences examined to determine the interspecific evolutionary relationships and modes of speciation in abalone will provide species-specific DNA sequences that can be employed to identify illegally harvested abalone products to the species level, even after processing has precluded morphological identification. Highly polymorphic nuclear genetic markers such as RAPDs (randomly amplified polymorphic DNA) and mini- and micro-satellite loci being developed for examination of intraspecific population structures may enable stock identification of confiscated abalone in species having areas open and closed to harvest. This report provides a review and evaluation of the application of biochemical and molecular genetic markers in the delineation of species and population structure in abalone and the use of these genetic markers for formulating and enforcing conservation measures for abalone species.

Abalone species identification

A common problem in the enforcement of species-wide harvest prohibition is the continued or newly established legal harvest of closely related species after harvest closure for the target species. This provides an effective "cover" for the illegal harvest and sale of the prohibited species. The seizure of organisms or tissues believed to have been illegally harvested often takes place after all morphological features enabling species identification have been removed. Frequently, the product also has been processed in some manner (e.g., dried, canned, smoked, etc.) and deliberately mislabeled. These circumstances have resulted in development of genetic methods for the "forensic" species identification of tissues from a variety of terrestrial (Borgo et al. 1996; Miller et al. 1995; Murray et al. 1995) and aquatic (Bartlett and Davidson 1991; Powell et al. 1995; Palumbi and Cipriano 1998) organisms for use in enforcement, including one for an abalone species, *H. midae* (Sweijd et al. 1998).

Genetic loci of most use in species identification are those that exhibit fixed differences (reciprocal monophyly) among the closely related species of interest, yet possess only a single fixed genotype within each species. In the development of a species identification tool to be used for enforcement, samples from across the species' ranges must be examined to ensure that the full range of inter- and intra-specific variability has been characterized and a variety of tissue sources must be tested to ensure development of a robust assay method that will provide results even from processed tissues. The need to examine the genetic markers in processed tissues precludes the use of allozyme loci for this task, because tissue processing generally destroys the enzyme function on which the assay depends. Both mitochondrial and nuclear DNA markers have been employed for species identification of fresh and processed tissues. Rapidly evolving regions of the mitochondrial genome, identified early in evolutionary studies on a wide variety of organisms, have been employed in many species identification methods (Bartlett and Davidson 1991; Murray et al. 1995; Borgo et al. 1996). However, the mitochondrial genome is typically maternally inherited and therefore misidentifies hybrids to the maternal parent species (Palumbi and Cipriano 1998).

Table 1. Studies of genetic variation among abalone species.

	Gene	Species *N*	Nucleotide divergence, %	Intraspecific variation, %	Source
Mitochondrial	mtCO1	12	3.7–64.9	0.4	Metz et al. 1998
Intron	Lysin	3	2.7–6.9	+	Metz et al. 1998
	G protein	3	6.2–8.1	1.4	Metz et al. 1998
Nuclear exons	Lysin	20	4.5–48.6[a]	<1.0	Lee et al. 1995
		3	13.7–24.1	0	Metz et al. 1998
	18-kDa protein	5	—	—	Vacquier et al. 1997
		3	81.3–92.8	0	Metz et al. 1998

Note: *N* is the number of species examined.
[a]Calculated by author from sequences deposited in GenBank.

Table 2. Studies of genetic variation among abalone populations.

Species	Loci	Alleles	Heterozygosity	F_{st}	Source
Allozyme					
discus hannai	10	—	12.5	—	Fujio et al. 1983
rubra	12	2–11	14	0.022	Brown 1991
laevigata	13	—	19.5	0.014	Brown and Murray 1992
rufescens	4	2–5	—	—	Gaffney et al. 1996
fulgens	7	2	12	0.036	G. Zuniga, per. comm.
Microsatellite					
rufescens	1	21	47	—	Kirby et al. 1998
rubra	3	8–41	19–38[a]	0.077	Huang et al. 2000
Minisatellite					
rubra	2	4–13	65	0.016	Huang et al. 2000
RAPD					
rubra	6	—	—	0.086	Huang et al. 2000
laevigata	3	—	—	0.223	Hanna et al. 2000

Note: The number of loci examined and range of alleles observed at individual loci is given. The degree of population subdivision, F_{ST}, observed in each study is indicated.
[a] Heterozygote deficiencies present at all loci.

Copies of nuclear loci are inherited from both parents and therefore genotypes at species-specific nuclear loci can be used to identify with high probability both purebred organisms and their hybrid combinations. However, nuclear genes, subjected to less genetic drift than the mitochondrial genome, typically evolve more slowly (Vawter and Brown 1986; Avise 1994) and become stochastically species-specific at a correspondingly slower rate. One solution has been to include intron (noncoding nuclear DNA interspersed with coding DNA) sequences in the comparison of interspecific sequence divergence among closely related species. The noncoding sequences, relatively unconstrained by selection, often display size or sequence differentiation among closely related species that is reciprocally monophyletic. Interspecific intron variation has been used in species identification techniques recently developed for Pacific salmon (McKay et al. 1997; Withler et al. 1997) and whales (Palumbi and Cipriano 1998).

Another approach is to target nuclear loci involved with the establishment of reproductive isolation in the evolution of related species for which positive selection may be expected to both accelerate interspecific differentiation and maintain intraspecific monomorphism (Liou and Price 1994; Vacquier et al. 1997). For abalone, positive selection has led to extremely high levels of interspecific divergence for sperm fertilization and egg receptor genes (Lee et al. 1995; Metz et al. 1998; Swanson and Vacquier 1998), associated with little intraspecific polymorphism (Metz et al. 1998) (Table 1). Levels of divergence among Pacific rim abalone species observed at these reproductive loci exceeded those observed at a mitochondrial gene and nuclear intron sequences (Metz et al. 1998).

For abalone, the gene encoding the lysin sperm protein (the protein that creates a hole in the vitteline membrane of the ovum during fertilization) displays the required characteristics of interspecific divergence and intraspecific conservation of DNA sequence (Vacquier et al. 1990; Shaw et al. 1994; Vacquier et al. 1997; Metz et al. 1998) for a species identification tool. Sweijd et al. (1998) developed PCR primers to the lysin DNA sequence to amplify exon (coding DNA) and intron portions of the gene that differed among *H. midae* (a commercially exploited South African species) and two other species as which *H. midae* might be misrepresented, the recreationally harvested South African *H. spadicea* and the commercially harvested Australian *H. rubra*. The lysin sequences amplified in each species differed in both intron length and exon DNA sequence, providing species identification not only by the size of the DNA fragment amplified but also by the RFLP (restriction fragment length polymorphism) patterns resulting from digestion of the DNA with various restriction enzymes. Because sizes of the amplified DNA products in the three abalone species were relatively large (700–1300 basepairs in length), amplification of the entire sequence was not always successful from processed (dried or canned) tissue. Thus, species-specific primers that only amplified a short portion of the lysin sequence in the target species were also developed for *H. midae* and *H. spadicea* for use on poor quality DNA samples (Sweijd et al. 1998). The degree of sequence variation in the haliotid lysin and other rapidly evolving reproductive loci (Lee et al. 1995; Metz et al. 1998; Swanson and Vacquier 1998), especially once the intron sequences are characterized completely, will likely enable development of a comprehensive species identification methodology for protected abalone species on a worldwide basis, similar to ones developed for whales (Baker et al. 1996; Palumbi and Cipriano 1998).

Population structure

The opportunities and predilection for dispersal over long distances among marine fish and invertebrate species afforded by their pelagic life history stages fueled expectations that genetic subdivision into populations or stocks in these species would be nonexistent or minimal in comparison with levels observed in terrestrial animals and freshwater and anadromous fishes (Waples 1998). Levels of differentiation among local populations of marine fish and shellfish over large geographic regions often are low (Burton 1983; Graves 1998; Shaklee and Bentzen 1998), but some marine species exhibit significant population structure over small geographic regions in spite of the apparent potential for long-distance dispersal (Palumbi 1996; Shaklee and Bentzen 1998). Studies of population structure generally are based on the measurement of allele frequencies at neutral genetic loci in geographically distinct samples of an organism. The allele frequencies are used to estimate F_{ST}, the proportion of the total genetic diversity that is apportioned among populations. Waples (1998) presented an illuminating review of the pitfalls associated with using genetic markers to estimate and interpret population structure parameters for species with high levels of gene flow among local populations. Estimation of F_{ST} for high gene flow species generally can be improved by surveying a large number of loci, random sampling of large numbers of adult individuals, and testing the stability of observed genetic differences over time. Stability in the geographic pattern of genetic differentiation over time increases confidence that a biologically meaningful population structure has been detected even when levels of differentiation are low. In contrast, a high level of temporal variation in allele frequencies indicates that true population structure has not been elucidated even when differentiation among geographic samples is great (Waples 1998).

Allozyme and mini- or micro-satellite loci are theoretically equally informative in the determination of population structure (Waples 1998), although in practice microsatellite loci often detect fine-scale population structure in cases where allozymes do not (Hughes and Queller 1993; Bentzen et al. 1996; Huang et al. 2000), largely as a result of their greater polymorphism. In addition, variation at allozyme loci may sometimes result from natural selection as well as, or instead of, gene flow. Examples of selective forces operating on allozymes seem disproportionately common for invertebrate species, including abalone (Koehn et al. 1980; Wilkins et al. 1980; Fujino et al. 1984; Karl and Avise 1992). The greater polymorphism of the noncoding microsatellite markers also makes them generally more useful than allozymes in stock identification, provided that the microsatellite loci applied to this task are not excessively polymorphic (Smouse and Chevillon 1998). Differing microsatellite allele frequencies among abalone populations in different areas may enable identification of the probable geographic origin of abalone tissue samples suspected to have been illegally harvested. For seized samples of abalone species for which harvest is legal only in some regions, mathematical mixture models can be used to estimate contributions from potential source populations. Hanna et al. (2000) identified Western Australia as the probable source region of illegally harvested greenlip abalone (*H. laevigata*), using principal component analysis of RAPD data from the confiscated individuals in comparison with five baseline samples from three regions.

For abalone, genetic surveys of population structure were initially based on allozyme loci, but nuclear DNA markers are being developed and applied. Allozyme variation was examined in the Pacific abalone *H. discus hannai* in Japan (reviewed in Fujino 1992) and in two closely related Australian species, *H. rubra* and *H laevigata* (Brown 1991; Brown and Murray 1992; Shepherd and Brown 1993). Small surveys of *H. rufescens* (Gaffney et al. 1996) and *H. fulgens* (G. Zuniga, Escuela Nacional de Ciencias Biologicas-IPN, Prolongacion de Carpio y Plan de Ayala s/n, Mexico D.F. 11340, personal communication) were conducted in California and the central portion of Baja California in Mexico, respectively. For wild abalone, significant differences in allele frequencies among sampling sites were observed in all of these studies, although the magnitude and geographic pattern of differentiation apparently differed among species (Table 2). The application of RAPDs, mini- and micro-satellites to analysis of structure in *H. rubra*, and RAPDs in *H. laevigata*, revealed greater polymorphism within populations and differentiation among populations (Hanna et al. 2000; Huang et al. 2000) than did allozyme analysis (Table 2).

The observed genetic differentiation indicated that abalone species in general are not structured as large panmictic populations homogenized by gene flow due to extensive larval dispersal. The continuously distributed *H. rubra* in southern Australia exhibited genetic differentiation at both allozyme and nuclear DNA loci that was well explained by isolation-by-distance models, although the scale of population differentiation differed depending upon the type of marker applied. Even for the allozyme data, which indicated a general "neighborhood size" of 500 km for *H. rubra*, genetic heterogeneity was sometimes detected on a much smaller geographic scale. Greater microspatial genetic heterogeneity was observed for *H. laevigata*, a species with a patchier distribution than *H. rubra*, in both the allozyme and RAPD data sets. Whereas the allozyme study indicated some isolation-by-distance for *H. laevigata* (Shepherd and Brown 1993), the limited RAPD data set provided evidence of much stronger isolation-by-distance among the states of Western Australia, South Australia, and Victoria. The genetic differentiation at RAPD loci among states allowed the probable identification of confiscated abalone to source state (Hanna et al. 2000). No evidence for isolation-by-distance was observed in the allozyme study of *H. fulgens*, but this study consisted of limited sampling of five sites within a small geographic region (<100 km of coastline).

Heterozygote deficits at allozyme loci were observed within samples for three haliotid species (Brown and Murray 1992; G. Zuniga, Escuela Nacional de Ciencias Biologicas-IPN, Prolongacion de Carpio y Plan de Ayala s/n, Mexico D.F. 11340, personal communication), although statistical significance would be low if corrections for multiple tests were conducted. Moreover, the heterozygote deficiencies were not strongly consistent at individual loci (indicative of null alleles or natural selection) or within individual samples (indicative of inbreeding or of including multiple populations within samples, the Wahlund effect). The tendency for homozygote excess at allozyme loci in *H. fulgens* was attributed to recent inbreeding at sampling locations, due to reduced abundances resulting from heavy exploitation. For *H. rubra*, very large heterozygote deficits were observed in pooled-population samples of *H. rubra* at all three microsatellite loci examined (Huang et al. 2000). However, heterozygote deficits were not observed at two minisatellite loci examined in the same study. Microsatellite allele identification may have been problematic or the microsatellite loci may have been better indicators of the Wahlund effect (due to pooled samples) or of localized inbreeding in *H. rubra* than were the minisatellite loci. Detection of either inbreeding or the Wahlund effect with the relatively small samples of this study would indicate a great restriction of gene flow over small distances (60–70 km) in this species.

The microspatial variation observed for both *H. fulgens* and *H. laevigata*, and possibly *H. rubra*, is consistent with the possibility of a "sweepstakes-style" reproduction for marine invertebrates (Hedgecock 1994). According to this hypothesis, even in populations of high abundance only a small fraction of mature adults effectively contribute to reproduction in each generation because of a limited window of oceanographic conditions compatible with successful spawning (Hedgecock 1994). The resulting large variance in individual reproductive success should lead to more rapid genetic drift (changing allele frequencies) within abalone populations and to much smaller estimates of effective population size than expected from an examination of actual census numbers of abalone within populations.

Examination of inbreeding levels, sweepstakes-style reproduction, and other models of population structure in abalone will require assessment of spatial and temporal variability in allele frequencies with genetic markers more polymorphic than allozyme loci. To this end, the development of mini- and micro-satellite loci for abalone species shows promise for improving resolution of population structure (Hanna et al. 2000; Huang et al. 2000). The use of these markers will also help overcome the requirement for sampling large amounts of tissue for allozyme analysis, as the tissue sample required for DNA extraction is small (a single abalone tentacle will suffice) and can be stored in ethanol before use (Kirby et al. 1998; Metz et al. 1998).

Abalone stock rebuilding

In general, relatively low levels of genetic differentiation have been observed among local populations of abalone (Brown and Murray 1992; Shepherd and Brown 1993; Gaffney et al. 1996; Metz et al. 1998). However, indications of fine-scale geographic structure are emerging from investigations in which molecular genetic markers have been applied (Hanna et al. 2000; Huang et al. 2000). Historical gene flow in the form of larval dispersal may have been been high enough to prevent strong genetic differentiation at neutral loci and therefore likely sufficiently great to preclude strong adaptive differentiation among geographically proximate populations. Nevertheless, numerous studies have indicated that recruitment in abalone populations was primarily from local sources, with immigration from distant sources having much less impact on population abundance (Shepherd and Brown 1993; Tegner 1993).

The apparent contradiction between mainly local recruitment and little genetic differentiation among local populations in some abalone species may reflect the different magnitudes of migration among local populations required to result in genetic homogenization (average exchange among populations of only one or a few individuals per generation) and to influence population abundance (e.g., the recovery of a depleted population may require hundreds or thousands of immigrants over several generations). Thus, although abalone populations occupying adjacent sites in a discontinuous habitat may be genetically linked, their population dynamics are likely independent. Genetic similarity among local populations cannot therefore be interpreted to mean that, once harvest is curtailed, populations depleted by fishing will be replenished from nearby populations in a time frame of interest to human managers.

Recruitment in at least some abalone species is subject to partial or complete failure at low adult densities. This phenomenon is termed the Allee effect due to an original description of reduced reproductive potential caused by adult scarcity (Allee et al. 1949). Thus, depleted abalone populations may face "double jeopardy." Even after exploitation ceases, both local and long-distance recruitment may be insufficient to enable a population to rebuild or even maintain its existing low level of abundance. This expectation is borne out by observations of nonexistent or very slow recovery of abalone populations even after all legal exploitation ceases (Tegner 1993; Davis et al. 1998; Campbell 2000). A complicating factor, however, in interpreting the biological significance of nonrecovery of abalone populations is the frequent establishment or persistence of illegal fisheries after fishery closures are imposed (Tegner 1993; Daniels and Floren 1998). The high market value of abalone fuels fishery efforts even when adult abundances are low. An exception to the general observation of local recruitment may be some "sink" populations of abalone, in which local recruitment is insufficient for population maintenance and sustained existence is dependent upon larval immigration from nearby source populations (Shepherd and Brown 1993). Such populations would be expected to be especially vulnerable to depletion under conditions of direct exploitation, indirect exploitation of the source population, or both.

The lack of evidence for strong intraspecific genetic differentiation at neutral loci and a likely corresponding lack of adaptation at loci under selection, over small- to medium-scale geographic regions, suggests that transplantation within similar geographic confines may be a viable option for rebuilding some depleted abalone stocks. The numbers of individuals required for a successful transplant are likely to be large as minimum viable population sizes for abalone may be at least a thousand individuals (Shepherd and Brown 1993). Initial transplant or "seeding" efforts with a variety of species in California, New Zealand, and Japan have provided variable recovery rates for the transplanted juveniles (Saito 1984; Tegner and Butler 1985; Tegner and Butler 1989; Henderson et al. 1988; Schiel 1993; Kojima 1995; Gaffney et al. 1996; Rogers-Bennett and Pearse 1998). In some cases, successful contribution to existing populations was observed (Schiel 1993; Kojima 1995; Gaffney et al. 1996), but there has been no evidence of seeded individuals contributing to succeeding generations. In one case, the failure of long-term survival of transplanted *H. fulgens* was attributed to illegal harvesting (Tegner 1993). Thus, as with all enhancement efforts, first addressing the factors that caused the initial population decline is necessary before enhancement is undertaken.

Shaklee and Bentzen (1998) review the reasons why the genetic population structure of a marine organism should be well understood before enhancement is undertaken. They also review the role of genetic markers in evaluating the success of enhancement programs. Given the current limited understanding of genetic structure in abalone and the lack of success of most abalone transplants, the take-home message is that conservation of existing wild populations is likely to be more effective both biologically and economically than enhancement measures undertaken after population depletion (Rogers-Bennett and Pearse 1998; Shaklee and Bentzen 1998).

Forensic stock identification

The polymorphic genetic markers used to determine intraspecific population structure in organisms also provide useful tools for determining the source of conspecific individuals of unknown origin. Typically, the intraspecific genetic heterogeneity among populations is much less than the interspecific genetic differences that can be used to identify an individual sample to species. Rather than the unique alleles at a few loci that differentiate species, it is frequency differences of shared alleles over many loci that distinguish populations. Because not every individual from every population carries equally "informative" (i.e., stock-specific) alleles, the assignment of individuals to source population (e.g., by jackknife discriminant analysis) is much more difficult than the assignment of individuals to species.

The difficulty of assignment of individuals to population increases with increasing numbers of populations and decreasing genetic distances among populations (Smouse and Chevillon 1998). Although classification of individuals of unknown origin to probable source population has been performed for some organisms (Waser and Strobeck 1998), the accuracy of the results has rarely been verified. A simpler task is to determine the proportional contributions of various populations to a sample containing individuals from one or more unknown populations, commonly referred to as a "mixed-stock" sample in fisheries literature. This approach enables more accurate determination of the proportion of individuals from each source population (even if all individuals come from a single source population), but each individual cannot be assigned to a source population (Beacham et al. 1995; Galvin et al. 1995).

For marine invertebrate organisms such as abalone, the characteristically low level of genetic distance among populations hinders the identification of individuals to their source populations and increases the need to sample more individuals and more loci in order to correctly profile each population and subsequently carry out stock composition analysis on individuals of unknown origin. Smouse and Chevillon (1998) indicated that the best genetic markers for this purpose are a "modest number of independently segregating, multiallelic, codominant loci, each locus with a small number of alleles, with each allele in moderate frequency." Microsatellite loci, now relatively easy to isolate from genomic DNA libraries, are the most promising class of DNA markers to meet these requirements. Microsatellites also circumvent the potential problems associated with the use of loci under selection that would hinder the use of allozyme loci for stock identification in marine invertebrates. However, because of the high number of alleles at some microsatellite loci (>50) and the difficulties in sampling populations of marine organisms, care must be taken to develop microsatellite loci with limited allelic diversity for use in stock identification. Hanna et al. (2000) have demonstrated the potential for using nuclear loci for stock identification of illegally harvested abalone to the most probable source population (the one showing the highest genetic similarity with the confiscated samples) through principal component analysis of RAPD variation in five population samples and six individuals of *H. laevigata*.

The identification of individual abalone to source population with the accuracy required for use in legal proceedings will likely be difficult when the genetic baseline for a species contains representative samples from tens or hundreds of populations throughout the species range. Under these circumstances, genetic analysis of multiple individuals and the application of conditional maximum likelihood mixture models (Fournier et al 1984; Millar 1987; Pella and Milner 1987) to the genotypes observed in the seized samples may enable relatively accurate and precise estimation of the proportional contributions of each population to the mixed-stock sample of seized individuals. This technique has been widely used in stock composition estimation for mixed-stock samples of Pacific salmon species caught in ocean fisheries (Utter et al. 1987; Wood et al. 1989; Shaklee et al. 1990; Beacham and Wood 1999). The methodology is readily transferable to any organism whose population structure has been described with the appropriate genetic markers.

The most laborious aspect of this methodology is the initial survey of genetic variation in all populations that might contribute to any mixed-stock sample required to create the baseline for subsequent analysis of mixed-stock samples. The level of genetic differentiation observed in the initial survey and the degree of temporal stability in observed allele frequencies will provide information on realistic goals to set for the analysis of mixture samples to geographic origin (i.e., the geographic boundaries of individual populations to be considered in the analysis). If genetic identification of seized individuals is to be considered an important enforcement tool for an exploited abalone species in which some areas are closed to fishing, then the closed areas should correspond relatively well to genetically distinct abalone populations (whether local populations or metapopulations). If abalone belong to a single, primarily interbreeding population which extends across the boundaries separating areas open and closed to fishing, identification of seized individuals to that population of origin will be of no value in determining the legality of the harvest.

Microsatellite loci have been isolated from *H. rubra* (Huang and Hanna 1998; Huang et al. 2000) and *H. rufescens* (Kirby et al. 1998) and are apparently also abundant in the genome of *H. kamtschatkana* (K.M. Miller, Pacific Biological Station, Nanaimo, B.C. V9R 5K6, personal communication). The most common dinucleotide repeat in abalone microsatellites is apparently the GT(CA) motif (Huang and Hanna 1998; Kirby et al. 1998) that is also most common in mammalian organisms (Rafalski et al. 1996). Abalone microsatellites with tri- and tetra-nucleotide core motifs are also relatively abundant

(K.M. Miller, personal communication). Preliminary surveys have indicated that the microsatellite loci developed thus far reveal population differentiation (Huang et al. 2000; Kirby et al. 1998) and that nonlethal sampling of abalone will be possible in surveys conducted for population characterization (Kirby et al. 1998). Microsatellite loci that can be amplified from more than one species (without an occurrence of null or nonamplifying alleles) would facilitate rapid and inexpensive delineation of population structure and estimation of stock composition for mixed-stock samples of abalone. All three microsatellite loci isolated from *H. rubra* amplified products in at least one other of the additional 14 haliotid species surveyed (Huang and Hanna 1998). However, for two of the loci successful amplification was limited to other Australian species and the third locus was amplified from only Australian and two South Korean species. The likelihood of widespread cross-species amplification of haliotid microsatellite loci remains an open question.

Marine protected areas

The establishment of marine protected areas (MPAs), with varying degrees of restriction on human access, harvest, and environmental disruption, is being considered as a conservation tool for marine organisms, including abalone, in a variety of jurisdictions worldwide (Dugan and Davis 1993; Shepherd and Brown 1993; Keeley 1997; Anonymous 1998). For depleted abalone populations, the hope is that successful recruitment from outside the MPA will lead to reestablishment of viable populations within the protected area (Tegner 1993). For healthy abalone populations, the hope is that MPAs will conserve biodiversity and provide surplus larval production that will recruit to the fishery resource outside the protected area (Davis 1989; Wallace 1999).

In order for a MPA to fulfill twin goals of population preservation and surplus larval production for abalone, the area must encompass one or more self-sustaining abalone populations of sufficient size to withstand both the large fluctuations in reproductive success characteristic of marine shellfish and current levels of environmental change. Surveys of abalone population structure with polymorphic genetic markers, such as microsatellite markers, will provide information on the geographic extent of a local or metapopulation within haliotid species (e.g., Shepherd and Brown 1993) useful for the delineation of MPAs for conservation. However, due to the independence of haliotid population genetics and dynamics discussed above, evaluation of the likelihood of providing surplus recruitment to populations outside the MPA would also require studies of larval dispersal and recruitment. Abalone may generally exhibit metapopulation structure (Shepherd and Brown 1993; Huang et al. 2000), but little is known of the population dynamics involved. Metapopulations may be of the "source-sink" type in which gene flow is basically unidirectional from large source local populations to ephemeral sink local populations (Pulliam 1988; Pulliam and Danielson 1991) or of the "balanced exchange" type, in which gene flow is bidirectional and migration rates are inversely proportional to local population size (McPeek and Holt 1992; Doncaster et al. 1997).

Obviously, the enclosure of sink abalone populations within a MPA will not only preclude larval dispersal from the MPA to surrounding exploited populations, but harvest of the related source populations might lead to loss of the protected populations even in the absence of direct fishing pressure. Conservation of a "balanced exchange" abalone metapopulation in a MPA would require that a sufficient number of productive local populations be included within the MPA boundaries to maintain dispersal patterns within the MPA and possibly provide recruitment to outside populations. Shepherd and Brown (1993) described a metapopulation of *H. laevigata* that was partially (21% of abalone habitat) encompassed within a marine reserve but underwent a drastic decline in abundance between 1970 and 1990. They concluded that light fishing on the unprotected portion of the metapopulation, combined with environmentally induced recruitment failure in several years, led to the stock collapse and that protection of a larger segment of the metapopulation from harvest might have improved the ability of the abalone population to withstand environmental adversity.

Tegner (1993) concluded that the fishery closure along several kilometres of shoreline in southern California would not bring about a recovery of *H. fulgens* or *H. corrugata* in the near future because of their limited larval dispersal whereas the more varied dispersal opportunities afforded *H. rufescens* by an extended spawning period might lead to a more rapid recovery for this species. The differing distribution of genetic variation observed for the sympatric *H. rubra* and *H. laevigata* (Shepherd and Brown 1993; Hanna et al. 2000; Huang et al. 2000) indicated that patterns of larval dispersal, and hence both gene flow and recruitment to nearby populations, also differed between them. Both species-specific and site-specific patterns of larval dispersal (Shepherd and Brown 1993) have implications for the number, size, and spacing of MPAs that will most benefit the individual species. Assumptions about larval recruitment into or out of MPAs require testing.

Success of MPAs for the rebuilding of depleted abalone populations or protecting healthy populations that may provide surplus production to surrounding areas also depends heavily on the ability to enforce the fishing closures. Tegner (1993) suggests that the ability to enforce harvest restrictions is a critical requirement for the success of any abalone conservation efforts based on protected areas. For many abalone species, the simultaneous existence of areas open and closed to harvest, often in close proximity, exacerbates difficulties in the enforcement of fishery closures. Use of polymorphic genetic markers in population surveys will not only assist in biologically effective delineation of MPAs but may allow forensic identification of the source population(s) of abalone allegedly harvested from MPAs through maximum likelihood estimation of stock composition as described above. For abalone species in which molecular genetic stock identification is considered an important enforcement tool, a genetic survey of population structure should be conducted before establishment of areas closed to harvest. This would allow determination of the geographic scale of abalone populations to which mixed-stock samples could be reliably ascribed and thus enable designation of closed areas from which illegally harvested abalone could be readily identified.

Acknowledgments

I gratefully acknowledge the provision of unpublished manuscripts by Drs. Peter Hanna, Neville Sweijd, Gerardo Zuniga, and their co-authors. Dr. Kristi Miller critically reviewed the manuscript and calculated the nucleotide divergence levels on lysin gene sequences retrieved from GenBank. Drs. Ron Burton and Peter Hanna provided comments and information that improved this manuscript.

References

Allee, W.C., Emerson, A.E., Park, O., and Schmidt, K.P. 1949. Principles of animal ecology. Saunder Publishing Co., Philadelphia, Penn. 837 p.

Anonymous. 1998. Marine protected areas: a strategy for Canada's Pacific Coast. Department of Fisheries and Oceans, Vancouver, B.C. and the British Columbia Land Use Co-ordination Office, Victoria, B.C. 28 p.

Avise, J.C. 1994. Molecular markers, natural history, and evolution. Chapman and Hall, New York.

Baker, C.S., Cipriano, F., and Palumbi, S.R. 1996. Molecular genetic identification of whale and dolphin products from commercial markets in Korea and Japan. Mol. Ecol. **5**: 671–685.

Bartlett, S.E., and Davidson, W.E. 1991. Identification of *Thunnus* tuna species by polymerase chain reaction and direct sequence analysis of their mitochondrial cytochrome *b* genes. Can. J. Fish. Aquat. Sci. **48**: 309–317.

Beacham, T.D., and Wood, C.C. 1999. Application of microsatellite DNA variation to estimation of stock composition and escapement of Nass River sockeye salmon (*Oncorhynchus nerka*). Can. J. Fish. Aquat. Sci. **56**: 1–14.

Beacham, T.D., Withler, R.E., and Wood, C.C. 1995. Stock identification of sockeye salmon by means of minisatellite DNA variation. N. Am. J. Fish. Manage. **15**: 249–265.

Bentzen, P., Taggart, C.T., Ruzzante, D.E., and Cook, D. 1996. Microsatellite polymorphism and the population structure of Atlantic cod (*Gadus morhua*) in the northwest Atlantic. Can. J. Fish. Aquat. Sci. **53**: 2706–2721.

Borgo, R., Souty-Grosset, C., Bouchon, D., and Gomot, L. 1996. PCR-RFLP analysis of mitochondrial DNA for identification of snail meat species. J. Food Sci. **61**: 1–4.

Brown, L.D. 1991. Genetic variation and population structure in the blacklip abalone, *Haliotis rubra*. Aust. J. Mar. Freshw. Res. **42**: 47–90.

Brown, L.D., and Murray, N.D. 1992. Population genetics, gene flow and stock structure in *Haliotis rubra* and *Haliotis laevigata*. *In* Abalone of the world: biology, fisheries and culture. *Edited by* S.A. Shepherd, M.J. Tegner, and S.A. Guzmán del Próo. Fishing News Books, Blackwell Science Ltd., Oxford, U.K. pp. 24–33.

Burton, R.S. 1983. Protein polymorphisms and genetic differentiation of marine invertebrate populations. Mar. Biol. Lett. **4**: 193–206.

Campbell, A. 1997. Possible criteria for re-opening the northern abalone (*Haliotis kamtschatkana*) fishery in British Columbia. Canadian Stock Assessment Secretariat Research Document 97/64. PSARC Secretariat, Pacific Biological Station, 3190 Hammond Bay Road, Nanaimo, B.C. V9R 5K6. 47 p.

Campbell, A. 2000. Review of northern abalone, *Haliotis kamtschatkana*, stock status in British Columbia. Can. Spec. Publ. Fish. Aquat. Sci. 130. pp. 41–50.

Daniels, R., and Floren, R. 1998. Poaching pressures on northern California's abalone fishery. J. Shellfish Res. **17**: 859–862.

Davis, G.E. 1989. Designated harvest refugia: the next stage of marine fishery management in California. Calif. Coop. Oceanic Fish. Invest. Rep. **30**: 53–58.

Davis, G.E., Richards, D.V., Haaker, P.L., and Parker, D.O. 1992. Abalone population declines and fishery management in southern California. *In* Abalone of the world: biology, fisheries and culture. *Edited by* S.A. Shepherd, M.J. Tegner, and S.A. Guzmán del Próo. Fishing News Books, Blackwell Science Ltd., Oxford, U.K. pp. 237–249.

Davis, G.E., Haaker, P.L., and Richards, D.V. 1998. The perilous condition of white abalone *Haliotis sorenseni*, Bartsch, 1940. J. Shellfish Res. **17**: 871–875.

Doncaster, C.P., Clobert, J., Doligez, B., Gustafsson, L., and Danchin, E. 1997. Balanced dispersal between spatially varying local populations: an alternative to the source-sink model. Am. Nat. **150**: 425–445.

Dugan, J.E., and Davis, G.E. 1993. The role of refugia in management of coastal fisheries. Can. J. Fish. Aquat. Sci. **50**: 2040–2052.

Fournier, D.A., Beacham, T.D., Riddell, B.E., and Busack, C.A. 1984. Estimating stock composition in mixed stock fisheries using morphometric, meristic, and electrophoretic characteristics. Can. J. Fish. Aquat. Sci. **41**: 400–408.

Fujino, K. 1992. Review of genetics and stock management of the Pacific abalone. *In* Abalone of the world: biology, fisheries and culture. *Edited by* S.A. Shepherd, M.J. Tegner, and S.A. Guzmán del Próo. Fishing News Books, Blackwell Science Ltd., Oxford, U.K. pp. 491–503.

Fujino, K., Sasaki, K., and Okumura, S.-I. 1984. Probable involvement of thermostability variations of enzymes in mechanisms of occurrence of deficient abalone. Bull. Jpn. Soc. Sci. Fish. **50**: 597–601.

Fujio, Y., Yamanaka, R., and Smith, P.J. 1983. Genetic variation in marine mollusks. Bull. Jpn. Soc. Sci. Fish. **49**: 1809–1817.

Gaffney, P.M., Rubin, V., Hedgecock, D., Powers, D.A., Morris, G., and Hereford, L. 1996. Genetic effects of artificial propagation: signals from wild and hatchery populations of red abalone in California. Aquaculture, **143**: 257–266.

Galvin, P.T., McKinnell, S., Taggart, J.B., Ferguson, A., O'Farrell, M., and Cross, T.F. 1995. Genetic stock identification of Atlantic salmon (*Salmo salar* L.) using single locus minisatellite DNA profiles. J. Fish. Biol. **47** (Suppl. A): 186–199.

Graves, J.E. 1998. Molecular insights into the population structures of cosmopolitan marine fishes. J. Hered. **89**: 427–437.

Hanna, P.J., Chai, Z., Cummins, S., and Huang, B. 2000. Applications of molecular biology to abalone fisheries and aquaculture. J. Med. Appl. Malacol. (In press).

Hedgecock, D. 1994. Does variance in reproductive success limit effective population sizes of marine organisms? *In* Genetics and evolution of aquatic organisms. *Edited by* A.R. Beaumont. Chapman and Hall, London, U.K. pp. 122–134.

Henderson, K.C., Parker, D.O., and Haaker, P.L. 1988. The survival and growth of transplanted adult pink abalone, *Haliotis corrugata*, at Santa Catalina Island. Calif. Fish Game, **74**: 82–86.

Huang, B.X., and Hanna, P.J. 1998. Identification of three polymorphic microsatellite loci in blacklip abalone, *Haliotis rubra* (Leach), and detection in other abalone species. J. Shellfish Res. **17**: 795–799.

Huang, B.X., Hanna, P.J., and Peakall, R. 2000. Analysis of population genetic structure of blacklip abalone (*Haliotis rubra*) using RAPD, minisatellite and microsatellite markers. Mar. Biol. (In press).

Hughes, C.R., and Queller, D.C. 1993. Detection of highly polymorphic microsatellite loci in a species with little allozyme polymorphism. Mol. Ecol. **2**: 131–137.

Karl, S.A., and Avise, J.C. 1992. Balancing selection at allozyme loci in oysters: implications from nuclear RFLPs. Science (Washington, D.C.), **256**: 100–102.

Keeley, F. 1997. Assembly Bill 1241: Sea life recovery and management Act. California Assembly, Sacramento, Calif. 30 p.

Kirby, V.L., Villa, R., and Powers, D.A. 1998. Identification of microsatellites in the California red abalone, *Haliotis rufescens*. J. Shellfish Res. **17**: 801–804.

Koehn, R.K., Newell, R.I.E., and Immermann, F. 1980. Maintenance of an aminopeptidase allele frequency cline by natural selection. Proc. Natl. Acad. Sci. U.S.A. **77**: 5385–5389.

Kojima, H. 1995. Evaluation of abalone stock enhancement through the release of hatchery-reared seeds. Mar. Freshw. Res. **46**: 689–695.

Lee, Y.-H., Ota, T., and Vacquier, V.D. 1995. Positive selection is a general phenomenon in the evolution of abalone sperm lysin. Mol. Biol. Evol. **12**: 231–238.

Liou, L.W., and Price, T.D. 1994. Speciation by reinforcement of premating isolation. Evolution, **48**: 1451–1459.

McKay, S.J., Smith, M.J., and Devlin, R.H. 1997. Polymerase chain reaction-based species identification of salmon and coastal trout in British Columbia. Mol. Mar. Biol. Biotechnol. **6**: 131–140.

McPeek, M.A., and Holt, R.D. 1992. The evolution of dispersal in spatially and temporally varying environments. Am. Nat. **140**: 1010–1027.

McShane, P.E. 1992. The early life history of abalone: a review. *In* Abalone of the world: biology, fisheries and culture. *Edited by* S.A. Shepherd, M.J. Tegner, and S.A. Guzmán del Próo. Fishing News Books, Blackwell Science Ltd., Oxford, U.K. pp. 120–138.

Metz, E.C., Robles-Sikisaka, R., and Vacquier, V.D. 1998. Nonsynonymous substitution in abalone sperm fertilization genes exceeds substitution in introns and mitochondrial DNA. Proc. Natl. Acad. Sci. U.S.A. **95**: 10676–10681.

Millar, R.B. 1987. Maximum likelihood estimation of mixed stock fishery composition. Can. J. Fish. Aquat. Sci. **44**: 583–590.

Miller, I., Gutleb C., Kranz, A., and Gemeiner, M. 1995. Forensics on wild animals — differentiation between otter and pheasant blood using electrophoretic methods. Electrophoresis, **16**: 865–868.

Moritz, C. 1994. Defining "evolutionarily significant units" for conservation. TREE, **9**: 373–375.

Murray, B.W., McClymont, R.A., and Strobeck, C. 1995. Forensic identification of ungulate species using restriction digests of PCR-amplified mitochondrial-DNA. J. Forensic Sci. **40**: 943–951.

Palumbi, S.R. 1996. Microspatial genetic structure and speciation in marine taxa with high dispersal abilities. *In* Molecular zoology. *Edited by* J.D. Ferraris and S.R. Palumbi. Wiley-Liss, New York. pp. 101–117.

Palumbi, S.R., and Cipriano, F. 1998. Species identification using genetic tools: the value of nuclear and mitochondrial gene sequences in whale conservation. J. Hered. **89**: 459–464.

Pella, J.J., and Milner, G.B. 1987. Use of genetic marks in stock composition analysis. *In* Population genetics and fishery management. *Edited by* N. Ryman and F.M. Utter. University of Washington Press, Seattle, Wash. pp. 247–276.

Powell, M.L., Meade, M.E., and Watts, S.A. 1995. Identification of six commercially cultured crayfish species using total protein separation by isoelectric focusing. J. Crustacean Biol. **15**: 763–770.

Prince, J.D., and Shepherd, S.A. 1992. The Australian abalone fisheries and their management. *In* Abalone of the world: biology, fisheries and culture. *Edited by* S.A. Shepherd, M.J. Tegner, and S.A. Guzmán del Próo. Fishing News Books, Blackwell Science Ltd., Oxford, U.K. pp. 407–426.

Prince, J.D., Sellers, T.L., Ford, W.B., and Talbot, S.R. 1987. Experimental evidence for limited dispersal of haliotid larvae (genus *Haliotis*: Mollusca: Gastropoda). J. Exp. Mar. Biol. Ecol. **106**: 243–263.

Prince, J.D., Sellers, T.L., Ford, W.B., and Talbot, S.R. 1988. Confirmation of a relationship between the localized abundance of breeding stock and recruitment for *Haliotis rubra* Leach (Mollusca: Gastropoda). J. Exp. Mar. Biol. Ecol. **122**: 91–104.

Pulliam, H.R. 1988. Sources, sinks, and population regulation. Am. Nat. **132**: 652–661.

Pulliam, H.R., and Danielson, B.J. 1991. Sources, sinks, and habitat selection — a landscape perspective on population dynamics. Am. Nat. **137**: S50–S66.

Rafalski, J.A., Vogel, J.M., Morgante, M., Powell, W., Andre, C., and Tingey, S.V. 1996. Generating and using DNA markers in plants. *In* Nonmammalian genomic analysis — a practical guide. *Edited by* B. Birren and E. Lai. Academic Press, Sydney, Aust. pp. 75–134.

Rogers-Bennett, L., and Pearse, J.S. 1998. Experimental seeding of hatchery-reared juvenile red abalone in northern California. J. Shellfish Res. **17**: 877–880.

Saito, K. 1984. Ocean ranching of abalones and scallops in northern Japan. Aquaculture, **39**: 361–373.

Schiel, D.R. 1993. Experimental evaluation of commercial-scale enhancement of abalone *Haliotis iris* populations in New Zealand. Mar. Ecol. Prog. Ser. **97**: 167–181.

Shaklee, J.B., and Bentzen, P. 1998. Genetic identification of stocks of marine fish and shellfish. Bull. Mar. Sci. **62**: 589–621.

Shaklee, J.B., Busack, C., Marshall, A., Miller, M., and Phelps, S.R. 1990. The electrophoretic analysis of mixed-stock fisheries of Pacific salmon. *In* Isozymes: structure, function, and use in biology and medicine. *Edited by* Z.-I. Ogita and C.L. Markert. Prog. Clin. Biol. Res. **344**: 235–262.

Shaw, A., Lee, Y-H., Stout, C.D., and Vacquier, V.D. 1994. The species-specificity and structure of abalone sperm lysin. Dev. Biol. **5**: 209–215.

Shepherd, S.A., and Brown, L.D. 1993. What is an abalone stock: implications for the role of refugia in conservation. Can. J. Fish. Aquat. Sci. **50**: 2001–2009.

Shepherd, S.A., and Partington, D. 1995. Studies on Southern Australian abalone (genus *Haliotis*). XVI. Recruitment, habitat, and stock relations. *In* Progress in abalone fisheries research. *Edited by* S.A. Shepherd, R.W. Day, and A.J. Butler. Mar. Freshw. Res. **46**: 669–680.

Shepherd, S.A., Lowe, D., and Partington, D. 1992. Studies on Southern Australian abalone (genus *Haliotis*). XIII. Larval dispersal and recruitment. J. Exp. Mar. Biol. Ecol. **164**: 247–260.

Shepherd, S.A., Turrubiates-Morales, J.R., and Hall, K. 1998. Decline of the abalone fishery at La Natividad, Mexico: overfishing or climate change? J. Shellfish Res. **17**: 839–846.

Smouse, P.E., and Chevillon, C. 1998. Analytical aspects of population-specific DNA fingerprinting for individuals. J. Hered. **89**: 143–150.

Swanson, W.J., and Vacquier, V.D. 1998. Concerted evolution in an egg receptor for a rapidly evolving abalone sperm protein. Science (Washington, D.C.), **281**: 710–712.

Sweijd, N.A., Bowie, R.C.K., Lopata, A.L., Marinaki, A.M., Harley, E.H., and Cook, P.A. 1998. A PCR technique for forensic, species-level identification of abalone tissue. J. Shellfish Res. **17**: 889–895.

Tarr, R.J.Q., Williams, P.V.G., and Mackenzie, A.J. 1996. Abalone, sea urchins, and rock lobster: a possible ecological shift may affect traditional fisheries. S. Afr. J. Mar. Sci. **17**: 319–323.

Tegner, M.J. 1993. Southern California abalones: can stocks be rebuilt using marine harvest refugia. Can. J. Fish. Aquat. Sci. **50**: 2010–2018.

Tegner, M.J., and Butler, R.A. 1985. The survival and mortality of seeded native red abalones, *Haliotis rufescens*, on the Palos Verdes Peninsula. Calif. Fish Game, **71**: 150–163.

Tegner, M.J., and Butler, R.A. 1989. Abalone seeding. *In* Handbook of culture of abalone and other marine gastropods. *Edited by* K.O. Hahn. CRC Press, Boca Raton, Fla. pp. 157–182.

Utter, F.M., Teel, D.J., Milner, G.B., and McIsaac, D. 1987. Genetic estimates of stock compositions of 1983 chinook salmon harvests off the Washington coast and the Columbia River. Fish. Bull. **85**: 13–23.

Vacquier, V.D., Carner, K.R., and Stout, C.D. 1990. Species-specific sequences of abalone lysin, the perm protein that creates a hole in the egg envelope. Proc. Natl. Acad. Sci. **87**: 5792–5796.

Vacquier, V.D., Swanson, W.J., and Lee, Y-H. 1997. Positive Darwinian selection on two homologous fertilization proteins: what is the selective pressure driving their divergence? J. Mol. Evol. **44** (Suppl. 1): s15–s22.

Vawter, L., and Brown, W.M. 1986. Nuclear and mitochondrial DNA comparisons reveal extreme rate variation in the molecular clock. Science (Washington, D.C.), **234**: 194–196.

Wallace, S.S. 1999. Evaluating the effects of three forms of marine reserve on northern abalone populations in British Columbia, Canada. Conserv. Biol. **13**: 882–887.

Waples, R.S. 1998. Separating the wheat form the chaff: patterns of genetic differentiation in high gene flow species. J. Hered. **89**: 438–450.

Waser, P.M., and Strobeck, C. 1998. Genetic signatures of interpopulation dispersal. TREE, **13**: 43–44.

Wilkins, N.P., Fujino, K., and Sasaki, K. 1980. Genetic studies on the Pacific abalone IV. Thermostability difference among phosphoglucomutase variants. Bull. Jpn. Soc. Sci. Fish. **46**: 549–553.

Withler, R.E., Beacham, T.D., Ming, T.J., and Miller, K.M. 1997. Species identification of Pacific salmon by means of a major histocompatibility complex gene. N. Am. J. Fish. Manage. **17**: 929–938.

Wood, C.C., Rutherford, D.T., and McKinnell, S. 1989. Identification of sockeye salmon (*Oncorhynchus nerka*) stocks in mixed-stock fisheries in British Columbia and southeast Alaska using biological markers. Can. J. Fish. Aquat. Sci. **46**: 2108–2120.

Infectious diseases of abalone (*Haliotis* spp.) and risks associated with transplantation

Susan M. Bower

Abstract: Infectious diseases of abalone (*Haliotis* spp.) can be grouped into three categories. Category 1 consists of pathogens that have caused severe disease and mortality: (*i*) *Vibrio fluvialis II* caused pustule disease in abalone cultured in Dalian, China; (*ii*) *Labyrinthuloides haliotidis* decimated young abalone cultured in British Columbia; (*iii*) *Perkinsus olseni* affected wild populations in South Australia; (*iv*) an introduced sabellid polycheate has become a pest in culture facilities in California, (*v*) withering foot syndrome presumed to be caused by a gastrointestinal rickettsia-like procaryote affected wild *H. cracherodii* in California; and (*vi*) amyotrophia, of presumed viral etiology, affected *H. discus* during seed production in Japan. Category 2 includes parasites of lesser concern that either are benign (e.g., coccidia in the kidney of abalone in California) or have special life-cycle requirements (e.g., helminths). Nevertheless, the risk of these parasites becoming a scourge to naive animals in a new habitat is unpredictable. Category 3 includes organisms that are detrimental only under certain situations: (*i*) ubiquitous opportunistic fungi and bacteria can cause disease under inadequate culture conditions; and (*ii*) shell-boring organisms (e.g., *Cliona* spp. and *Polydora* spp.) require nutrient-rich water for proliferation detrimental to abalone. However, some of these organisms (e.g., specific fungi and *Boccardia knoxi*) may be harmful regardless of conditions and thus deserving of classification in Category 1. Additional new pathogens are likely to be encountered as increased attention is focused on highly valued abalone. To protect wild stocks and to avoid curtailing development of abalone culture, disease issues must be considered and precautions taken to avoid accidentally introducing foreign pathogens when transplanting abalone.

Résumé : Les pathogènes qui infectent les ormeaux (*Haliotis* spp.) peuvent être groupés en trois catégories. La première catégorie comprend les pathogènes qui causent des maladies graves, mortelles dans certains cas : (*i*) *Vibrio fluvialis II* a causé une pustulose chez des ormeaux cultivés à Dalian, en Chine; (*ii*) *Labyrinthuloides haliotidis* a décimé de jeunes ormeaux cultivés en Colombie-Britannique; (*iii*) *Perkinsus olseni* a affecté des populations sauvages dans le sud de l'Australie; (*iv*) un polychète introduit de la famille des Sabellidés est devenu un ravageur dans des installations d'aquiculture en Californie; (*v*) un procaryote rickettsoïde gastrointestinal aurait semble-t-il provoqué le syndrome de dépérissement du pied chez des *H. cracherodii* sauvages en Californie; et (*vi*) un virus, croit-on, serait à l'origine de l'amyotrophie qui a affecté *H. discus* dans une culture durant la production du naissain au Japon. La deuxième catégorie comprend des parasites moins préoccupants qui sont bénins (p. ex. une coccidie présente dans le rein d'ormeaux en Californie) ou qui ont besoin de conditions particulières pour pouvoir compléter leur cycle vital (p. ex. des helminthes). Néanmoins, il n'est pas impossible que ces parasites puissent devenir un fléau pour des ormeaux transplantés dans un nouvel habitat, bien qu'on ne puisse en établir le risque. La troisième catégorie comprend des organismes qui ne sont nuisibles que dans certaines conditions : (*i*) des champignons et bactéries opportunistes ubiquistes peuvent provoquer des maladies dans des conditions de culture inadéquates; et (*ii*) les organismes qui percent les coquilles (p. ex. *Cliona* spp et *Polydora* spp.) ne peuvent proliférer au point de nuire aux ormeaux que dans des eaux riches en éléments nutritifs. Cependant, certains des organismes de cette catégorie (p. ex. certains champignons et *Boccardia knoxi*) peuvent être nuisibles peu importe les conditions et devraient donc entrer dans la première catégorie. On trouvera probablement de nouveaux pathogènes vu l'intérêt de plus en plus marqué pour les ormeaux de grande valeur. Pour protéger les stocks sauvages et éviter que le développement de la culture des ormeaux ne soit freiné, on doit porter attention aux maladies et prendre des précautions pour empêcher l'introduction accidentelle de pathogènes étrangers lors de la transplantation des ormeaux.

[Traduit par la Rédaction]

S.M. Bower. Fisheries and Oceans Canada, Pacific Biological Station, Nanaimo, BC V9R 5K6, Canada.
e-mail: BowerS@dfo-mpo.gc.ca

Correct citation: Bower, S.M. 2000. Infectious diseases of abalone (*Haliotis* spp.) and risks associated with transplantation. *In* Workshop on Rebuilding Abalone Stocks in British Columbia. *Edited by* A. Campbell. Can. Spec. Publ. Fish. Aquat. Sci. 130. pp. 111–122.

Introduction

In comparison to other shellfish that are more extensively and intensively cultured, very little is known about the diseases of abalone. For example, the synopsis by Bower et al. (1994) lists many more diseases and parasites in other groups of molluscs (i.e., oysters, mussels, clams, or scallops) than in abalone. Despite the limited research conducted on abalone pathogens, several infectious diseases have been encountered. This paper provides a literature review of the infectious diseases of abalone. In doing so, the diseases are grouped into three categories according to the consequences that they have imposed on wild or cultured abalone stocks. The first category includes pathogens that have caused severe problems in abalone in various parts of the world. The second category includes parasites that are benign or have special life-cycle requirements that may prohibit their becoming established in another location. The third category consists of organisms that are usually ubiquitous and benign but can have a serious impact under certain environmental conditions. The potential risk that each pathogen could impose on abalone if inadvertently introduced into a new area is assessed. Also, methods of prevention or control that have been applied are mentioned.

Category 1

This category includes six pathogens that have caused severe disease in either wild or cultured stocks of abalone.

Vibrio fluvialis II

Pustule or blister disease was encountered in abalone (*Haliotis discus hannai*) from several abalone hatcheries and in wild stocks in the vicinity of Dalian, China. The blister-like lesions on the foot were associated with mortalities approaching 50–60% among cultured abalone (3–8 cm shell length, SL) in at least three facilities (Liu et al. 1995; Li et al. 1998). Lesions comprised an intense haemocytic infiltration and the connective tissue and muscle fibres in the center of the lesions were usually dissolved and denatured (Li et al. 1997*b*). The lesions first formed on the surface of the abalone foot and tended to migrate into the tissue as the disease progressed. In advanced stages of the disease, only haemocytes and bacteria remained near the center of the lesions (Li et al. 1997*b*).

Stained smears from well-developed lesions revealed abundant Gram-negative, short, rod-shaped bacteria. The bacteria were most prolific in cultures of tryptone – beef broth – agar medium made up with 20–30 parts per thousand seawater at 37°C and pH of 6–11 (Li at al. 1998). The cultured bacteria had the following characteristics: glucose fermentation with the production of gas, reduction of nitrate to nitrite, a positive Voges-proskauer test, hydrolization of arginine but not amylum or gelatin, fermentation of arabinose but not xylose or inositol, oxidase and catalase positive, and sensitivity to vibriostatic agent 0/129 (150 μg) (Nie et al. 1995). Examination of the ultrastructure revealed a bacterium with a thin cellular wall, a clear central zone with many glycogenosomes and free ribosomes in the plasma and a single polar flagellum. The bacterium was identified as *Vibrio fluvialis II* and cultured forms caused lesions, identical to those in production abalone, in 100% of the experimental abalone exposed by inoculation (Lui et al. 1995).

There are no known methods of controlling this pathogen (Li et al. 1998). Three strains of *V. fluvialis II* that were isolated from diseased abalone had varying resistance to 18 different antibiotics. The mechanism of drug resistance was related to gene mutations in the bacterium (Li et al. 1996). Results of other experiments suggest that the survival of juvenile and adult abalone could be improved by exposing abalone (via oral ingestion or injection) to a formalin treated culture of *V. fluvialis II* and the serum of inoculated abalone had a higher agglutinin titre against the bacterium (Li et al. 1997*a*).

To date, this abalone disease is known to occur only in China. However, *V. fluvialis II* is present in aquatic, particularly estuarine, environments throughout the world (Lee et al. 1981). Although the conditions in the abalone culture facilities in China may have predisposed the abalone to infection, the possibility of the existence of a strain modified for pathogenicity to abalone can not be discounted. Thus, to avoid the risk of introducing this pathogen to other culture facilities or natural stocks, only animals certified to be free of infection should be considered for transplantation from areas where the disease occurs. In addition, imported animals must be held in quarantine and assayed for cryptic or subclinical infections prior to release into the new environment.

Labyrinthuloides haliotidis

Labyrinthuloides haliotidis was pathogenic to small juvenile northern abalone (*Haliotis kamtschatkana*) and small juvenile red abalone (*Haliotis rufescens*) and was involved in the demise of the only attempt at abalone culture in British Columbia, Canada (Bower 1987*a*). This parasite is a thraustochytrid (achlorophyllous, eukaryotic protists). Thraustochytrids have often been grouped with the lower fungi but were included in the subkingdom Protozoa, phylum Labyrinthomorpha by Levine et al. (1980). *Labyrinthuloides haliotidis* has only been observed in small abalone (<0.5 cm SL). Within 2 weeks of first being detected in a raceway, over 90% of the 100 000 small abalone in that raceway succumbed to infection and the disease quickly spread between raceways. The high mortality caused by *L. haliotidis* was one of the reasons that this particular abalone culture facility is no longer in operation.

The vegetative stage of *L. haliotidis* is spheroid (5–9 mm in diameter) and has a surface organelle called a sagenogenetosome which is unique to the phylum Labyrinthomorpha. Several sagenogenetosomes were scattered on the surface of each parasite forming openings in the thin laminated cell wall and each produced the ectoplasmic membrane net upon which the parasite moved (0.3 ± 0.16 $mm \cdot min^{-1}$) and obtained nutrients. Following removal from a source of nutrients (i.e., placement in sterile seawater), the vegetative cell underwent synchronous multiple fission to form a zoosporoblast (6–10 mm in diameter) containing about 10 zoospores. The uninucleate, slightly oval, motile, biflagellated zoospores (4–6 mm long and 2.5–4.5 mm wide) escaped through a rupture in the zoosporoblast wall. The two flagella were laterally attached to the ovoid zoospore. The anterior flagellum (9–15 mm in length) had a brush of mastigonemes along one side (about a third of the circumference) and the posterior flagellum (5–10 mm in length) was naked and had a tapered tip (Bower 1987*a*). The flagella were shed when the zoospore contacted a hard surface or after about 24 h of active swimming in seawater. The resulting cell was morphologically similar but slightly smaller than the vegetative stage and survived in sterile seawater at about 5°C for at least 2 years (Bower 1987*b*).

Vegetative stages that developed from zoospores were infective to small abalone. Within 4 h of contacting the host, sagenogenetosomes developed extracellular lytic activity that disrupted the plasmalemma layer of the host epithelial cells adjacent to the parasite, eventually lysing the host cell. By 24 h post exposure, the ectoplasmic net was well developed, allowing the parasite to move into and within the head and foot tissues of the abalone and dividing forms of the parasite were observed (Bower et al. 1989*a*). Within 10 days after exposure to about 10^4 parasites in 20 mL of seawater, about 90% of the abalone (<4.0 mm SL and 140 days of age) died with numerous parasites throughout the head and foot (Bower 1987*c*). Tissues of heavily infected abalone were slightly swollen with a loss of integrity. As dead abalone decomposed, vegetative stages released from the tissues developed into zoosporoblasts that produced zoospores within about 24–72 h. Parasites released from infected abalone were infective to other abalone on contact.

Prevalence and intensity of infection decreased and time to death increased as the abalone increased in age and size. Abalone greater than 1.5 cm SL could not be infected even when about 1.5×10^4 *L. haliotidis* were injected intramuscularly. The mechanism of defence against this parasite is not known. There was no indication of an inflammatory response in young susceptible abalone. Possibly the resistance of older abalone corresponded to the development of cellular or humeral defence mechanisms as the abalone matured (Bower 1987*c*).

Although alternate hosts have not been described, *L. haliotidis* can utilize diverse sources of nutrients but seems incapable of coexisting with bacteria. Small juvenile Japanese scallops (*Patinopecten yessoensis*) and juvenile Pacific oysters (*Crassostrea gigas*) both less than 8 months of age were resistant to infection. However, two of the oysters with badly cracked shells became infected suggesting that *L. haliotidis* was capable of utilizing oyster tissue as nutrients for growth and multiplication if it was able to gain access to the soft tissues of the oyster (Bower 1987*c*).

Axenic cultures of *L. haliotidis* grew well on several different liquid media and agar-based solid media. An *in vitro* life cycle could be produced by alternately placing the parasite in minimum essential medium with 10% foetal calf serum (where rapid production of vegetative stages occurred through binary fission) and sterile seawater (where the vegetative stages transformed into zoosporoblasts and zoospores were produced) (Bower 1987*b*). Cultured *L. haliotidis* were infective to small abalone (Bower 1987*c*; Bower et al. 1989*a*). The vegetative stage also grew on pine pollen (*Pinus contorta*) in seawater but failed to produce zoosporoblasts and zoospores (Bower 1987*b*).

The spheroid parasite is readily observed with light microscopy (100 times magnification) in the head and foot of the small abalone squashed in seawater between a glass slide and cover slip. *Labyrinthuloides haliotidis* is also evident in histological sections prepared using routine procedures. However, due to the morphological similarities of *L. haliotidis* to other thraustochytrids, identification of the parasite outside the host is almost impossible. A direct fluorescent antibody technique showed promise in facilitating the detection of this parasite (Bower et al. 1989*b*). However, this technique has not been fully tested to verify its specificity.

The source of infection in the abalone hatchery was not established. Employing sanitary techniques could prevent transmission of the disease between raceways. The parasite was destroyed in 20 min when exposed to 25 $mg \cdot L^{-1}$ of chlorine in seawater. The fungicide cyclohexamide at 1–2 $mg \cdot L^{-1}$ for 23 h per day on five consecutive days cured infected abalone. However, this treatment had the following disadvantages: (*i*) being carcinogenic to people, (*ii*) being detrimental to diatoms upon which the abalone fed, (*iii*) being ineffective against nongrowing but infective zoospores such that reinfection occurred within 2–3 weeks following treatment, and (*iv*) inducing resistant forms (as few as three successive treatments resulted in the production of forms requiring twice the amount of cyclohexamide for lethal affects) (Bower 1989). Ozone treatment of incoming water may only be efficacious if ozone exposure is greater than 0.97 mg ozone per litre for 25 min (Bower et al. 1989*c*). Considerable precautions should be implemented to prevent entry of *L. haliotidis* into a facility because this parasite is (*i*) so devastating to abalone <5 mm SL and (*ii*) virtually impossible to control in an abalone culture facility. These precautions include the physical separation of broodstock from the progeny as soon as possible following spawning, the disinfecting of fertilized eggs, and the rearing of the progeny in treated (disinfected) water until they are >5 mm SL.

The impact of *L. haliotidis* on wild abalone stocks and its geographical range are unknown because abalone, of the size that are susceptible to infection, are too tiny to be found in the wild. Apart from the occurrence of this parasite in an abalone culture facility in Saanich Inlet and in a subsidiary hatchery in Sooke on Vancouver Island, British Columbia, the natural distribution of this parasite is unknown. This parasite is very eclectic in nutrient requirements and can survive on virtually nothing for considerable periods of time. Thus, once introduced into a new area *L. haliotidis* may never leave. To prevent the extension of the natural range of this parasite, transplantation of abalone between geographical areas for stock rehabilitation and culture purposes should be avoided.

Perkinsus olseni

Perkinsus olseni proliferates in the foot and mantle of *Haliotis rubra, Haliotis laevigata, Haliotis cyclobates*, and *Haliotis scalaris* in South Australia (Lester and Davis 1981; O'Donoghue et al. 1991). A species of *Perkinsus*, that is believed to be *P. olseni*, occurs in various molluscs from the Great Barrier Reef but was not detected in abalone from that area. *Perkinsus* sp. has not been reported in molluscs from Tasmania (Goggin and Lester 1995). In two outbreaks of *P. olseni* in abalone culture facilities in Australia, about 30–40% mortality occurred among *H. laevigata* of 3–4 cm SL. In both cases *H. rubra* taken from an infected site had been introduced to the facility prior to the outbreaks (Goggin and Lester 1995).

This parasite was also associated with substantial losses of *H. laevigata* from the Gulf of St. Vincent, South Australia (Lester 1986; Lester et al. 1990; O'Donoghue et al. 1991). After the epizootic had passed, two attempts were made to repopulate the area by transplanting adult *H. laevigata*. Eight months after the first attempt, many of the transplanted abalone had died and 15 of 20 survivors examined were heavily infected with *P. olseni*. Only the second attempt, conducted a year after the first attempt and three years following the epizootic event, appeared to be successful (Goggin and Lester 1995).

Perkinsus olseni belongs to a group of protozoa that inhabits at least 67 species of marine molluscs from warm temperate to tropical waters around the world (Perkins 1996). The most studied member of the group is *Perkinsus marinus* that infects American oysters (*Crassostrea virginica*) south from about Damariscotta River, Maine, along the east coast of North and Central America (Ford 1996). *P. marinus* has been associated with the devastation of oyster stocks in Chesapeake Bay, USA (Burreson and Ragone Calvo 1996). Another species, *Perkinsus atlanticus*, was described from diseased clams along the coast of southern Europe (Azevedo 1989). However, the relationship between *P. atlanticus* and *P. olseni* is now in question. Molecular studies (nucleotide sequence of the internal transcribed spacers (ITS) in the ribosomal gene cluster (rDNA)) indicated that *P. olseni* occurs in many species of molluscs from Australia and is homologous to *P. atlanticus* from Portugal (Goggin 1994). *Perkinsus olseni* from abalone was experimentally transmitted in the laboratory and was highly infectious to a range of molluscs including three lamellibranchs from Australian waters: pearl oysters (*Pinctada sugillata*), blood cockles (*Anadara trapezia*), and rock oysters (*Saccostrea commercialis*). Also, *Perkinsus* spp. (that may or may not be *P. olseni*) isolated from other molluscs were infectious to abalone (Lester et al. 1990; Goggin et al. 1989).

Abalone infected with *P. olseni* often produce pustules (spherical brown abscesses up to 8 mm in diameter containing a caseous creamy-brown deposit) in the adductor muscle and mantle. Lesions on the surface of the abalone appear as hemispherical cream or brown swellings and pustules greater than 3 mm in diameter contain purulent material of degrading haemocytes, parasites, and debris (O'Donoghue et al. 1991). *Haliotis rubra* and *H. laevigata* with pustules in the meat have reduced market value (Lester and Davis 1981).

The most sensitive technique for detecting infection is the examination of abalone tissues incubated in Fluid Thioglycollate Medium according to the procedures described by Ray (1966) for *P. marinus*. After 3–14 days of incubation in Fluid Thioglycollate Medium at 25°C, followed by staining with Lugol's iodine, the prezoosporangia of *P. olseni* appear blue-black in colour and 56–94 μm in diameter. The parasite can also be detected by histological examination of abalone tissues (Lester and Davis 1981). Three stages are found in the connective tissues of the host: (*i*) immature trophozoites (2–3 μm in diameter), (*ii*) mature trophozoites (3–18 μm in diameter) with large vacuoles (up to 10 μm diameter with some containing a weakly eosinophilic vaculoplast) commonly called the "signet-ring" stage, and (*iii*) tomonts (dividing cells, 12–35 μm in diameter) containing 2–32 developing immature trophozoites. Abalone will respond to infection by an accumulation of haemocytes that eventually develops into a pustule. Parasites that do not die within the necrotic pustules develop further into the large prezoosporangium (60–95 μm in diameter). Unfortunately, neither the thioglycollate technique described by Ray (1966) nor histological examination is specific for *P. olseni*. Future studies involving both the developments of molecular assays and host specificity investigations will be required to differentiate between the species of *Perkinsus*.

Transmission of this parasite occurs directly between individual molluscs. Prezoosporangia that escape from necrotic pustules or decaying dead abalone undergo further development to zoosporangia in seawater. Within nine days at 20°C and three days at 28°C, hundreds of motile, biflagellated zoospores (about 3 by 5 μm) exit from the zoosporangium. The zoospores are infective to abalone as well as other molluscs (Goggin et al. 1989).

There are no known methods of controlling *P. olseni* in wild abalone stocks. In *H. rubra* and *H. laevigata*, stress such as high temperature (e.g., 20°C) may predisposes the abalone to disease. However, infected *H. rubra* but not *H. laevigata* appear to contain and possibly eliminate the infection at 15°C and during the winter (Lester and Davis 1981; Goggin and Lester 1995). Disease outbreaks in abalone culture facilities were controlled by isolating infected tanks, removal of infected abalone, followed by washing equipment in fresh water (Goggin and Lester 1995). Prezoosporangia and trophozoites can tolerate a wide range of salinities and temperatures and when enclosed in tissue were also resistant to chlorine. Also, because this parasite can survive freezing (–60°C) in abalone tissue for 197 days, its potential for being spread is high in relation to processing plants (Goggin et al. 1990). Thus, abalone from areas with records of the disease should not be imported into areas with no record of *P. olseni* unless suitable precautions are taken not to return potentially contaminated materials to the marine environment.

Sabellid polycheate

An undescribed species of sabellid polychaete, of the subfamily Fabracidae, possibly in the genus *Pseudofabricia* or perhaps from an unknown genus (Oakes and Fields 1996), seriously impacted abalone culture in California during the 1990s (McBride 1998). This sabellid is not native to California and was probably introduced with cultured abalone from South Africa that were not quarantined on arrival in California. By late 1990, the sabellid occured in most abalone culture facilities (including an offshore production facility) in California, USA, and in Baja California, Mexico. Concerns about the spread of this colonizer have interfered with the reseeding plans for rehabilitation of overutilized abalone populations (Hauser 1997). Although this polychaete colonized test molluscs held adjacent to an effluent outlet from an affected abalone hatchery, the polychaete has not yet been observed in wild abalone stocks in California (Shields et al. 1998). This sabellid also caused problems in South African abalone farms among *Haliotis midae*, especially when high abalone stocking densities were combined with poor hygiene and marginal water quality in culture tanks (Cook 1998; Ruck and Cook 1998, 1999).

Haliotis rufescens and other species of abalone as well as other species of marine gastropods including snails and limpets can be infested by this polychaete. Although not directly life threatening to abalone, this sabellid polychaete has become a serious pest of abalone in aquaculture facilities in California (Oakes and Fields 1996). Detrimental effects include significantly reduced growth rates, grossly deformed shells with the absence of respiratory aperture (gill pore) formation on the shells' leading edge, increased mortality due to the inability of the abalone to right themselves when dislodged from their substrate, and reduced marketability.

Adult sabellid polychaetes are about 4–5 mm in length with a branchial crown, eight thoracic segments, and three abdominal segments. The branchial crown (about 1 mm in diameter) may be observed along the leading edge of the abalone shell when the abalone is submerged in seawater. The first thoracic segment has an anterior dorsal collar that is bifurcated mid-dorsally by the faecal groove. This sabellid is hermaphroditic and usually broods 6–10 orange eggs at the posterior end of the tube. The orange ovoid eggs (about 0.25 mm long) can sometimes be seen at the end of the tubes from the interior side of the abalone shell. After hatching, the larvae crawl out the opening of the tube (Oakes and Fields 1996).

The actively crawling benthic larvae, about 1 mm in length, lack a branchial crown. Larvae spread from one abalone to another when the abalone come in contact with each other but, there is evidence to suggest that the larvae may also be disbursed short distances via the water column or along the substrate. The larval sabellid attaches to the shell at the growing edge and covers itself with a mucus tube forming a thin membrane. As the abalone secretes new shell it covers the developing juvenile sabellid with nacre but the polychaete keeps the area around its anterior end open to the environment. When the abalone attempts to form the prismatic layer of the shell in the vicinity of the sabellid, the polychaete interferes with the process.

A fast growing abalone can encapsulate a small number of sabellids and extend its shell beyond them. Slower growing abalone appear to be more susceptible to heavy infections. Heavy infestations (as observed in some abalone facilities) result in a shell with a thickened leading edge that is very fragile and porous, due to the honeycombed affect caused by high polychaete populations. The shell grows downward instead of outward (as in normal abalone) giving infected shells a characteristic dome shape. In an abalone culture facility, infestations can rapidly build up to a damaging level within 60 days, indicating a brief generation time for this polychaete (Oakes and Fields 1996).

The ability of the sabellid to withdraw into its tube makes it very difficult to eliminate without harming the abalone. Even if the adult hermaphrodite polychaete could be destroyed, there are often many fertilized eggs left in the tubes that may be able to hatch later. Lower water temperatures result in a decrease in polychaete reproduction but will also reduce abalone growth. Leighton (1998) claimed that high temperatures (48 h exposure to 28.5°C) killed all stages of the sabellids with only minor mortalities among abalone species tolerant to these temperatures (i.e., *H. fulgens* and *H. corrugata*). However, other workers have observed that a small percentage of the sabellids can survive this high temperature treatment (C. Friedman, California Dept. Fish and Game, Bodega Marine Laboratory, P.O. Box 247, Bodega Bay, CA 94923, personal communication). Nevertheless, this treatment is not suitable for abalone species that have upper lethal limits well below 28°C (i.e., *H. rufescens*). Oakes et al. (1995) proposed dipping the shells of affected abalone in molten wax to cause anoxia and death to sabellids within the plugged burrows. Shields et al. (1998) proposed that the application of lipid-walled microcapsules (LWMs) had potential for controlling sabellid infestations in abalone culture facilities. The use of LWMs takes advantage of the filter-feeding nature of the polychaetes while avoiding impact on the abalone which are benthic herbivores. Although the majority of the polychaetes readily consumed and digested the LWMs, the appropriate toxin(s) for incorporation into the LWMs to control polychaete populations with minimal impact on the abalone and environment has yet to be identified.

The impact of the disease on a culture facility can be reduced by aggressively culling juvenile abalone that are slow growing and removing abalone with signs of sabellid infestation, because only heavy infestations result in shell deformation. In addition, inspection of seed before sale, improved farm sanitation practices, and use of screens on farm discharges are being used to reduce spread of the sabellid between and within farms and to the environment (McBride 1998). Tanks can be disinfected with freshwater baths. Facilities free of this pest should not obtain abalone for stocking from facilities where this pest has been observed.

Withering foot syndrome

Withering foot syndrome is characterized as a withering or wasting of the abalone and is evident in abalone with a body mass that is smaller than normal relative to shell size. The disease is associated with mass mortalities of *Haliotis cracherodii*, but occasionally *H. rufescens* and *H. corrugata* are affected (Friedman 1996; Altstatt et al. 1996). Recently, a dramatic increase in the number of cultured *H. rufescens* with withering foot syndrome was noticed in conjunction with El Niño – Southern Oscillation (ENSO) elevated seawater temperatures (Moore et al. 1999). Although the etiological agent has not been specifically identified and is still under investigation, a rickettsiales-like infection of the digestive gland is thought to be the cause (Gardner et al. 1995). A similar-looking disease of unknown etiology has been reported from cultured *H. discus hannai* on the northern coast of China (Guo et al. 1999).

Diseased abalone and extreme population declines have been reported from the coast of California south of Point Conception since 1985 (Lafferty and Kuris 1993; Davis 1993; Kismohandaka et al. 1995). Mass mortalities of *H. cracherodii* were first noted in the California Channel Islands in 1986 (Haaker et al. 1992; Ruediger and VanBlaricom 1993; VanBlaricom et al. 1993). Mortalities progressively spread throughout the California Channel Islands causing population crashes on six of the eight Channel Islands by 1992 with up to 99% of the *H. cracherodii* lost at several sites (Haaker et al. 1992; Richards and Davis 1993). In 1993, the black abalone fishery in California was closed. North of Point Conception, disease and mortalities were initially limited to the immediate vicinity of a warm-water discharge of a thermal electric power plant in Diablo Cove, California (70 km north of Point Conception) (Steinbeck et al. 1992). However, sampling of locations between Diablo Cove and Point Conception that began in 1992 indicated that this disease is progressing northward from Point Conception (Altstatt et al. 1996).

This disease affects all sizes of abalone and causes lethargy, retraction of the visceral tissues, atrophying of the foot muscle (thereby adversely affecting the ability of the abalone to adhere to the substrate), and is lethal. Affected abalone were discoloured (pale) and weakened and the soft tissues were atrophied and nonresponsive to stimuli (Haaker et al. 1992). In the field, affected abalone can be detached from the substrate by hand and do not attempt to right themselves when turned upside down. Elevated temperatures accelerated disease progression and decreased survival (Lafferty and Kuris 1993; Tissot 1995; Friedman et al. 1997*a*; Moore et al. 1999). At 18–20°C, death usually occurred within one month of the appearance of the clinical signs (Friedman 1996).

Diseased abalone consumed 4.4 times less kelp, 1.2 times less oxygen, and excreted 3.8 time more ammonia per gram wet weight than did healthy abalone (Kismohandaka et al. 1993). Severe metabolic alterations were detected in abalone before visible atrophy of the foot occurred. Haemocyanin concentration in the blood decreased, glycogen in the foot muscle was depleted, haemocyte abundance was reduced, and haemocytes with abnormal morphology increased in wasted abalone (Friedman 1996; Shields et al. 1996). In addition, haemocytes were more chemotactically active but the capability of the stimulated cells to engulf and destroy foreign particles appeared to be compromised and may contribute to mortality associated with the disease (Friedman et al. 1999). In samples examined by histology, severe foot muscle fiber depletion was observed suggesting that diseased abalone may be using foot muscle protein as a source of energy (Kismohandaka et al. 1995). However, no pathogens were found in the muscle or blood tissues. Initially, heavy infections of coccidia in the kidney (see below) were thought to cause the disease but a correlation between coccidial infection and withering syndrome was not found (Steinbeck et al. 1992; VanBlaricom et al. 1993; Kuris et al. 1994; Friedman et al. 1997*a*).

Gardner et al. (1995) proposed that extensive infections of Gram-negative intracellular prokaryotes in the epithelium of the intestinal tract, especially in the enzymes secreting cells of the digestive diverticula, caused the disease. The prokaryotes had morphological characteristics of the order Rickettsiales. They were rod-shaped and ribosome-rich with trilaminar cell walls and were accumulated into intracellular colonies within epithelial cells. Infection of the digestive diverticula was accompanied by a loss of digestive enzyme granules from epithelial cells and apparently a metaplasia of enzyme secretory cells to cells morphologically similar to epithelial cells lining the gut (Gardner et al. 1995). Thus, heavily infected abalone were speculated to undergo muscle tissue catabolism resulting in the withering disease. Initial field studies found no correlation between either the presence or intensity of infection and the physical condition of the abalone (VanBlaricom et al. 1993; Friedman 1996). However, subsequent investigations in the laboratory suggested a possible relationship between the presence of infection and reduced survival in *H. cracherodii* (Friedman et al. 1997*a*). Also, Moore et al. (1999) reported that 70 *H. rufescens* from four culture facilities had a highly significant correlation between the rickettsiales-like procaryote infection intensity and withering syndrome clinical signs. Nevertheless, further research on the role of the rickettsiales-like procaryote in withering syndrome is required.

To prevent the spread of this disease it is essential that the identity of the etiological agent be confirmed followed by the development of sensitive and specific diagnostic tools. The use of such tools could curtail the spread of the disease via aquaculture and restocking activities by assuring that animals proposed for relocation are free of infection. Also, animals other then abalone could be screened for infection to verify that alternate “carrier” hosts do not occur. In affected culture facilities, the severity of the disease may be curtailed if water temperatures could be reduced to about 15°C or less (Moore et al. 1999).

The recovery of black abalone populations affected by mass mortalities from withering foot syndrome seems to be closely linked with temperature. Oceanographic factors that result in elevated seawater temperatures (i.e., ENSO) had a strong negative impact on the recovery of black abalone populations in southern California (Tissot 1995). These elevated temperatures were also associated with a dramatic increase in the number of red abalone with withering foot syndrome in culture facilities in California (Moore et al. 1999). Despite the devastation caused to black abalone populations, a few large, old individuals can still be found and some small juveniles have been seen (Haaker 1997). Also, the research of Tissot (1995) suggests that black abalone populations in southern California may recover with the subsidence of ENSO oceanographic conditions.

Amyotrophia

Abalone with amyotrophia develop muscle atrophy in the mantle and foot, which impedes feeding and adhesion to the substrate, followed by impaired shell growth and mortality. Epizootic mass mortalities attributable to this disease have been observed in juvenile Japanese black abalone (*H. discus discus*) during seed production and subsequent nursery stages at several facilities in Japan since early 1980. Amyotrophia in cultured abalone has now been detected throughout western Japan (Nakatsugawa et al. 1999). Disease usually occurs among juvenile abalone reared at 18–20°C. Initially the enzo abalone (*H. discus hannai*) was thought to be resistant to the disease. However, when the major species for seed production was changed to the enzo abalone, this species was also found to be vulnerable to amyotrophia (Nakatsugawa et al. 1999).

Moribund and dead abalone with amyotrophia have many tumors and muscle atrophy near the nerve trunk of the pedal ganglia and their transverse commissures. The nuclei of tumor cells were contracted and the centers of some tumors were necrotic (Nakatsugawa et al. 1999). No pathogens were detected during histopathological examinations. However, virus-like particles (about 100 nm in diameter) were observed in the cytoplasm of cells near the nerve trunk of diseased *H. discus discus* examined with an electron microscope (Otsu and Sasaki 1997). Virus-like particles of similar size and retrovirus-like morphology were also isolated from diseased *H. discus discus* in primary cultures of abalone haemocytes (Nakatsugawa et al. 1999). Although amyotrophia could be experimentally transmitted between abalone with a filtrate (0.22 μm membrane filter) from diseased *H. discus discus*, the isolates from the primary culture of abalone haemocytes did not reproduce the disease (Nakatsugawa et al. 1999). Thus, further research is required to confirm the etiology of amyotrophia in Japanese abalone and to find mechanisms of preventing or controlling the disease.

Category 2

This category includes parasites that are thought to be benign in their current hosts and geographic range and other parasites that have special life-cycle requirements that may prohibit their becoming established in another location. Various protozoa are thought to exemplify the first scenario and larval nematodes and trematodes that complete their life cycles in specific vertebrates are included in the latter scenario.

Ciliates

Numerous species of ciliates in the orders Thigmotrichida, Peritrichida, Heterotrichida, and Hypotrichida have been reported to live in close association with marine molluscs (Lauckner 1983). Botes et al. (1998) reported vast numbers of ciliophorans of the genus *Mantoscyphidia* attached to the gills of *H. midae* and *Haliotidis spadicea* in southern Africa with no apparent adverse affects.

Kidney coccidia

Margolisiella (=*Pseudoklossia*) *haliotis* (see Desser and Bower (1997) for an explanation of the systematics) is commonly found (prevalence of about 69%) in the kidneys of abalone, *H. cracherodii, H. rufescens, H. corrugata, Haliotis fulgens, Haliotis walallensis,* and *H. kamtschatkana* along the coast of California (Friedman et al. 1995; Friedman et al. 1997*a*). This parasite has not yet been observed in natural stocks of *H. kamtschatkana* in British Columbia or Washington State (Friedman 1991). However, vegetative stages of the parasite were observed in 2 of 40 *H. rufescens* subsampled from about a thousand animals illegally imported from California into barrel culture in Bamfield, British Columbia, in 1991. In experimental trials in California, the coccidian was directly transmitted from *H. rufescens* to *H. kamtschatkana* after 10.5 months of cohabitation and all *H. kamtschatkana* were infected within 17 months (Friedman et al. 1993). It is not known if the parasite has now become established in British Columbia.

This parasite was initially implicated in black abalone (*H. cracherodii*) mass mortalities associated with withering foot syndrome (Steinbeck et al. 1992; Haaker et al. 1992). However, it is no longer considered to be the cause of this disease and is now thought to be benign in nature despite the heavy infections that have been observed (Friedman 1991; Kuris et al. 1994; Friedman et al. 1997*a*).

All stages in the life cycle of *M. haliotis*, including vegetative multiplication (merogony with development of meronts containing merozoites) and sexual reproduction (gametogony with development of microgametes, macrogametes, and oocysts containing sporocysts with sporozoites), occur within the epithelial cells of the abalone kidney (Friedman et al. 1995). Infection is initiated when an elongate sporozoite (about 7 μm × 2.5 μm) penetrates a kidney epithelial cell and develops by ectomerogony to form a meront (about 13 μm in diameter) containing about 28 merozoites and a single residuum. After an unknown number of generations of merogony (vegetative multiplication), elongate meozoites (about 10 × 3 μm) that infect a kidney epithelial cell will develop into either a microgamont (about 20 μm in diameter) containing about 40 microgametocytes or into a single macrogametocyte (about 25 μm in diameter). Following fertilization by a microgametocyte, the macrogametocyte develops into an oocyst (about 31 μm in diameter) containing 6–12 sporocysts and a residuum. Each sporocyst contains two sporozoites (about 7 μm long and 2.5 μm wide) and a sporocyst residuum. The oocysts rupture the host cell and are believed to exit the abalone via the nephridial ducts to initiate the infection in another abalone making this parasite homoxenous (complete life cycle in one host).

Infected kidney epithelial cells become extremely hypertrophied and heavy infections appear to cause serious kidney damage. However, *M. haliotis* elicits no haemocytic response suggesting that it is not recognized as an invader by the abalone host (Friedman et al. 1995). Despite the heavy natural and experimental infections that have been observed in the field and as a result of laboratory experiments, no change in the condition of the abalone or mortalities have been observed (Friedman et al. 1993).

Nematode parasitism

Larval stages of the nematode *Echinocephalus pseudouncinatus* (superfamily Spiruroidea, family Gnathostomidae) parasitize the foot of *H. corrugata* and *H. fulgens* from southern California and the Gulf of California, Mexico (Millemann 1951). In enzootic locations, larval *E. pseudouncinatus* were also reported from several species of sea urchins (Pearse and Timm 1971). Abalone and sea urchins serve as intermediate hosts for this nematode, which is thought to develop to adults in elasmobranch fish, specifically, the horned shark (*Heterodontus francisci*) and the bat stingray (*Myliobatis californicus*) (Millemann 1963). Ingestion of living larval worms from an intermediate host may cause *larval migrans* in humans, a situation where the larval nematode does not develop into an adult but penetrates through the gut wall and wanders aimlessly in the body before being overcome by the person's internal defence mechanisms.

In abalone, a blister is associated with the burrowing and encystment of the larval nematode in the abalone foot. These blisters apparently weaken and decrease the efficacy of the foot muscle as a hold-fast organ making infected abalone easy to remove from the rock substrate (Millemann 1951). The larval nematode (18–21 mm long) within the blister-like cyst has a head-bulb holding 6–8 rows of hooks with 30–50 hooks per row (Millemann 1963).

There are no known methods of preventing or controlling this disease. However, elasmobranch fish infected with the adult stage of this parasite must occur in the vicinity of the abalone in order to initiate the infection in abalone.

Trematode parasitism

Abalone can serve as an intermediate host for trematodes. Because only incompletely developed juvenile stages of the parasites occur in abalone, the specific identities of the trematodes involved are not known. However, the prevalence of infection appears to be low and encystment of the parasites in the foot musculature or gonad does not appear to affect the survival of the abalone host but may cause castration (Harrison and Grant 1971; Shepherd and Breen 1992). The trematodes involved require development to the adult stage in a vertebrate host to complete the life cycle before the parasite can be transmitted to other abalone. With many trematodes, only specific species of vertebrates can serve as suitable final hosts. This requirement limits the potential of trematodes to cause a problem in another habitat or within an abalone culture facility.

Category 3

This category includes organisms that have a serious impact only under environmental conditions that are stressful to abalone and (or) facilitate enhanced multiplication of the organism.

Ubiquitous opportunistic organisms

Included in this group are organisms such as fungi and bacteria that are normally present in the marine environment but can become problematic if given the opportunity. Numerous species of fungus can invade the tissues of abalone held under suboptimal culture conditions. For example, fungal mycelium were found in tubercle-like swellings (up to 5 mm in length) on the mantle and foot of abalone (*Haliotis seiboldii*) held in a live-storage facility in Japan (Hatai 1982). Friedman et al. (1997*b*) isolated a fungus (provisionally identified as a member of the Class Deuteromycotina) from lesions on the inner shell surface of *Haliotis iris* and *Haliotis australis* from Jacks Bay, on the South Island of New Zealand (see section on shell-boring organisms below). The pathogenicity of this fungus and the possible relationship with environmental conditions remains to be investigated.

Bacterial species in the genus *Vibrio* are notorious opportunists of cultured larvae and juvenile shellfish and the disease they cause is called vibriosis. Vibriosis has been documented among larval and juvenile abalone (e.g., *H. rufescens* and *H. kamtschatkana*) in hatcheries (Elston and Lockwood 1983). *Vibrio* spp. are usually not considered a problem for abalone larval culture because the larval period is relatively short and stringent sanitary practices are effective in avoiding potential problems. However, if vibriosis develops in cultures of juvenile abalone, the problem can be serious and costly (Ebert and Houk 1989). Recently, Reuter and McOrist (1999) reported mortality due to *Vibrio harveyi* among juvenile *H. rubra* from a wild harvest that were being cultured in large open tank marine systems in southern Australia.

In cultured abalone, vibriosis is a systemic infection in the soft tissues that results in tissue necrosis (due to production of exotoxin by the bacteria) and death. Histological examination of diseased abalone revealed the presence of rod-shaped bacteria (usually slightly curved) within the tissues (Elston and Lockwood 1983). The *Vibrio* spp. can also be isolated and cultured from diseased abalone on several bacterial culture media including TCBS agar plates especially designed for the culture of *Vibrio* spp.

Vibrio spp. are ubiquitous; hence eradication of the etiological agent of vibriosis is impossible. However, vibriosis appears to be directly related to poor husbandry and other stressful conditions that predispose the abalone to infection. Sources of infection are broodstock, food supply, surfaces utilized by abalone, incoming seawater, and air-borne contaminants (Lizárraga-Partida et al. 1998). In order to alleviate the problem, the source of infection should be determined by culturing bacteria from the above possible sources followed by thorough cleaning of the likely sources. Culturing larvae under optimal conditions (i.e., suitable temperature and salinity for the abalone species) and the use of sterilized seawater (e.g., irradiated with ultraviolet light) can reduce the development of vibriosis. Batches of abalone containing many infected individuals should be destroyed in an approved manner followed by disinfection of all containers and equipment in contact with the infected stock. Vibriosis can also be avoided by limiting the exposure of cultured abalone to physical and chemical stresses (Elston and Lockwood 1983).

Shell-boring organisms

Various species of shell-boring organisms colonize the shells of living and dead molluscs, including abalone. Many species of the boring sponge *Cliona* spp. inhabit abalone shells around the world. Extensive inhabitation can cause some shells to thicken (up to 5 times that of uninfected shells) but the vast network of tunnels weakens the shell causing the abalone to be more vulnerable to predation (Forster 1967; Hansen 1970; Shepherd 1973). However, infestation rates vary markedly between species, abalone age, habitat, and place, and in most situations infestations are benign (Shepherd and Breen 1992).

This is also true for the boring polychaetes in the family Sipionidae where most infestations are innocuous and usually of low intensity with burrows being confined to the shell. However, in southern Japan, the flesh weight of *H. diversicolor aquatilis* decreased significantly when infested with more than ten *Polydora* sp. per shell (Kojima and Imajima 1982). In South Africa, *Polydora* spp. occur in high numbers on some farms and cause severe shell damage and stress to some *H. midae* by penetrating the mantle cavity (Ruck and Cook 1999). In one inlet in British Columbia (Port Neville), the shells of wild *H. kamtschatkana* were extensively riddled with a labyrinth of *Polydora* (possibly *P. limicola, P. ligni*, and (or) *P. websteri*) burrows resulting in fragile shells that cracked easily. The shells also had internal deformities such as nacre-covered nodules and patches of dark discoloration (Bower and Horne, unpublished). Another spionid polychaete, *Boccardia knoxi,* is proving to be a pest to abalone culture in Australia (R.J.G. Lester, Dept. of Microbiology and Parasitology, University of Queensland, Brisbane, Queensland 4072, Australia, personal communication). As more information on the pathogenicity of these polychaetes becomes available, they may warrant upgrading to Category 1 diseases.

High prevalence and intensity of infestation by shell borers have been associated with increased levels of organic pollution (Anger 1977; Rice and Simon 1980). This association was certainly true for the situation in British Columbia where poor water circulation caused by the narrow inlet and nutrient-rich water resulting from logging activities probably induced abnormal proliferation of *Polydora*. The toxoglossan snail *Oenopota levidensis* (native to British Columbia and resides subtidally in the same zones as abalone) is a predator of *Polydora* spp. and may have the potential of being used as a method of biological control in abalone culture facilities. Lleonart (1999) found that air drying (about 4 h at 21°C and about 60% humidity) of abalone from open water culture facilities in Tasmania significantly reduced infestation by both *B. knoxi* and *Polydora hoplura* with no apparent detrimental affects on the abalone. Also, preliminary field exposure results indicated that settlement by both these polychaetes occurred during a short dispersive period making farm management practices feasible for avoiding problems (Lleonart 1999).

Recently, a fungus was reported from shell lesions in *H. iris, H. australis,* and *Haliotis virginea virginea* in southern New Zealand (Friedman et al. 1997*b*; Grindley et al. 1998). The lesions were blisters of conchiolin containing fungal hyphae and occasionally nacreous material that formed on the inner shell surface near the shell apex. Lesions in some severely affected abalone extended into the foot-muscle attachment site. In some cases, the external surface of the shell apex was crumbling and caving in. Although fungal hyphae were consistently found within the conchiolin and shell matrix, none were observed in the adjacent soft tissue of the abalone and abalone condition was apparently not affected. To date, abalone from other parts of the world have not been assayed for shell mycosis. Thus, the specificity and global distribution of this disease is unknown. Until further research has determined the identity, distribution, and potential impact of this pathogen, caution is required when transporting diseased animals out of enzootic areas.

Discussion

As the culture of the various species of abalone increases globally, additional infectious diseases that affect abalone productivity are likely to be encountered. An inadvertent introduction of any of these pathogens into a new location or culture facility could result in severe consequences to cultured abalone, wild stocks, and possibly even to other organisms in the recipient environment. Also, the risk of a supposedly benign parasite turning into a scourge to naive animals in a new habitat is often unpredictable and should be avoided. In most cases, the potential of a parasite (even if not pathogenic in its enzootic habitat) for causing a problem under different environmental conditions is not known. However, the approach of trying to prevent a problem from developing is far better than trying to find a solution.

As more examples of catastrophes resulting from the indiscriminate transplantation of shellfish come to light, international agencies are expending additional effort in trying to reduce the risks of shellfish transfers while attempting not to curtail international trade. For example, the Office International des Epizooties (O.I.E.), the World Animal Health Organization, has set up an International Aquatic Animal Health Code to facilitate international trade in aquatic animals and products. This Code provides detailed definitions of minimum health guarantees to be required of trading partners in order to avoid the risk of spreading aquatic animal diseases. In addition, each year a Working Group on Pathology and Diseases of Marine Organisms (WGPDMO) within the Mariculture Committee of the International Council for the Exploration of the Sea (ICES) convenes to compile and analyze aquatic animal disease information and formulate recommendations to ICES Member Countries.

Guidelines for the transplantation of aquatic organisms have been drawn up by ICES (Sindermann 1988*a*, 1988*b*). Essentially, the imported individuals should be brood stock from a reliable source known to be free of disease. This imported brood stock should be held in quarantine and never released into the new environment. If the F1 progeny experience low mortalities and are found to be free of known disease on microscopic examination, they may be released from quarantine and used for culture purposes. Any deviation from these strict guidelines will increase the risk of introducing diseases that may be debilitating to aquaculture and the wild harvest industry. Of course, regardless of the precautions taken, there is no guarantee that a parasite enzootic and not pathogenic for native animals may occur in the recipient environment and be pathogenic for the introduced stocks.

Many countries have adopted shellfish health protection regulations. The Directorate-General for Agriculture of the Commission of the European Communities has developed European Economic Community legislation on aquatic animal health. This legislation is a directive that defines the animal health conditions governing the marketing of aquatic animals and products and includes provisions for the conservation of species. In the United States of America, many of the states enforce shellfish disease control programs for regulating the introduction of shellfish to prevent the spread of infectious diseases, harmful pests, and predators In Canada, effort is underway to develop shellfish health protection regulations. Incorporated into these regulations will be procedures that can be implemented to avoid the accidental introduction of pathogens. Each province in Canada and most developed countries have an introductions and transfers committee that issues permits or licences for the importation of shellfish intended for culture or long-term maintenance. These committees also have access to resources for the evaluation of import requests. Information from these resources can be utilized to avoid unnecessary risks to the stocks already resident in an area.

References

Altstatt, J.M., Ambrose, R.F., Engle, J.M., Haaker, P.L., Lafferty, K.D., and Raimondi, P.T. 1996. Recent declines of black abalone *Haliotis cracherodii* on the mainland coast of central California. Mar. Ecol. Prog. Ser. **142**: 185–192.

Anger, K. 1977. Benthic invertebrates as indicators of organic pollution in the western Baltic Sea. Int. Revue Ges. Hydrobiol. **62**: 245–254.

Azevedo, C. 1989. Fine structure of *Perkinsus atlanticus* n. sp. (Apicomplexa, Perkinsea) parasite of the clam *Ruditapes decussatus* from Portugal. J. Parasitol. **75**: 627–635.

Botes, H., Basson, L., and Van As, J.G. 1998. Ectosymbiotic ellobiophryid peritrichs associated with abalone. Proc. Microscopy Soc. S. Africa, **28**: 75.

Bower, S.M. 1987*a*. *Labyrinthuloides haliotidis* n. sp. (Protozoa: Labyrinthomorpha), a pathogenic parasite of small juvenile abalone in a British Columbia mariculture facility. Can. J. Zool. **65**: 1996–2007.

Bower, S.M. 1987*b*. Artifical culture of *Labyrinthuloides haliotidis* (Protozoa: Labyrinthomorpha), a pathogenic parasite of abalone. Can. J. Zool. **65**: 2013–2020.

Bower, S.M. 1987*c*. Pathogenicity and host specificity of *Labyrinthuloides haliotidis* (Protozoa: Labyrinthomorpha), a parasite of juvenile abalone. Can. J. Zool. **65**: 2008–2012.

Bower, S.M. 1989. Disinfectants and therapeutic agents for controlling *Labyrinthuloides haliotidis* (Protozoa: Labyrinthomorpha), an abalone pathogen. Aquaculture, **78**: 207–215.

Bower, S.M., McGladdery, S.E., and Price, I.M. 1994. Synopsis of infectious diseases and parasites of commercially exploited shellfish. Ann. Rev. Fish Dis. **4**: 1–199.

Bower, S.M., McLean, N., and Whitaker, D.J. 1989*a*. Mechanism of infection by *Labyrinthuloides haliotidis* (Protozoa: Labyrinthomorpha), a parasite of abalone (*Haliotis kamtschatkana*) (Mollusca: Gastropoda). J. Invert. Pathol. **53**: 401–409.

Bower, S.M., Whitaker, D.J., and Elston, R.A. 1989*b*. Detection of the abalone parasite *Labyrinthuloides haliotidis* by a direct fluorescent antibody technique. J. Invert. Pathol. **53**: 281–283.

Bower, S.M., Whitaker, D.J., and Voltolina, D. 1989*c*. Resistance to ozone of zoospores of the thraustochytrid abalone parasite, *Labyrinthuloides haliotidis* (Protozoa: Labyrinthomorpha). Aquaculture, **78**: 147–152.

Burreson, E.M., and Ragone Calvo, L.M. 1996. Epizootiology of *Perkinsus marinus* disease of oysters in Chesapeake Bay, with emphasis on data since 1985. J. Shellfish Res. **15**: 17–34.

Cook, P. 1998. The current status of abalone farming in South Africa. J. Shellfish Res. **17**: 601–602.

Davis, G.E. 1993. Mysterious demise of southern California black abalone, *Haliotis cracherodii* Leach, 1814. J. Shellfish Res. **12**: 183–184.

Desser, S.S., and Bower, S.M. 1997. Margolisiella kabatai gen. et sp. n. (Apicomplexa: Eimeriidae), a parasite of native littleneck clams, Protothaca staminea, from British Columbia, Canada, with a taxonomic revision of the coccidian parasites of bivalves (Mollusca: Bivalvia). Folia Parasitol. **44**: 241–247.

Ebert, E.E., and Houk, J.L. 1989. Abalone cultivation methods used at the California Department of Fish and Game's marine resources laboratory. *In* Handbook of culture of abalone and other marine gastropods. *Edited by* K.O. Hahn. CRC Press, Inc., Boca Raton, Fla. pp. 239–254.

Elston, R.A., and Lockwood, G.S. 1983. Pathogenesis of vibriosis in cultured juvenile red abalone, *Haliotis rufescens* Swainson. J. Fish Dis. **6**: 111–128.

Ford, S.E. 1996. Range extension by the oyster parasite *Perkinsus marinus* into the northeastern United States: response to climate change? J. Shellfish Res. **15**: 45–56.

Forster, G.R. 1967. The growth of *Haliotis tuberculata*: results of tagging experiments in Guernsey. J. Mar. Biol. Assoc. U.K. **47**: 287–300.

Friedman, C.S. 1991. Coccidiosis of California abalone, *Haliotis* spp. (Abstract) J. Shellfish Res. **10**: 236.

Friedman, C.S. 1996. Update on abalone withering syndrome. Alolkoy, Publ. Channel Islands. Natl. Mar. Sanctuary, **9**: 9.

Friedman, C.S., Roberts, W., Kismohandaka, G., and Hedrick, R.P. 1993. Transmissibility of a coccidian parasite of abalone, *Haliotis* spp. J. Shellfish Res. **12**: 201–205.

Friedman, C.S., Gardner, G.R., Hedrick, R.P., Stephenson, M., Cawthorn, R.J., and Upton, S.J. 1995. *Pseudoklossia haliotis* sp.n. (Apicomplexa) from the kidney of California abalone, *Haliotis* spp. (Mollusca). J. Invert. Pathol. **66**: 33–38.

Friedman, C.S., Thomson, M., Chun, C., Haaker, P., and Hedrick, R.P. 1997*a*. Withering syndrome of the black abalone, *Haliotis cracherodii* (Leach): water temperature, food availability, and parasites as possible causes. J. Shellfish Res. **16**: 403–411.

Friedman, C.S., Grindley, R., and Keogh, J.A. 1997*b*. Isolation of a fungus from shell lesions of New Zealand abalone, *Haliotis iris* Martyn and *H. australis* Gmelin. Moll. Res. **18**: 313–324.

Friedman, C.S., Robbins, T., Jacobsen, J.L., and Shields, J.D. 1999. Examination of the cellular immune response of black abalone, *Haliotis cracherodii*, with and without withering syndrome. (Abstract) J. Shellfish Res. **18**: 322.

Gardner, G.R., Harshbarger, J.C., Lake, J.L., Sawyer, T.K., Price, K.L., Stephenson, M.D., Haaker, P.L., and Togstad, H.A. 1995. Association of prokaryotes with symptomatic appearance of withering syndrome in black abalone *Haliotis cracherodii*. J. Invert. Pathol. **66**: 111–120.

Goggin, C.L. 1994. Variation in the two internal transcribed spacers and 5.8S ribosomal RNA from five isolates of the marine parasite *Perkinsus* (Protista, Apicomplexa). Mol. Biochem. Parasitol. **65**: 179–182.

Goggin, C.L., and Lester, R.J.G. 1995. *Perkinsus*, a protistan parasite of abalone in Australia: a review. Mar. Fish. Res. **46**: 639–646.

Goggin, C.L., Sewell, K.B., and Lester, K.B. 1989. Cross-infection experiments with Australian *Perkinsus* species. Dis. Aquat. Org. **7**: 55–59.

Goggin, C.L., Sewell, K.B., and Lester, R.J. 1990. Tolerances of *Perkinsus* spp. (Protozoa, Apicomplexa) to temperature, chlorine and salinity. J. Shellfish Res. **9**: 145–148.

Grindley, R.M., Keogh, J.A., and Friedman, C.S. 1998. Shell lesions in New Zealand *Haliotis* spp. (Mollusca, Gastropoda). J. Shellfish Res. **17**: 805–811.

Guo, X., Ford, S.E., and Zhang, F. 1999. Molluscan aquaculture in China. J. Shellfish Res. **18**: 19–31.

Haaker, P.L. 1997. Abalone at the Channel Islands. Alolkoy, Publ. Channel Islands Natl. Mar. Sanctuary, **10**: 4.

Haaker, P.L., Parker, D.O., Togstad, H., Richards, D.V., Davis, G.E., and Friedman, C.S. 1992. Mass mortality and withering foot syndrome in black abalone, *Haliotis cracherodii*, in California. *In* Abalone of the world: biology, fisheries and culture. Proceedings of the 1st International Symposium on Abalone. *Edited by* S.A. Shepherd, M.J. Tegner, and S.A. Guzmán del Próo. Fishing News Books, Blackwell Science Ltd., Oxford, U.K. pp. 214–224.

Hansen, J.C. 1970. Commensal activity as a function of age in two species of California abalone (Mollusca: Gastropoda). Veliger, **13**: 90–94.

Harrison, A.J., and Grant, J.F. 1971. Progress in abalone research. Tasmanian Fish. Res. **5**: 1–10.

Hatai, K. 1982. On the fungus *Haliphthorous milfordensis* isolated from temporarily held abalone (*Haliotis sieboldii*). Fish Pathol. (Tokyo) **17**: 199–204.

Hauser, H. 1997. The future of abalone. Alolkoy, Publ. Channel Islands Natl. Mar. Sanctuary, **10**: 5.

Kismohandaka, G., Friedman, C.S., Roberts, W., and Hedrick, R.P. 1993. Investigation of physiological parameters of black abalone with withering syndrome. (Abstract) J. Shellfish Res. **12**: 131–132.

Kismohandaka, G., Roberts, W., Hedrick, R.P., and Friedman, C.S. 1995. Physiological alterations of the black abalone, *Haliotis cracherodii* Leach, with withering syndrome (Abstract). J. Shellfish Res. **14**: 269–270.

Kojima, H., and Imajima, M. 1982. Burrowing polychaetes in the shells of the abalone *Haliotis diversicolor aquatilis* chiefly on the species of *Polydora*. Bull. Jpn. Soc. Sci. Fish. **48**: 31–35.

Kuris, A.M., Schmitt, R.J., and Hedrick, R. 1994. Abalone wasting disease: role of coccidian parasites and environmental factors. California Sea Grant Biennial Report of Completed Projects 1990–92. California Sea Grant College, La Jolla, Calif. pp. 120–123.

Lafferty, K.D., and Kuris, A.M. 1993. Mass mortality of abalone *Haliotis cracherodii* on the California Channel Islands: tests of epidemiological hypotheses. Mar. Ecol. Prog. Ser. **96**: 239–248.

Lauckner, G. 1983. Diseases of Mollusca: Bivalvia. *In* Diseases of marine animals. Volume II: Introduction, Bivalvia to Scaphopoda. *Edited by* O. Kinne. Biologische Anstalt Helgoland, Hamburg. pp. 477–961.

Lee, J.V., Shread, P., Furniss, A.L., and Bryant, T.N. 1981. Taxonomy and description of *Vibrio fluvialis* sp. nov. (synonym Group F vibrios, Group EF6). J. Appl. Bacteriol. **50**: 73–94.

Leighton, D.L. 1998. Control of sabellid infestations in green and pink abalones, *Haliotis fulgens* and *H. corrugata*, by exposure to elevated water temperatures. J. Shellfish Res. **17**: 701–705.

Lester, R.J.G. 1986. Abalone die-back caused by protozoan infection? Aust. Fish. **45**: 26–27.

Lester, R.J.G., and Davis, G.H.G. 1981. A new *Perkinsus* species (Apicomplexa, Perkinsea) from the abalone *Haliotis ruber*. J. Invert. Pathol. **37**: 181–187.

Lester, R.J.G., Goggin, C.L., and Sewell, K.B. 1990. *Perkinsus* in Australia. *In* Pathology in marine science. *Edited by* F.O. Perkins and T.C. Cheng. Academic Press Inc., San Diego. Calif. pp. 189–199.

Levine, N.D., Corliss, J.O., Cox, F.E.G., Deroux, G., Grain, J., Honigberg, B.M., Leedale, G.F., Loeblich, A.R., Lom, J., Lynn, D., Merinfeld, E.G., Page, F.C., Poljansky, G., Sprague, V., Vavra, J., and Wallace, F.G. 1980. A newly revised classification of the Protozoa. J. Protozool. **27**: 37–58.

Li, T., Ding, M., Song, X., Xiang, J., and Liu, R. 1996. Preliminary studies on the mechanism of *Vibrio fluvialis II* resistance to antibiotics. Oceanol. Limnol. Sin. **27**: 637–645.

Li, T., Ding, M., Xiang, J., and Liu, R. 1997*a*. Immunological studies on *Haliotis discus hannai* with *Vibrio fluvialia II*. Oceanol. Limnol. Sin. **28**: 27–32.

Li, T.-W., Zhang, J., Ding, M.-J., Wang, S.-J., Shi, P., Xiang, J.-H., and Liu, R.-Y. 1997*b*. Histology and ultrastructure of pustule disease in abalone *Haliotis discus hannai* (Mollusca: Gastropoda). Acta Zool. Sin. **43**: 238–242.

Li, T., Ding, M., Zhang, J., Xiang, J., and Liu, R. 1998. Studies on the pustule disease of abalone (*Haliotis discus hannai* Ino) on the Dalian coast. J. Shellfish Res. **17**: 707–711.

Liu, J., Nie, L., Li, T., Ding, M., Song, X., and Zhao, R. 1995. A study on pustule disease of *Haliotis discus*. J. Fish. Sci. (China), **2**: 78–84.

Lizárraga-Partida, M.L., Anguiano-Beltrán, C., Searcy-Bernal R., and Vázquez-Moreno, E. 1998. Bacterial water quality in abalone farms of Baja California. J. Shellfish Res. **17**: 689–692.

Lleonart, M. 1999. 1. Mud worm treatment trials and 2. Initial report: abalone mud worm field studies in infected and progression patterns. *In* Abalone aquaculture sub-program. Proceedings of the 6th Annual Abalone Aquaculture Workshop. World Aquaculture Society Conference, Sydney, April 1999. Fisheries Research & Development Corporation, Deakin, Australia. pp. 1–14.

McBride, S.C. 1998. Current status of abalone aquaculture in the Californias. J. Shellfish Res. **17**: 593–600.

Millemann, R.E. 1951. *Echinocephalus pseudouncinatus* n.sp., a nematode parasite of the abalone. J. Parasit. **37**: 435–439.

Millemann, R.E. 1963. Studies on the taxonomy and life history of Echinocephalid worms (Nematoda: Spiruroidea) with a complete description of *Echinocephalus pseudouncinatus* Millemann, 1951. J. Parasitol. **49**: 754–764.

Moore, J.D., Robbins, T.T., and Friedman, C.S. 1999. Withering syndrome in farmed red abalone, *Haliotis rufescens*. (Abstract) J. Shellfish Res. **18**: 319.

Nakatsugawa, T., Nagai, T., Hiya, K., Nishizawa, T., and Muroga, K. 1999. A virus isolated from juvenile Japanese black abalone *Nordotis discus discus* affected with amyotrophia. Dis. Aquat. Org. **36**: 159–161.

Nie, L., Liu, J., and Li, T. 1995. Physiological and biochemical characteristics of the pathogen of pustule disease of *H. discus*. Chin. J. Microecol. **7**: 33–36.

Oakes, F.R., and Fields, R.C. 1996. Infestation of *Haliotis rufescens* shells by a sabellid polychaete. Aquaculture, **140**: 139–143.

Oakes, F.R., Fields, R.C., Arthur, P.F., and Trevelyan, G.A. 1995. The use of wax to control the shell parasites of red abalone, *Haliotis rufescens*. (Abstract). J. Shellfish Res. **14**: 273.

O'Donoghue, P.J., Phillips, P.H., and Shepherd, S.A. 1991. *Perkinsus* (Protozoa: Apicomplexa) infections in abalone from South Australian waters. Trans. Royal Soc. S. Aust. **115**: 77–82.

Otsu, R., and Sasaki, K. 1997. Virus-like particles detected from juvenile abalones (*Nordotis discus discus*) reared with an epizootic fatal wasting disease. J. Invertebr. Pathol. **70**: 167–168.

Pearse, J.S., and Timm, R.W. 1971. Juvenile nematodes (*Echinocephalus pseudouncinatus*) in the gonad of sea urchins (*Centrostephanus coronatus*) and their effect on host gametogenesis. Biol. Bull. (Woods Hole), **104**: 95–103.

Perkins, F.O. 1996. The structure of *Perkinsus marinus* (Mackin, Owen and Collier, 1950) Levine, 1978 with comments on taxonomy and phylogeny of *Perkinsus* spp. J. Shellfish Res. **15**: 67–87.

Ray, S.M. 1966. A review of the culture method for detecting *Dermocystidium marinum*, with suggested modifications and precautions. Proc. Natl. Shellfish Assoc. **54**: 55–69.

Reuter, R.E., and McOrist, S. 1999. Mortality due to *Vibrio harveyi* in farmed blacklip abalone, *Notohaliotis ruber*. *In* Book of abstracts. *Edited by* C.L. Browdy and R. Fletcher (co-program chairs). The Annual International Conference and Exposition of the World Aquaculture Society, 26 April – 2 May, 1999, Sydney, Australia. p. 629.

Rice, S.A., and Simon, J.L. 1980. Intraspecific variation in the pollution indicator polychaete *Polydora ligni* (Spionidae). Ophelia, **19**: 79–115.

Richards, D.V., and Davis, G.E. 1993. Early warnings of modern population collapse in black abalone *Haliotis cracherodii*, Leach, 1814 at the California Channel Islands. J. Shellfish Res. **12**: 189–194.

Ruck, K.R., and Cook, P.A. 1998. Sabellid infestations in the shells of South African molluscs: implications for abalone mariculture. J. Shellfish Res. **17**: 693–699.

Ruck, K.R., and Cook, P.A. 1999. Polychaete worms: a threat to abalone farming? *In* Book of abstracts. *Edited by* C.L. Browdy and R. Fletcher (co-program chairs). The Annual International Conference and Exposition of the World Aquaculture Society, 26 April – 2 May, 1999, Sydney, Australia. World Aquaculture Society, Baton Rouge, La. p. 658.

Ruediger, J.L., and VanBlaricom, G.R. 1993. Abalone withering syndrome at San Nicolas Island, California. (Abstract) J. Shellfish Res. **12**: 132.

Shepherd, S.A. 1973. Studies on southern Australian abalone (genus *Haliotis*) 1. Ecology of five sympatric species. Aust. J. Mar. Freshw. Res. **24**: 217–257.

Shepherd, S.A., and Breen, P.A. 1992. Mortality in abalone: its estimation, variability and causes. *In* Abalone of the world: biology, fisheries and culture. Proceedings of the 1st International Symposium on Abalone. *Edited by* S.A. Shepherd, M.J. Tegner, and S.A. Guzmán del Próo. Fishing News Books, Blackwell Science Ltd., Oxford, U.K. pp. 276–304.

Shields, J.D., Perkins, F.O., and Friedman, C.S. 1996. Hematological pathology of wasting syndrome in black abalone. (Abstract) J. Shellfish Res. **15**: 498.

Shields, J.D., Buchal, M.A., and Friedman, C.S. 1998. Microencapsulation as a potential control technique against sabellid worms in abalone culture. J. Shellfish Res. **17**: 79–83.

Sindermann, C.J. 1988*a*. Role of the International Council for the Exploration of the Sea (ICES) in matters concerned with transfers and introductions of marine organisms (Abstract). J. Shellfish Res. **7**: 560.

Sindermann, C.J. 1988*b*. Disease problems created by introduced species. *In* Developments in aquaculture and fisheries science. Volume 17. Disease diagnosis and control in North American marine aquaculture. *Edited by* C.J. Sindermann and D.V. Lightner. Elsevier, Amsterdam. pp. 394–398.

Steinbeck, J.R., Groff, J.M., Friedman, C.S., McDowell, T., and Hedrick, R.P. 1992. Investigations into a mortality among populations of the California black abalone, *Haliotis cracherodii*, on the central coast of California, USA. *In* Abalone of the world: biology, fisheries and culture. Proceedings of the 1st International Symposium on Abalone. *Edited by* S.A. Shepherd, M.J. Tegner, and S.A. Guzmán del Próo. Fishing News Books, Blackwell Science Ltd., Oxford, U.K. pp. 203–213.

Tissot, B.N. 1995. Recruitment, growth, and survivorship of black abalone on Santa Cruz Island following mass mortality. Bull. S. Calif. Acad. Sci. **94**: 179–189.

VanBlaricom, G.R., Ruediger, J.L., Friedman, C.S., Woodard, D.D., and Hedrick, R.P. 1993. Discovery of withering syndrome among black abalone *Haliotis cracherodii* Leach, 1814, populations at San Nicolas Island, California. J. Shellfish Res. **12**: 185–188.

The effects of sea otters (*Enhydra lutris*) on abalone (*Haliotis* spp.) populations

Jane Watson

Abstract: Historically sea otters (*Enhydra lutris*) were common in coastal regions of the North Pacific. They were hunted to near extinction from the mid-1700s until 1911 but were reintroduced to portions of their historic range in the 1960s and 1970s. Sea otters prey on invertebrates and limit the size and abundance of their prey. Extirpation of the sea otter eliminated its role as a predator and resulted in dense populations of invertebrates, including abalone (*Haliotis* spp.). Abundant abalone populations, along with developing markets, resulted in the growth of commercial abalone fisheries. However, sea otters and abalone fisheries are mutually exclusive. In areas occupied by otters, abalone occur at low but stable densities and are restricted to crevices. The co-existence of otters and abalone appears to depend on habitat type and the life history characteristics of abalone. Even in the absence of sea otters, commercial abalone fisheries along the west coast of North America have collapsed. This paper reviews the status of sea otter populations, the effects of sea otters on abalone, and the history of abalone fisheries within the historic range of sea otters.

Résumé : Par le passé, les loutres de mer (*Enhydra lutris*) étaient nombreuses dans les régions côtières du Pacifique Nord. Entre le milieu du XVIIIe siècle et 1911, elles ont presque été chassées jusqu'à l'extinction, mais on les a réintroduites dans des parties de leur aire de répartition d'origine dans les années 60 et 70. La loutre de mer se nourrit d'invertébrés et limite ainsi la taille et l'abondance de ses proies. Sa disparition avait éliminé son rôle de prédateur et provoqué l'augmentation de la densité des populations d'invertébrés, dont les ormeaux (*Haliotis* spp.). L'abondance des populations d'ormeaux, ainsi que les marchés en développement, ont entraîné la croissance de la pêche commerciale de cette espèce. Toutefois, les loutres de mer et les ormeaux sont mutuellement exclusifs. Dans les endroits occupés par les loutres, les densités d'ormeaux sont faibles mais stables, et leur présence se limite aux fissures des rochers. La coexistence des loutres et des ormeaux semble dépendre du type d'habitat et des caractéristiques historiques du mode de vie des ormeaux. Même en l'absence des loutres de mer, la pêche commerciale à l'ormeau sur la côte Ouest de l'Amérique du Nord a chuté. La présente étude revoit l'état des populations de loutres de mer, les effets qu'elles ont sur les ormeaux et l'histoire de la pêche à l'ormeau à l'intérieur de l'aire de répartition historique des loutres de mer.

[Traduit par la Rédaction]

Introduction

The sea otter (*Enhydra lutris*) is a small marine carnivore, which preys upon benthic invertebrates and, in some cases, slow-moving demersal fishes (Estes and Palmisano 1974; Estes et al. 1982). Sea otters once were distributed from the Kurile Islands along coastal areas of the North Pacific to cold, upwelled regions of Baja California, Mexico (Kenyon 1969). Although the exact size of the otter population before the arrival of Europeans is unknown, estimates range from 150 000 to 300 000 animals (Kenyon 1969; Johnson 1982). Midden remains indicate that aboriginal people exploited sea otters and likely reduced otter abundance around village sites (Simenstad et al. 1978; Walker 1982). Large-scale, commercial exploitation of sea otters off North America began in the mid-1700s. This intense, maritime fur trade, which lasted until sea otters were protected in 1911, may have reduced the sea otter population to fewer than 1000 individuals located in 12 remnant populations (Kenyon 1969).

In the absence of exploitation, sea otter populations have grown and spread throughout much of their historic range (Estes 1990). Expansion of the sea otter's range was aided in the late 1960s and 1970s by a series of reintroductions. Sea otters were captured from Amchitka Island and Prince William Sound and were translocated to southeast Alaska, British Columbia, Washington State, and Oregon. With the exception of the reintroduction to Oregon, all of these translocation efforts were successful and resulted in well-established sea otter populations (Jameson et al. 1982). In 1987, sea otters were captured from the central coast of California and translocated to San Nicolas Island, one of the eight Channel Islands, located in the southern California Bight (Benz 1996).

Before the mid-1800s sea otters probably limited the size, abundance, and distribution of many invertebrate populations (Johnson 1982; Estes and VanBlaricom 1985; Watson and Smith 1996). Overexploitation of sea otters eliminated their role as predators and, with reduced predation, dense populations of large invertebrates resulted. These abundant invertebrate populations, along with developing markets, contributed to the growth of commercial and recreational

J. Watson. Malaspina University-College, 900 Fifth St., Nanaimo, BC V9R 5S5, Canada, and
Fisheries Centre, 2204 Main Mall, University of British Columbia, Vancouver, BC V6T 1Z4, Canada.
Correct citation: Watson, J. 2000. The effects of sea otters (*Enhydra lutris*) on abalone (*Haliotis* spp.) populations. *In* Workshop on Rebuilding Abalone Stocks in British Columbia. *Edited by* A. Campbell. Can. Spec. Publ. Fish. Aquat. Sci. 130. pp. 123–132.

fisheries (Estes and VanBlaricom 1985; Watson and Smith 1996).

Sea otters forage in both rocky and soft-bottom areas, preying on a wide variety of invertebrates. Foraging occurs to a depth of about 100 m (Newby 1975), although most foraging occurs in depths of <40 m (Estes et al. 1981). Sea otters have high metabolic rates and consume 20–30% of their body weight in prey each day (Costa 1978). Sea otters are very effective predators and limit the abundance of their prey (reviewed in Riedman and Estes 1990). By limiting the abundance of herbivorous invertebrates, particularly sea urchins, sea otters reduce grazing pressure, which enhances the abundance of fleshy algae, especially kelp (Estes and Duggins 1995). This increase in kelp can in turn enhance nearshore productivity (Duggins et al. 1989), increase fish recruitment (Bodkin 1986, 1988), and alter patterns of invertebrate recruitment (Duggins et al. 1990). Although the effects of sea otter foraging on community composition are relatively well understood north of Point Conception, California, the role that sea otters played historically in structuring communities south of Point Conception is thought to be less clear by some. The presence of large kelp beds, in the absence of sea otters, suggests that factors other than sea otters also affect community structure in this area (Foster and Schiel 1988).

Conflicts between sea otters and invertebrate fishers have become increasingly common as sea otter populations increase and reoccupy their historic range (Estes and VanBlaricom 1985; Wendell et al. 1986; Estes 1990; Wendell 1994; Watson and Smith 1996). There is general agreement that many commercial shellfisheries will not coexist with sea otters. In many cases, the decline of commercially harvestable densities of invertebrates has coincided with the range expansion of sea otters (Wendell et al. 1986; Wendell 1994). However, there is controversy over the actual effects of sea otters on some shellfisheries, largely because it is often difficult to distinguish between the effects of sea otter predation and human harvest. Even in the absence of sea otters, examples of sustainable, high-yield, commercial invertebrate fisheries are surprisingly rare (Estes et al. 1993). For a fishery to be sustainable, recruitment to the exploited portion of the stock must replace the harvest (Ludwig et al. 1993). Data supporting stock-recruitment models for invertebrate populations are rare (Estes and VanBlaricom 1985) and many marine invertebrate species appear to be subject to episodic and irregular recruitment or recruitment governed by environmental or geographic features rather than standing stock (Ebert and Russell 1988; Ebert et al. 1994).

Sea otters have been held responsible for increasing the rate of decline or eliminating commercial abalone fisheries in central California (Wendell 1994). However, the history of abalone fisheries along the west coast of North America makes it clear that a variety of factors, including sea otter predation, have been responsible (Estes and VanBlaricom 1985; Watson and Smith 1996). Sea otters cannot be held responsible for the collapse of abalone fisheries in British Columbia, southeast Alaska, or Southern California where abalone fisheries collapsed in the absence of significant sea otter predation (Sloan and Breen 1988; Davis et al. 1996; papers in Campbell 2000*a*).

In this paper, I review the status and known effects of sea otters on abalone populations from an ecological perspective. To do this, I review the population biology and ecology of sea otters throughout their North American range where abalone and sea otters overlap and examine how sea otter populations in each of these areas are known to interact with abalone stocks. Commercial abalone fisheries and sea otters appear to be mutually exclusive. However, the effects of sea otters on abalone species that have co-existed on an evolutionary time scale (e.g., Estes and Steinberg 1988) are different than those of commercial exploitation and may provide valuable insights into the biology of exploited abalone populations.

The population ecology of sea otters

Sea otter population dynamics vary geographically (Estes et al. 1996). Prior to 1990, sea otter populations in the western Aleutians, southeast Alaska, British Columbia, and Washington State were increasing at 17–20%, which is near the maximum estimated rate of increase (r_{max}). These high rates of increase probably occurred because sea otters were protected and permitted to recover, food resources were not limiting, and sea otters have few natural predators (but see Estes et al. 1998). Population growth in sea otters has traditionally been thought to be density-dependent, regulated by the availability of food (Kenyon 1969); population growth rates decline as populations approach equilibrium, which varies depending on habitat (Estes 1990). In areas where unoccupied habitat is available, sea otter mortality rates appear low. Sea otters expand into new areas hence equilibrium density and population growth are achieved through dispersal (Lubina and Levin 1988; Estes 1990; Watson et al. 1996). In most areas, range expansion is by males expanding into new areas apparently as resources become limiting. Females expand into the areas vacated by males (Loughlin 1980; Garshelis et al. 1984; Wendell et al. 1986).

In south central Alaska, sea otters give birth as early as three years of age and by age five all females have given birth (Bodkin et al. 1993). Females give birth at yearly intervals and live to an age of at least 15 years (Riedman and Estes 1990). The age of first reproduction and the probability of females giving birth appear to be constant, regardless of the growth rate of the population (Jameson and Johnson 1993; Bodkin et al. 1993; Monson and Degange 1995). Thus, mortality rates in populations that are growing are low, whereas mortality rates in populations near or at equilibrium density are high, especially among newly weaned pups and juveniles (Estes 1990). However, Estes et al. (1998) report a precipitous decline in sea otter populations in the central and western Aleutian Islands, apparently brought about by killer whale predation on sea otters. A substantial decline in pinniped abundance appears to have caused killer whales to switch from their preferred prey of pinnipeds to sea otters. Sea otter populations in this region are presently thought to be regulated by predation rather than by density-dependent factors.

Status of sea otter populations

Sea otter population in California

The sea otter population in California has grown slower

than more northern sea otter populations. From 1983 to 1994, spring survey results indicated that the population was increasing about 5–6% per year. However, from 1994 to 1996, the average growth rate was –2% per year (Estes et al. 1996). This low rate of population growth appears to be explained by increased mortality from a variety of sources including infectious diseases, trauma, emaciation, and miscellaneous causes (Thomas and Cole 1996).

Sea otters in California are recognized as a subspecies of sea otter distinct from their northern cogeners (Wilson et al. 1991; Anderson et al. 1996) and as such are classified as threatened under the U.S. Endangered Species Act and as depleted under the Marine Mammal Protection Act. The California sea otter is also fully protected under California State Law. The range of the California sea otter presently extends from Point Conception to San Francisco Bay. In 1998, the sea otter population size was estimated to have 2114 animals (J. Estes, U.S. Geological Survey, University of California, Santa Cruz, CA 95064, personal communication).

Oil spills are considered a major threat to the California sea otter population. To reduce this risk, attempts were made to establish a breeding population of sea otters outside the existing range of the California sea otter. Consequently, from 1987 to 1990, 139 sea otters were translocated from the coast of mainland California to San Nicolas Island, 70 miles west of Los Angeles (Benz 1996). In an effort to ensure sufficient shellfish resources for commercial and recreational exploitation, the conditions of the translocation included the restriction of the mainland sea otter population to north of Point Conception and containment of the translocated sea otter population to a "translocation zone" around San Nicolas Island. Legislation requires that sea otters found south of Point Conception and outside the translocation zone be relocated to north of Point Conception or back to San Nicolas Island (Benz 1996).

Maintaining separate zones for sea otters and shellfish resources, or zonal management, does not appear to work. In 1998, a group of male sea otters moved south of Point Conception. To date, no attempts have been made to relocate the sea otters that move seasonally south of Point Conception. Even with zonal management, it is not clear that the exclusion of sea otters will guarantee a persistent commercial or recreational abalone fishery (VanBlaricom 1996).

Sea otter population in Washington

In 1969 and 1970, 59 Alaskan sea otters were reintroduced to Washington State (Jameson et al. 1986). In 1997, 502 sea otters were counted between Destruction Island and Neah Bay. The population has increased at a rate of about 11.4% per year since regular surveys were started in 1989. Most range expansion has occurred in the northern portion of the range. There is evidence, however, that the rate of increase may be declining (R. Jameson, U.S. Geological Survey, Oregon State University, Corvallis, Oreg., personal communication). This expanding population has recently come into conflict with a commercial fishery for red sea urchins (*Strongylocentrotus franciscanus*). There is no commercial harvest of abalone in Washington State (Watson and Smith 1996).

Sea otter population in British Columbia

A total of 89 Alaskan sea otters from Amchitka Island and Prince William Sound were reintroduced to British Columbia in a series of three translocations in 1969, 1970, and 1972 (Bigg and McAskie 1978). Since reintroduction, the British Columbian sea otter population has increased at a rate of about 18.6% per year (Watson et al. 1996) and in 1998 included an estimated 2500 animals located in two disjunct areas. Approximately 2000 sea otters occurred along the west coast of Vancouver Island, from Cape Scott in the north to Estevan Point in the south, although extralimital sightings were increasingly common along both the inside and outside waters of Vancouver Island. A second group of sea otters, off the central portion of the British Columbia coast, included approximately 500 animals in 1998. The success of the translocations to British Columbia may in part be due to the fact that sea otters introduced to Washington State dispersed north, augmenting the numbers released into British Columbian waters.

In British Columbia, the sea otter is classified as threatened and is protected under terms of the Federal Fisheries Act and the Provincial Wildlife Act (Watson et al. 1996).

Sea otter population in southeast Alaska

Sea otters were reintroduced to several sites in southeast Alaska from 1965 to 1969. The results of surveys conducted between 1994 and 1995 suggested that there were 8707 (±6292) sea otters between Cape Yakataga and the British Columbia border. The principal causes of mortality in this population include incidental take in fisheries, native harvest, predation, and food depletion. (C. Gorbics, U.S. Fish and Wildlife Service, Anchorage AK 99503, personal communication). From 1975 to 1987, the growth rate of the sea otter population in southeast Alaska was estimated at 17.6% per year (Estes 1990), although some estimates of population growth were lower (D. Woodby, Alaska Fish and Game, Douglas, Alaska 99824-0020, personal communication).

The biology and ecology of abalone

Abalone, in the genus *Haliotis*, generally occur in shallow water, on rock substrate, near or within algal beds, where they graze on drift or attached algae as well as diatoms. They tend to be slow-growing, long-lived, and have unpredictable recruitment (Tegner et al. 1996). Abalone are dioecious, broadcast spawners with a lecithotrophic larvae. Spawning appears to occur in all months but peaks in the spring and summer, with enormous variability both within and between species in timing of reproduction (Sloan and Breen 1988). As with many broadcast spawners abalone appear to require high concentrations of sperm to increase fertilization success. Some species aggregate to spawn (Breen and Adkins 1980; McShane 1995), suggesting that abalone density may affect fertilization success. The zone of effective fertilization has been reported to be from 1 m in some species (Davis et al. 1996), to slightly greater than 1 m in other species (McShane 1995).

The larvae are short-lived, spending 3–12 days in plankton, depending on species, latitude, and water temperature (Sloan and Breen 1988; McShane 1995). Abalone larval dispersal appears to be limited under most environmental conditions. They are weak swimmers and local water currents may determine dispersal. Studies indicate the dispersal of abalone may be limited to 0–10 km suggesting that stock replenishment by distant populations may not always be possible (Prince et al. 1987, 1988; Tegner and Butler 1985*a*; McShane et al. 1988; Shepherd et al. 1992; Tegner 1993; McShane 1995; Sasaki and Shepherd 1995; Tegner et al. 1996).

Larvae are thought to settle preferentially on crustose coralline algae, which is maintained by herbivore grazing, although laboratory studies indicate that settlement can occur on a variety of substrates (Sloan and Breen 1988). Chemicals present in coralline algae may induce metamorphosis (Morse and Morse 1984). After settlement, depending on the species, small abalone occupy open habitats for up to two months before apparently moving to cryptic habitats where they are less vulnerable to predation (Mottet 1978; Shepherd and Turner 1985; McShane 1996). Predation on recently settled abalone is thought to be high but is difficult to document (Tegner and Butler 1985*b*; Shepherd et al. 1992).

Juvenile abalone feed on diatoms and epibenthic bacteria (Ebert and Houk 1984). As they grow, abalone trap drift algae as their major source of food (Cox 1962; Tegner 1989). Along the west coast of North America, canopy-forming kelp species (*Nereocystis* and *Macrocystis*) seem to be the most important source of algal drift in the abalone diet (Cox 1962; Breen 1980; Breen 1986).

The cryptic habits of juvenile abalone may reduce predation, although octopus, panularid lobsters, crabs, fish, and predatory gastropods are known to prey on juveniles. Sea stars, fish, crab, and octopus prey on adult abalone; however, only sea otter and human exploitation appear to have a significant impact on the abundance and size of abalone populations (reviewed in Sloan and Breen 1988).

In addition to predation, widespread mass mortality of black abalone (*H. cracheroidii*) populations has reduced once-dense intertidal abalone populations by about 97% in southern California (Alstatt et al. 1996). This disease, known as withering foot syndrome, was initially confined to black abalone located in southern California. The disease however has spread north of Point Conception and there are anecdotal reports of withered red and pink abalone in southern California. This disease threatens recreational fisheries and mariculture operations and may affect the ability of abalone populations to recover from exploitation by additionally reducing density (VanBlaricom et al. 1993).

Abalone fisheries

Within the historic North American range of sea otters, there are eight species of abalone (Table 1). Populations of all eight species have supported recreational and subsistence fisheries and commercial fisheries have occurred in all areas except Washington and Oregon states. The effects of sea otters on abalone are impossible to evaluate without considering human exploitation.

Aboriginal exploitation of abalone occurred for at least 7000 years (Walker 1982). Abalone meat was consumed and the shells were used as fishing lures, in jewelry, and as an inlay for carvings (Stewart 1977). Shells from the red abalone (*H. rufescens*), which were deemed superior to pinto or northern abalone (*H. kamtschatkana*), were one of the items the Spanish used to trade for sea otter pelts with First Nations people along the west coast of Vancouver Island (Ogden 1941). In California, native Americans harvested abalone in large numbers. The size and abundance of abalone shells found in middens suggested that in many cases sea otters may not have limited the abundance and size of abalone near village sites (Walker 1982).

The first commercial fisheries for abalone began in the mid-1800s in California. Chinese fishers harvested abalone from the intertidal and shallow subtidal zones using poles and hooks. The product was dried or canned and shipped to Asia. As shallow stocks were depleted and regulations changed, the fishery became a diving fishery, dominated by Japanese fishers. With the development of advanced technology, including rubber suits and SCUBA, the fishery expanded to deeper water (Estes and VanBlaricom 1985).

Commercial abalone fisheries in California, British Columbia, and Alaska intensified as Asian markets developed in the 1970s (Estes and VanBlaricom 1985; Breen 1986). Commercial landings of abalone in California peaked in 1950 and then moderated at about 1.8 and 2.3 million kg per year until the late 1960s when they declined to 700 000 kg per year (Cicin-Sain et al. 1977). Early fisheries for abalone focused on red and pink abalone (*H. corrugata*), but as these stocks declined, landings of less desirable species, black (*H. cracherodii*), green (*H. fulgens*), and white (*H. sorensii*) abalone increased. Abalone harvest was completely prohibited in southern California in 1997 due to low stocks (Davis et al. 1996). Recreational harvest of red abalone by free diving, north of San Francisco, remains as the only abalone fishery open in California. By restricting this fishery to free diving, both access and effort are limited. This management strategy has resulted in sustainable harvests (Pollard 1992), although poaching and a spreading sea otter population are thought to threaten the recreational fishery (Karpov et al. 2000).

In British Columbia, landings of *H. kamtschatkana* peaked in 1977 at 474 000 kg and severe catch limits were imposed to reduce harvest and protect stocks (Breen 1986). However in 1990, with abalone stocks severely depleted, all harvest of abalone was prohibited (Campbell 2000*b*). Likewise, in Alaska, the harvest of northern abalone increased to about 172 000 kg in 1979 (Woodby et al. 2000) and catches were reduced by quota and seasonal closures in an effort to protect dwindling stocks (Breen 1980; Watson and Smith 1996). In 1995, the Alaskan annual harvest was 6000 kg and subsequently the fishery was closed (Woodby et al. 2000).

In all cases, commercial fisheries for abalone in North America can be characterized by rapid increases in landings, followed by precipitous declines, and in most cases closure of commercial harvest (Tegner et al. 1996; Karpov et al. 2000). The reason for the collapse of these abalone fisheries is not entirely clear and may vary between geographic area, but overharvesting, poor or episodic recruitment, disease, and predation by natural predators all appear to have played a role. In contrast, very limited access and few natural pred-

Table 1. Scientific and common names and range of the eight abalone species that occur within the North American range of the sea otter (adapted from Estes and VanBlaricom 1985).

Abalone species		Range	
		From north	To south
Pinto	*H. kamtschatkana*	SE Alaska	Pt. Sur, California
Flat	*H. walallensis*	British Columbia	La Jolla, California
Red	*H. rufescens*	Sunset Bay, Oregon	Bahia Tortuga, Baja California, Mexico
Black	*H. cracheroidii*	Coos Bay, Oregon	Cabo San Lucas, Baja California, Mexico
Pink	*H. corrugata*	Pt. Conception, California	Bahia Tortuga, Baja California, Mexico
Green	*H. fulgens*	Pt. Conception, California	Bahia Magdalena, Baja California, Mexico
White	*H. sorensenii*	Pt. Conception, California	Bahia Tortuga, Baja California, Mexico
Threaded	*H. assimilis*	Pt. Conception, California	Bahia Tortuga, Baja California, Mexico

ators characterize sustainable commercial abalone fisheries such as those found in Australia (Shepherd et al. 2000).

Sea otter and abalone interactions

California

Sea otters are known to affect the abundance, size, and distribution of abalone in California (McLean 1962; Ebert 1968; Lowry and Pearse 1973; Cooper et al. 1977; Hines and Pearse 1982; Pollard 1992). Fisher (1939) first predicted that competition between the abalone fishery and sea otters was inevitable. In California, abalone found in areas occupied by sea otters are restricted to refugial habitats, such as crevices, which are not accessible to foraging otters and are less abundant and smaller than in similar areas without sea otters (Pollard 1992). Before the commercial fur trade, most abalone were probably cryptic and restricted to habitats inaccessible to foraging sea otters, except perhaps in the intertidal zone, where abalone may have been less vulnerable to otter predation because of daily exposure and because the intertidal zone was close to village sites from which sea otters were hunted (Ebert 1968; Walker 1982).

A series of studies conducted in California in the late 1950s and 1960s confirmed Fisher's earlier predictions. Abalone were reported to comprise between 2 and 63% of the sea otters' diet depending upon the availability of abalone or how long sea otters had resided in the area (Fisher 1939; Boolootian 1965; Hall and Schaller 1964; Ebert 1968; Wild and Ames 1974). As abalone and sea urchins became less abundant, sea otters switched to species such as mussels and sea stars (Ebert 1968). Ebert (1968) found that off Pico Creek, central California, an area that had been recently occupied by sea otters, abalone composed up to 63% of a sea otter's diet. He noted that initially sea otters harvested one abalone per two dives, but within three months otters were diving nine to ten times before surfacing with an abalone. Abalone caught early on were usually brought to the surface intact, whereas abalone caught in the latter period of his observations invariably showed signs of having been beaten with a rock (Ebert 1968).

McLean (1962), working off the central coast of California, first reported that sea otter foraging had reduced the effects of urchin grazing, increased kelp abundance, and restricted abalone and urchins to crevices. The first quantitative studies of the effects of sea otters on abalone came from diving surveys conducted by the California Department of Fish and Game in the Point Estero area of central California. Between 1965 and 1967, red abalone densities ranged from 0.075 to 0.085 m^{-2}. Once sea otters had reoccupied the area, abalone density dropped to about 0.01 m^{-2} (Wild and Ames 1974; Wendell 1994). Abalone density stabilized at about 0.007 m^{-2}, which is similar to densities reported for some regions long occupied by sea otters (Ebert 1968; Wendell 1994), although abalone density in areas occupied by otters varied with habitat type or crevice availability (North 1965). Pollard (1992) reported that areas with refugial habitat (crevices) could support small but stable abalone populations even in the presence of sea otter foraging.

Abalone populations also declined in areas outside of the sea otters' range. These declines were attributed to overharvesting (Tegner et al. 1996) or disease (Alstatt et al. 1996). Wendell (1994) suggested that abalone fishing could be disregarded as the cause of the decline in abalone density in the Point Estero region because the decline occurred over all size-classes and occurred from north to south as sea otters reoccupied the area.

Lowry and Pearse (1973) examined the effects of sea otter foraging on abalone at Hopkin's Marine Reserve in Monterey Bay. Sea otters returned to the area in the mid-1960s. Lowry and Pearse (1973) reported that in the presence of sea otters, nearly all abalone and sea urchins occurred within crevices. However, they also noted that where algal drift was abundant, urchins and abalone often occurred in crevices and acted as passive detritivores instead of grazing in the open. Despite the presence of sea otters in the area for over ten years, Lowry and Pearse (1973) reported abalone densities (*H. rufescens* and *H. walallensis*) of up to 0.21 m^{-2} with a wide size distribution indicating a broad age range. Unfortunately, there were no quantitative estimates of abalone density before the arrival of sea otters (Lowry and Pearse 1973).

Cooper et al. (1977) found that density and size frequency distribution of abalone in the same area had remained con-

Fig. 1. Annual changes in abalone abundance at two permanently marked sites located off the northwest coast of Vancouver Island from 1988 to 1998. Sea otters have occupied the Gull Island areas since about 1972 and the Noname Island area since about 1984. Points are means; bars are ±1 SE (N = 5).

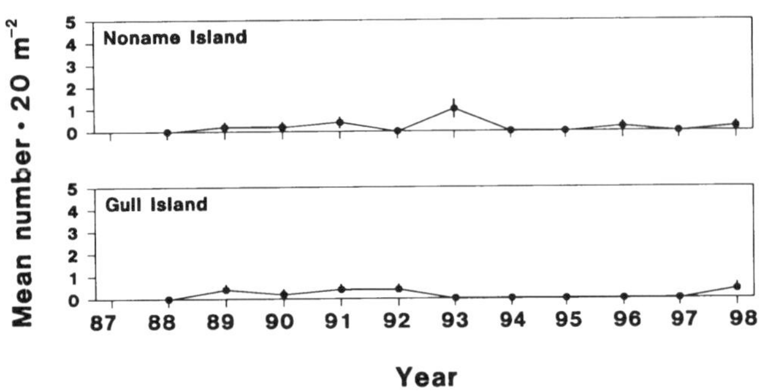

Fig. 2. Annual changes in abalone abundance at two permanently marked sites located off the southwest coast of Vancouver Island from 1987 to 1998. Sea otters have been absent from this area since their extirpation. Points are means; bars are ±1 SE (N = 5).

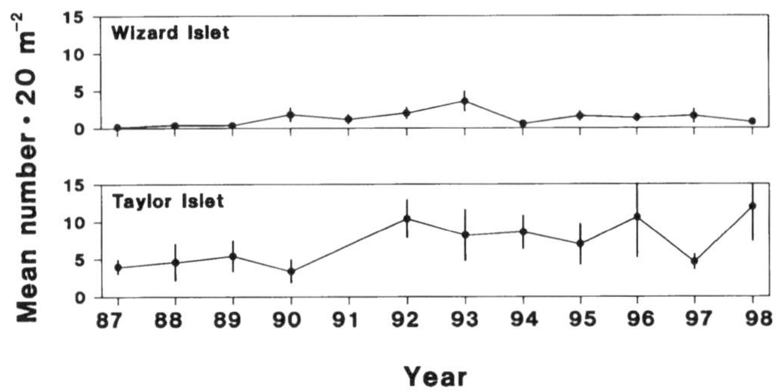

stant during the intervening four-year period. They proposed that the abalone population was stable, having achieved a refuge from sea otter predation. They further suggested that the stable size distribution indicated that recruitment was constant or at least balanced with mortality. Hines and Pearse (1982) mapped and measured live abalone in Hopkin's Marine Reserve. They documented the disappearance of the mapped abalone and collected and measured otter-predated abalone shells. They found that all abalone living in Hopkin's Reserve were in crevices, in densities not significantly different from those reported in earlier studies. They concluded, based on the production of abalone shells, that the crevice population of abalone underwent a complete turnover in three years with rapid growth and recruitment offsetting mortality. They did, however, notice a decline in the abundance of *H. walallensis*. Subsequent unpublished student studies conducted in the late 1980s and early 1990s support the idea that, despite continued predation by sea otters, the abalone population continued to be stable with respect to size and density (J. Pearse, University of California at Santa Cruz, Santa Cruz, CA 95064, personal communication).

The effects of sea otter predation on severely depleted abalone stocks is not known. In southern California, white abalone populations have declined to near extinction and may be at a density below which they can successfully recover on their own (Davis et al. 1998). Likewise black abalone populations have declined precipitously from withering disease (Alstatt et al. 1996). Removal of individuals from these limited populations may threaten the species (Tegner et al. 1996). Sea otter predation may represent a serious threat to abalone populations that have become rare and restricted in distribution (Davis 2000).

Washington State

The effect of sea otters on abalone has not been described in Washington State, although Kvitek et al. (1989) noted the absence of abalone at sites recently occupied by sea otters that they surveyed off the Olympic Peninsula.

British Columbia

In British Columbia, sea otters are known to prey on *H. kamtschatkana*, reducing the density, size, and distribution of abalone, compared to areas without sea otters (Breen et al. 1982; Watson 1993).

The effects of sea otters on nearshore community structure have been studied continuously off northwestern Vancouver Island since 1987 (Watson 1993). The direct effects of sea otter predation on community structure were observed by establishing permanently marked sites at three locations prior to the arrival of sea otters. Changes in algal and invertebrate abundance were documented as sea otters moved into the three sites. The changes associated with the arrival of sea otters were compared to changes at control sites established at two locations without sea otters and at two locations continuously with sea otters. Northern abalone was among the species monitored.

At each site, algal and invertebrate abundance was monitored annually in five 10 × 2 m belt transects established at depths of 8–10 m below MLW. The density of abalone at the two sites continuously occupied by sea otters (Gull and No Name islands) remained consistently below 1 abalone·20 m^{-2} from 1988 to 1998 (Fig. 1). During the same time, abalone abundance at sites without sea otters varied from a mean of 3.6 ± 1.3 (±1 SE) to 0.2 ± 0.2 abalone·20 m^{-2} at Wizard Islet, and 10.6 ± 5.3 to 4.0 ± 0.9 abalone·20 m^{-2} at Taylor Islet (Fig. 2). At all the sites reoccupied by sea otters during the study, abalone abundance was variable among sites before the arrival of sea otters (Fig. 3). At Union Island and Kamils Anchorage, mean abalone abundance was ≤1 abalone·20 m^{-2} both prior to and following the arrival of sea otters in 1988 and 1989, respectively (Fig. 3). In contrast, at Kyuquot Bay, abalone abundance dropped gradually from a mean of 19.4 ± 5.0 abalone·20 m^{-2} in 1988 to <1 abalone·20 m^{-2} in 1995. Although sea otters arrived in Kyuquot Bay in 1988, they initially foraged intermittently and declines in invertebrate prey were gradual (Fig. 3). As sea otters removed invertebrate herbivores from all sites, algal abundance increased making the finding of abalone difficult. Abalone at all five sites with sea otters occurred in crevices or in areas inaccessible to foraging sea otters.

Southeast Alaska

Rosenthal and Barilotti (1973) first reported the effects of sea otter foraging in southeast Alaska. They reported that sea urchins (*S. franciscanus*) made up the major portion of

Fig. 3. Annual changes in abalone abundance, from 1987 to 1998, at three permanently marked sites, located off the northwest coast of Vancouver Island, that were reoccupied by sea otters. Sea otters have reoccupied the Union Island site since 1990, the Kamils Anchorage site since 1988, and Kyuquot Bay since 1988. Points are means; bars are ±1 SE (N = 5).

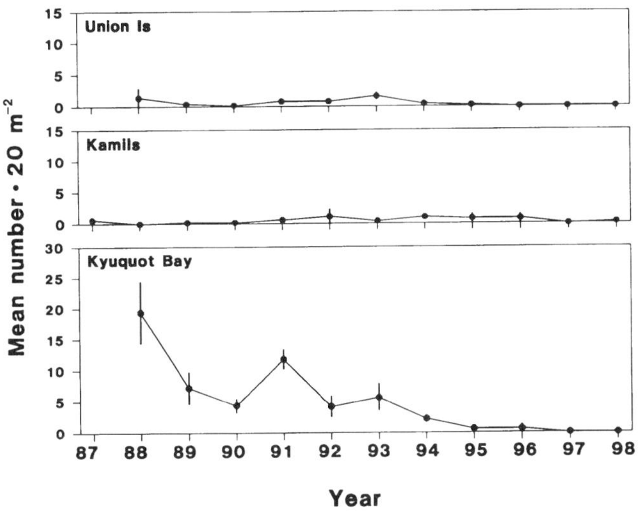

the sea otters' diet but that abalone were also a noticeable component. The effects of sea otters on *H. kamtschatkana* in southeast Alaska will likely be similar to those seen in British Columbia and California (Pitcher 1989). Woodby (2000) predicts that sea otters will limit fishing opportunities for abalone in southeast Alaska.

Ecological effects of sea otter foraging on abalone

The effects that sea otters and human harvesters have on abalone populations differ in several aspects. Although sea otters have no regard for size limits, quotas, or area closures, as abalone become rare, sea otters are energetically constrained and appear to switch to alternate prey (Ebert 1968). Even if human harvesters respect size limits and quotas, as the abundance of abalone declines, abalone becomes more valuable, making it worthwhile to pursue a rare species. Furthermore, human harvesters using specially designed tools have the ability to remove abalone from areas where sea otters cannot. The combination of sea otter predation and intense human harvest or heavy human harvest alone may prove too much for many abalone populations.

Sea otter predation restricts abalone to crevices, concentrating individuals in refugial habitats. Thus, although overall abalone density declines, the remaining abalone occur in localized patches. This aggregated distribution may allow for good mixing of gametes and high fertilization success. Therefore, even in areas where the overall density of abalone is low, abalone may not be sperm limited (*sensu* Levitan and Petersen 1995). The abundance of kelp, created when sea otters remove major invertebrate herbivores, ensures a steady supply of drift algae. Consequently, abalone restricted to crevice habitats by sea otters can remain in such refugia, functioning as passive detritivores.

The surrounding kelp forest may also affect the dispersal of abalone larvae (Pollard 1992). With short-lived, weakly swimming larvae (Mottet 1978), abalone are thought to have short dispersal distances (McShane 1995, 1996). Kelp beds reduce water flow (Jackson and Winant 1983) and are known to entrain larvae (Duggins et al. 1990). As the weakly swimming abalone larvae leave the crevices, they may be entrained by the surrounding kelp and settle within the kelp forest, where patches of coralline algae are maintained by gastropod grazers. The potentially high fertilization success and possibly increased settlement rate may explain the stable abalone population observed at Hopkin's Reserve where constant recruitment appeared to offset mortality (Hines and Pearse 1982). The maintenance of such refugial populations may be dependent on high-relief substrates that offer abalone an escape from sea otter predation (Pollard 1992). Species such as *H. kamtschatkana* with a small size of first reproduction, which increases their crypsis, may further be buffered from sea otter predation.

A long evolutionary co-existence between abalone and sea otters has meant that abalone populations, albeit at low densities, have persisted in the face of intense sea otter predation. When sea otters were extirpated, abalone populations increased. In the absence of sea otters, intense grazing by sea urchins reduced the abundance of fleshy algae and consequently algal drift. In the absence of drift algae, abalone likely switched from passive detritivory to active grazing. Without the limitations of otter predation, abalone were able to grow to large sizes and persisted in open habitats. Without kelp forests, abalone larvae may have dispersed more widely, often to inappropriate habitats, reducing successful recruitment. South of Point Conception, California, other biological factors may regulate urchin populations and dense kelp forests persist even in the absence of sea otters (Tegner and Levin 1983; Cowen 1983). In this environment, abalone may have continued to function as passive detritivores and have had higher recruitment rates because larvae were entrained within kelp beds.

Commercial fisheries for abalone are unlikely to persist in the presence of sea otters. However, in the absence of commercial exploitation, abalone populations have clearly persisted in the presence of sea otters. Limited-access recreational fisheries (Tegner et al. 1992) or very restricted access commercial fisheries (Shepherd et al. 2000), in the absence of natural predation, appear to be sustainable suggesting that intense human harvesting has the greatest impact on abalone populations. The combined pressures of commercial harvesting and sea otter predation may be greater than most abalone populations can support. In areas where human exploitation has reduced abalone populations to critically low levels, sea otter predation may threaten abalone populations (Davis 2000). Although sea otters pose a threat to abalone fisheries and severely reduced abalone populations, the foraging activities of sea otters do not threaten abalone populations. The life-history characteristics of abalone species, when examined in the light of the foraging tactics of sea otters, suggest there is a co-evolutionary relationship between sea otters and abalone. Although sea otter foraging reduces the size and density of abalone, precluding human exploitation, increased kelp abundance and subsequent changes in hydrodynamic regimes may enhance abalone recruitment and allow abalone to remain in crevice refugia where they are not accessible to sea otters. Thus, in the absence of human exploitation, small patchy populations of abalone may occur in areas where sea otters forage (Pollard 1992).

The present limited geographic range of sea otters means that otters cannot be held responsible for the collapse of the abalone fishery in southern California, British Columbia, or southeast Alaska (Sloan and Breen 1988; Davis et al. 1996). In fact, in many areas where harvest bans have been implemented, abalone stocks continue to decline, apparently as a result of increased poaching due to the enhanced market value brought about by severe harvest restrictions (see papers in Campbell 2000*a*).

The conflict between sea otters and abalone harvesters seems inevitable. However, by examining abalone populations in areas exposed to sea otter foraging alone, biologists may be able to identify the life-history characteristics that permit abalone and their predators to co-exist. These insights may be important to fisheries managers attempting to manage abalone fisheries in a sustainable manner.

Acknowledgments

This manuscript was greatly improved by comments from Alan Campbell, Jim Estes, Peter Haaker, and Mia Tegner. Glenn VanBlaricom, John Pearse, and Graeme Ellis also contributed through many stimulating discussions. Support for sea otter research in British Columbia was provided by Fisheries and Oceans Canada, the Friends of the Ecological Reserves, and the Vancouver Aquarium Marine Science Centre.

References

Alstatt, J.M., Ambrose, R.E., Engle, J.M., Haaker, P.L., Lafferty, K.D., and Raimondi, P.T. 1996. Recent declines of black abalone *Haliotis cracheroidii* on the mainland coast of California. Mar. Ecol. Prog. Ser. **142**: 185–192.

Anderson, C.G., Gittleman, J.L., Koepfli, K.P., and Wayne, R.K. 1996. Sea otter systematics and conservation: which are critical subspecies? School of Natural Resources and Environment, The University of Michigan, Ann Arbor, Mich. Endangered Species Update, **13**: 6–10.

Benz, C. 1996. Evaluating attempts to reintroduce sea otters along the California coastline. School of Natural Resources and Environment, The University of Michigan, Ann Arbor, Mich. Endangered Species Update, **13**: 31–35.

Bigg, M.B., and Mackaskie, I.B. 1978. Sea otters re-established in British Columbia. J. Mammal. **59**: 874–876.

Bodkin, J.L. 1986. Fish assemblages in *Macrocystis* and *Nereocystis* forests off central California. U.S. Natl. Mar. Fish. Serv. Fish. Bull. **84**: 799–807.

Bodkin. J.L. 1988. Effects of kelp forest removal on associated fish assemblages in central California. J. Exp. Mar. Biol. Ecol. **117**: 227–238.

Bodkin, J.L., Mulcahy, D., and Lensink, C.J. 1993. Age-specific reproduction in female sea otters (*Enhydra lutris*) from south-central Alaska; analysis of reproductive tracts. Can. J. Zool. **71**: 1811–1815.

Boolootian, R.A. 1965. The sea otters and its effects upon the abalone resource. *In* Third progress report to the legislature 1965. Senate Permanent Factfinding Committee on Natural Resources. Reg. Session. Secretary. 1 Senate of California, Sacramento, Calif. pp. 129–144.

Breen, P.A. 1980. Measuring fishing intensity and annual production in the abalone fishery of British Columbia. Can. Tech. Rep. Fish. Aquat. Sci. 947. 49 p.

Breen, P.A. 1986. Management of the British Columbia fishery for northern abalone (*Haliotis kamtschatkana*). *In* North Pacific Workshop on Stock Assessment and Management of Invertebrates. *Edited by* G.S. Jamieson, and N. Bourne. Can. Spec. Publ. Fish. Aquat. Sci. 92. pp. 300–312.

Breen, P.A., and Adkins, B.E. 1980. Spawning in a British Columbia population of northern abalone, *Haliotis kamtschatkana*. Veliger, **23**: 177–179.

Breen, P.A., Carson, T.A., Foster, J.B., and Stewart, E.A. 1982. Changes in subtidal community structure with British Columbia sea otter transplants. Mar. Ecol. Prog. Ser. **7**: 13–20.

Campbell, A. (*Editor*). 2000*a*. Proceedings of the Workshop on Rebuilding Abalone Stocks in British Columbia. Can. Spec. Publ. Fish. Aquat. Sci. 130. 158 p.

Campbell, A. 2000*b*. Review of northern abalone, *Haliotis kamtschatkana*, stock status in British Columbia. *In* Proceedings of the Workshop on Rebuilding Abalone Stocks in British Columbia. *Edited by* A. Campbell. Can. Spec. Publ. Fish. Aquat. Sci. 130. pp. 41–50.

Cincin-Sain, B., Moore, J.E., and Wyner, A.J. 1977. Management approaches for marine fisheries: the case of the California abalone. California University Institute of Marine Resources Sea Grant Publication No. 54 and IMR 77-101. 233 p.

Cooper, J., Wieland, J., and Hines, A. 1977. Subtidal abalone populations in areas inhabited by sea otters. Veliger, **20**: 163–167.

Costa, D. 1978. The sea otter: its interaction with man. Oceans, **21**: 24–10.

Cowen, R.K. 1983. The effect of the sheephead (*Semicossyphus pulcher*) on red sea urchin (*Strongylocentrotus franciscanus*) populations: an experimental analysis. Oecologia, **58**: 249–255.

Cox, K.W. 1962. California abalones, Family Haliotidae. Calif. Dep. Fish. Game Fish. Bull. 188. 133 p.

Davis, G.E. 2000. Refugia-based strategies to restore and sustain abalone (*Haliotis* spp.) populations in Southern California. *In* Workshop on Rebuilding Abalone Stocks in British Columbia. *Edited by* A. Campbell. Can. Spec. Publ. Fish. Aquat. Sci. 130. pp. 133–138.

Davis, G.E., Haaker, P.L., and Richards, D.V. 1996. Status and trends of white abalone at the California Channel Islands. Trans. Am. Fish. Soc. **125**: 42–48.

Davis, G.E., Haaker, P.L., and Richards, D.V. 1998. The perlious condition of white abalone *Haliotis sorenseni*, Bartsch, 1940. J. Shellfish. Res. **17**: 871–875.

Duggins, D.O., Simenstad, S.A., and Estes J.A. 1989. Magnification of secondary production by kelp detritus in coastal marine ecosystems. Science (Washington, D.C.), **245**: 170–173.

Duggins, D.O. Eckman, J.E., and Sewell, A.T. 1990. Ecology of understory kelp environments II. Effects of kelps on recruitment of benthic invertebrates. J. Exp. Mar. Biol. Ecol. **143**: 27–45.

Ebert, E.E. 1968. A food habits study of the southern sea otter, *Enhydra lutris nereis*. Calif. Fish. Game, **54**: 33–42.

Ebert, E.E., and Houk, J.N. 1984. Elements and innovations in the culture of red abalone, *Haliotis rufescens*. Aquaculture, **39**: 375–392.

Ebert, T.A., and Russell, M.P. 1988. Latitudinal variation in size structure of the west coast purple sea urchin: a correlation with headlands. Limnol. Oceanogr. **33**: 286–294.

Ebert, T.A., Schroeter, S.S., Dixon, J.D., and Kalvas, P. 1994. Settlement patterns of red and purple sea urchins (*Strongylocentrotus franciscanus* and *S. purpuratus*) in California USA. Mar. Ecol. Prog. Ser. **111**: 41–52.

Estes, J.A. 1990. Growth and equilibrium in sea otter populations. J. Anim. Ecol. **59**: 385–401.

Estes, J.A., and Duggins, D.O. 1995. Sea otters and kelp forests in Alaska: generality and variation in a community ecological paradigm. Ecol. Monogr. **65**: 75–100.

Estes, J.A., and Palmisano, J.F. 1974. Sea otters; their role in structuring nearshore communities. Science (Washington, D.C.), **185**: 1058–1060.

Estes, J.A., and Steinberg, P.D. 1988. Predation, herbivory, and kelp evolution. Paleobiology, **14**: 19–36.

Estes, J.A., and VanBlaricom, G.R. 1985. Sea otters and shellfisheries. *In* Marine mammals and fisheries. *Edited by* J.R. Beddington, R.J.H. Beverton, and D.M. Lavigne. George Allen and Unwin, London. pp. 187–235.

Estes, J.A., Jameson, R.J., and Johnson, A.M. 1981. Food selection and some foraging tactics of sea otters. *In* Proceedings of the Worldwide Furbearer Conference. *Edited by* J.A. Chapman and D. Purley. 3–11 August 1980, Frostburg, Md. pp. 606–641.

Estes, J.A., Jameson, R.J., and Rhode, E.B. 1982. Activity and prey selection in the sea otter: influence of population status on community status. Am. Nat. **120**: 242–258.

Estes, J.A., Rathbun, G.B., and Vanblaricom, G.R. 1993. Paradigms for managing carnivores; the case of the sea otter. Symp. Zool. Soc. Lond. **65**: 307–320.

Estes, J.A., Doak, D.F., Bodkin, J.R., Jameson, R.J., Monson, D., Watt, J., and Tinker, M.T. 1996. Comparative demography of sea otter populations. School of Natural Resources and Environment, The University of Michigan, Ann Arbor, Mich. Endangered Species Update, **13**: 11–13.

Estes, J.A., Tinker. M.T., Williams, T.M., and Doak, D.F. 1998. Killer whale predation on sea otters; linking oceanic and nearshore ecosystems. Science (Washington, D.C.), **282**: 369–580.

Fisher, E.M. 1939. Habits of the southern sea otter. J. Mammal. **20**: 21–36.

Foster, M.S., and Schiel, D.R. 1988. Kelp communities and sea otters: keystone species or just a another brick in the wall? *In* The community ecology of sea otters. *Edited by* G.R.VanBlaricom and J.A. Estes. Ecological Studies Vol. 65. Springer Verlag, Heidelberg, Germany. pp. 92–115.

Garshelis, D.L., Johnson, A.M., and Garshelis, J.A. 1984. Social organization of sea otters in Prince William Sound, Alaska. Can. J. Zool. **62**: 2648–2658.

Hall, K.R.L., and Schaller, G.B. 1964. Tool using behavior of the California sea otter. J. Mammal. **45**: 287–298.

Hines, A H., and Pearse, J.S. 1982. Abalones, shells and sea otters: dynamics of prey populations in central California. Ecology, **63**: 1547–1560.

Jackson, G.A., and Winant, C.D. 1983. Effect of a kelp forest on coastal currents. Cont. Shelf. Res. **2**: 75–80.

Jameson, R.J., Kenyon, K.W., Johnson, A.M., and Wright, H.M. 1982. History and status of translocated sea otter populations in North America. Wildl. Soc. Bull. **10**: 100–107.

Jameson, R.J., Kenyon, K.W., Jeffries, S., and VanBlaricom, G.R. 1986. Status of a translocated sea otter population and its habitat in Washington. Murrelet, **67**: 84–87.

Jameson, R.J., and Johnson, A.M. 1993. Reproductive characteristics of female sea otters. Mar. Mamm. Sci. **9**: 156–167.

Johnson, A.M. 1982. Status of Alaska sea otter populations and developing conflicts with fisheries. Trans. N. Am. Wildl. Nat. Res. Conf. Wildlife Management Institute, Wash. D.C. pp. 293–299.

Karpov, K.A., Haaker, P., Taniguchi, I., and Rogers-Bennett, L. 2000. Serial depletion and the collapse of the California abalone (*Haliotis* spp.) fishery. *In* Workshop on Rebuilding Abalone Stocks in British Columbia. *Edited by* A. Campbell. Can. Spec. Publ. Fish. Aquat. Sci. 130. pp. 11–24.

Kenyon, K.W. 1969. The sea otter in the eastern Pacific Ocean. N. Am. Fauna, **68**: 1–352.

Kvitek, R.G., Shull, D., Canestro, D., Bowlby, E.C., and Troutman, B.L. 1989. Sea otter abundance and benthic prey communities in Washington State. Mar. Mamm. Sci. **5**: 266–280.

Levitan, D.R., and Petersen, C. 1995. Sperm limitation in the sea. TREE, **10**: 228–231.

Loughlin, T.R. 1980. Home range and territoriality of sea otters near Monterey, California. J. Wildl. Manage. **44**: 576–582.

Lowry, L.F., and Pearse, J.S. 1973. Abalones and sea urchins in an area inhabited by sea otters. Mar. Biol. **23**: 213–219.

Lubina, J.A., and Levin, S.A. 1988. The spread of a reinvading species: range expansion in the California sea otter. Am. Nat. **131**: 526–543.

Ludwig, D., Hilborn, R., and Walters, C. 1993. Uncertainty, resource exploitation, and conservation, lessons from history. Science (Washington, D.C.), **260**: 36.

McLean, J.H. 1962. Sublittoral ecology of kelp beds of the open coast near Carmel, California. Biol. Bull. **122**: 213–219.

McShane, P.E. 1995. Recruitment variations in abalone: its importance to fishery management. Mar. Freshw. Res. **46**: 555–570.

McShane, P.E. 1996. Recruitment processes in abalone (*Haliotis* spp.). *In* Proceedings of an international workshop on survival strategies in early life history stages of marine resources. *Edited by* Y. Watanabe, Y. Yamashita, and Y. Oozeki. Balkema Publishers, Brookfield, Vermont. pp. 315–325.

McShane, P.E., Black, K.P., and Smith, M.G. 1988. Recruitment processes in *Haliotis rubra* (Mollusca: Gastropoda) and regional hydrodynamics in southeastern Australia imply localized dispersal of larvae. J. Exp. Mar. Biol. Ecol. **124**: 175–203.

Monson, D., and Degange, D.R. 1995. Reproduction, preweaning survival and survival of adult sea otters at Kodiak Island, Alaska. Can. J. Zool. **72**: 1161–1169.

Morse, A.N.C., and Morse, D.E. 1984. Recruitment and metamorphosis of *Haliotis* larvae induced by molecules uniquely available at the surface of crustose red algae. J. Exp. Mar. Biol. Ecol. **75**: 191–215.

Mottet, M. 1978. A review of the fishery biology of abalones. Wash. Dep. Fish. Tech. Rep. 37. 81 p.

Newby, T.C. 1975. A sea otter (*Enhydra lutris*) food dive record. Murrelet, **56**: 19.

North, W.J. 1965. Urchin predation. *In* Kelp habitat improvement project. Annual Report, 1 Feb – 31 March 1965. California Institute of Technology, Pasadena, Calif. pp. 57–61.

Ogden, A. 1941. The California sea otter trade 1784–1848. University of California Press, Berkeley and Los Angeles, Calif. 251 p.

Pitcher, K.W. 1989. Studies of southeastern Alaska sea otter populations: distribution, abundance, structure, range expansion, and potential conflicts with shellfisheries. U.S. Fish Wildl. Coop. Agree. No. 14-16-0009-954.

Pollard, S. 1992. Red abalone, *Haliotis rufescens*; relative impacts of recreational fisheries and sea otter predation on the abundance, size frequency, and microhabitat distribution of red abalone populations in central and northern California. M.Sc. thesis, University of California Santa Cruz, Calif.

Prince, J.D., Sellers, T.L., Ford, W.B., and Talbot, S.R. 1987. Experimental evidence for limited dispersal of haliotid larvae (genus *Haliotis*; Mollusca: Gastropoda). J. Exp. Mar. Biol. Ecol. **106**: 243–263.

Prince, J.D., Sellers, T.L., Ford, W.B., and Talbot, S.R. 1988. Confirmation of a relationship between the localized abundance of breeding stock and recruitment for *Haliotis rubra* Leach (Mollusca: Gastropoda). J. Exp. Mar. Biol. Ecol. **122**: 91–104.

Riedman, M.L., and Estes, J.A. 1990. The sea otter (*Enhydra lutris*): behavior, ecology, and natural history. U.S. Fish. Wildl. Serv. Biol. Rep. 90. 126 p.

Rosenthal, R.J., and Barilotti, D.C. 1973. Feeding behaviour of transplanted sea otters and community interactions of Chichagoff Island, southeast Alaska. *In* Kelp habitat improvement project. Annual Report, 1 July 1972 – 30 June 1973. California Institute of Technology, Pasadena, Calif. pp. 74–88.

Sasaki, R., and Shepherd, S.A. 1995. Larval dispersal and recruitment of *H. discus* and *Tegula* spp. on Miyagi coasts Japan. Mar. Freshw. Res. **56**: 519–29.

Shepherd, S.A., and Turner, J.A. 1985. Studies on southern Australian abalone (genus *Haliotis*) VI. Habitat preference, abundance, and predators of juveniles. J. Exp. Mar. Biol. Ecol. **93**: 285–298.

Shepherd, S.A., Lowe, D., and Partington, D. 1992. Studies on southern Australian abalone (genus *Haliotis*) XIII: larval dispersal and recruitment. J. Exp. Mar. Biol. Ecol. **164**: 247–260.

Shepherd, S.A., Preece, P.A., and White, R.W.G. 2000. Tired nature's sweet restorer? Ecology of abalone stock enhancement in Australia. *In* Workshop on Rebuilding Abalone Stocks in British Columbia. *Edited by* A. Campbell. Can. Spec. Publ. Fish. Aquat. Sci. 130. pp. 84–97.

Simenstad, C.A., Estes, J.A., and Kenyon, K.W. 1978. Aleuts, sea otters and alternate stable state communities. Science (Washington, D.C.), **200**: 403–411.

Sloan, N.A., and Breen, P.A. 1988. Northern abalone, *Haliotis kamtschatkana*, in British Columbia: fisheries synopsis of life history information. Can. Spec. Publ. Fish. Aquat. Sci. 103. 46 p.

Stewart, H. 1977. Indian fishing. Early methods on the northwest coast. Douglas MacIntyre, Vancouver/Toronto. University of Washington Press, Seattle, Wash. 181 p.

Tegner, M.J. 1989. The California abalone fishery: production, ecological interactions and prospects for the future. *In* Scientific approaches to management of shellfish resources. *Edited by* J.F. Caddy. John Wiley & Sons, Inc., New York. pp. 401–420.

Tegner, M.J. 1993. Southern California abalones: can stocks be rebuilt using marine harvest refugia? Can. J. Fish. Aquat. Sci. **50**: 2010–2018.

Tegner, M.J., and Butler, R.A. 1985*a*. Drift tube study of the dispersal potential of green abalone (*Haliotis fulgens*) larvae in the southern California Bight: implications for recovery of depleted populations. Mar. Ecol. Prog. Ser. **26**: 73–84.

Tegner, M.J., and Butler, R.A. 1985*b*. The survival and mortality of seeded and native red abalones, *Haliotis rufescens*, on the Palos Verdes Peninsula. Calif. Fish. Game, **71**: 150–163.

Tegner, M.J., and Levin, L.A. 1983. Spiny lobsters and sea urchins: an analysis of a predator prey interaction. J. Exp. Mar. Biol. Ecol. **73**: 125–150.

Tegner, M., DeMartini, J., and Karpov, K. 1992. The California red abalone fishery: a case study in complexity. *In* Abalone of the world: biology, fisheries and culture. *Edited by* S.A. Shepherd, M.J. Tegner, and S. Guzmán del Próo. Fishing News Books, Blackwell Science Ltd., Oxford, U.K. pp. 370–383.

Tegner, M.J., Basch L.V., and Dayton, P.K. 1996. Near extinction of an exploited marine invertebrate. TREE, **11**: 271–310.

Thomas, N.J., and Cole, R.A. 1996. The risk of disease and threats to the wild population. School of Natural Resources and Environment, The University of Michigan, Ann Arbor, Mich. Endangered Species Update, **13**: 12:23–27.

VanBlaricom, G.R. 1996. Saving the sea otter population in California: contemporary problems and future pitfalls. School of Natural Resources and Environment, The University of Michigan, Ann Arbor, Mich. Endangered Species Update, **13**: 12:85–91.

VanBlaricom, G.R., Ruediger, J.L., Friendman, C.S., Woodward, D.D., and Hedrick, R.P. 1993. Discovery of withering syndrome among black abalone *Haliotis cracheroidii* Leach 1814 populations at San Nicolas Island, California. J. Shellfish Res. **12**: 165–188.

Walker, P.L. 1982. California Indians, sea otters and shellfish: the prehistoric record. *In* Social science perspectives on managing conflicts between marine mammals and fisheries. *Edited by* B. Cicin-Sain, P.M. Grifman, and J.B. Richards. Marine Policy Program, Marine Science Institute University of California at Santa Barbara and the University of California Cooperative Extension. pp. 17–31.

Watson, J.C. 1993. The effects of sea otter (*Enhydra lutris*) foraging on shallow rocky communities off northwestern Vancouver Island, British Columbia. Ph.D. dissertation, University of California at Santa Cruz. 169 p.

Watson, J.C., and Smith, T.G. 1996. The effects of sea otters on invertebrate fisheries in British Columbia: a review. *In* Invertebrate working papers. Reviewed by the Pacific Stock Assessment Review Committee (PSARC) in 1993 and 1994. *Edited by* C.M. Hand and B.J. Waddell. Can. Tech. Rep. Fish. Aquat. Sci. 2089. pp. 262–303.

Watson, J.C., Ellis, G.E., Smith, T.G., and Ford, J.K.B. 1996. Updated status of the sea otter, *Enhydra lutris* in Canada. Can. Field-Nat. **111**: 2:277–286.

Wendell, F.E. 1994. Relationship between sea otter range expansion and red abalone abundance and size distribution in central California. Calif. Fish. Game, **80**: 2:45–56.

Wendell, F.E., Hardy, R.A., Ames, J.A., and Burge, R.T. 1986. Temporal and spatial patterns in sea otter, *Enhydra lutris*, range expansion in the loss of the Pismo clam fisheries. Calif. Fish. Game, **72**: 197–212.

Wild, P.W., and Ames, J.A. 1974. A report on the sea otter. *Enhydra lutris,* L. in California. Calif. Dep. Fish Game Mar. Res. Tech. Rep. **20**: 1–93.

Wilson, D.E., Bogan, M.A., Brownell, R.A., Burdin, A.M., and Maminov, M.K. 1991. Geographic variation in sea otters *Enhydra lutris*. J. Mammal. **72**: 22–36.

Woodby, D., Larson, R., and Rumble, J. 2000. Decline of the Alaska abalone (*Haliotis* spp.) fishery and prospects for rebuilding. *In* Workshop on Rebuilding Abalone Stocks in British Columbia. *Edited by* A. Campbell. Can. Spec. Publ. Fish. Aquat. Sci. 130. pp. 25–31.

Refugia-based strategies to restore and sustain abalone (*Haliotis* spp.) populations in Southern California

Gary E. Davis

Abstract: Modern fishery management strategies, based on species biology and monitored with landings data, failed to sustain either exploited populations of abalone, *Haliotis* spp., or fishery landings in southern California. California law and policy require that the State restore depleted populations and explore new management strategies to sustain abalones and the fisheries they could produce. Restoration strategies will vary by species and by the causes of depletion. Fishing depleted white abalone, *H. sorenseni*, to critically low densities (~100 · ha^{-1}) in the 1970s. By the 1990s, natural mortality had further reduced the population density to 1 · ha^{-1}. Captive breeding and rearing, with adult outplanting, may be the only way to restore white abalone with sufficient certainty to avoid extinction. Fishing also depleted pink abalone, *H. corrugata*, but sufficient numbers of adults exist to artificially aggregate critical spawning densities (2000–4000 · ha^{-1}) in the wild with little risk of extinction at this time. A withering syndrome of uncertain etiology reduced isolated remnant populations of black abalone, *H. cracherodii*, to extremely low densities. Restoration of this species will require finding and identifying disease-resistant survivors to use for captive breeding and rearing. Persistence of red abalone, *H. rufescens*, in northern California and experimental success of refugia created by translocating adult pink abalone and green abalone, *H. fulgens*, in southern California, indicate that refugia-based management could sustain exploited populations and fisheries. Refugia-based strategies are being explored to restore and sustain southern California pink and white abalone populations.

Résumé : Les stratégies modernes de gestion des pêches, qui sont fondées sur la biologie des espèces et dont les résultats sont surveillés à l'aide des données de débarquements, n'ont pas réussi à assurer le maintien des populations exploitées d'ormeaux, *Haliotis* spp., ou des débarquements dans le sud de la Californie. En vertu de sa législation et de ses politiques, l'État de la Californie doit reconstituer les populations appauvries et élaborer de nouvelles stratégies de gestion visant à assurer le maintien des populations d'ormeaux et des pêches que cette ressource pourrait susciter. Les stratégies de reconstitution des populations varieront selon les espèces et les causes de déplétion. La pêche a fait chuter les densités d'ormeau blanc, *Haliotis sorenseni*, à des niveaux critiques (environ 100 individus par ha) dans les années 1970. Vers les années 1990, la densité des populations n'était plus que de 1 ormeau par ha en raison de la mortalité naturelle. La reproduction et l'élevage en captivité, suivis de transplantations d'adultes dans le milieu naturel, pourrait être le seul moyen suffisamment sûr de reconstituer les populations d'ormeau blanc pour éviter l'extinction de l'espèce. La pêche a aussi fait chuter les populations d'ormeau jaune, *H. corrugata*, mais les adultes sont assez nombreux pour qu'on puisse constituer artificiellement des densités de reproducteurs suffisantes (2000–4000 individus par ha) dans le milieu naturel avec peu de risque d'extinction pour le moment. Un syndrome de dépérissement de cause inconnue a réduit à des densités extrêmement faibles des populations vestigiales isolées d'ormeau noir, *H. cracherodii*. Pour reconstituer les populations de cette espèce, on devra trouver et identifier des survivants résistants à la maladie pour en assurer la reproduction et l'élevage en captivité. La persistance de l'ormeau rouge, *H. rufescens*, dans le nord de la Californie et le succès de refuges expérimentaux qu'on a établis en déplaçant des ormeaux jaunes et des ormeaux verts, *H. fulgens*, adultes dans le sud de la Californie, montrent qu'une gestion utilisant des refuges pourrait assurer le maintien des populations exploitées et des pêches. On examine diverses stratégies fondées sur l'utilisation de refuges qui pourraient permettre de reconstituer les populations d'ormeaux jaunes et d'ormeaux blancs du sud de la Californie et d'en assurer la durabilité.

[Traduit par la Rédaction]

G.E. Davis. U.S. National Park Service, 1901 Spinnaker Dr., Ventura, CA 93001-4354, U.S.A.
e-mail: gary_davis@nps.gov
Correct citation: Davis, G.E. 2000. Refugia-based strategies to restore and sustain abalone populations in Southern California. *In* Workshop on Rebuilding Abalone Stocks in British Columbia. *Edited by* A. Campbell. Can. Spec. Publ. Fish. Aquat. Sci. 130. pp. 133–138.

Introduction

Southern California's abalone populations were once mainstays of large, popular, and valuable recreational and commercial fisheries. They are no more. The white abalone, *Haliotis sorenseni*, is poised on the brink of extinction (Davis et al. 1998). Black abalone, *H. cracherodii*, appear similarly threatened, but for different reasons (Richards and

Davis 1993; Altstatt et al. 1996). The status of green abalone, *H. fulgens*, is tenuous at best, while only a few remnant aggregations of pink, *H. corrugata*, and red, *H. rufescens*, abalone survive at remote islands (Karpov et al. 2000). By 1998, all abalone fishing in California south of San Francisco was prohibited to protect the small remnant populations of once abundant, widespread species. The refugium-based recreational red abalone fishery north of San Francisco was the only viable abalone fishery in the state (Karpov et al. 1998). Whether these species will survive and support fisheries again in the rest of California is still an open question.

The State of California now faces a series of risky decisions and expensive actions. Depleted populations of these five species must be rebuilt and the disturbed ecosystems in which they resided must be restored before the fisheries can be restarted on a sustainable basis so people can once again enjoy the benefits of these public resources. California law and policy clearly reflect public willingness and desire both to restore depleted marine resources and to sustain them after restoration (Wilson and Wheeler 1997).

It is important to understand and acknowledge that legal, scientifically managed, fishing contributed to the collapse of these exploited populations while unexploited sympatric species exposed to the same environmental factors, such as pollution and El Niño events, did not collapse. Understanding what went wrong should facilitate restoration efforts and help avoid the same kinds of conservation failures and population collapses in the future.

Southern California abalone fishery management was an experiment conducted without benefit of controls to determine its efficacy. Traditional management measures (size, season, bag, and entry limits) and enhancement by outplanting failed to sustain the populations of these five species in southern California (Karpov et al. 2000; Tegner 2000). Learning from that experience and applying the knowledge to future endeavors is the essence of scientific, or adaptive, management. Objective analysis of this conservation failure may help move public attitudes past the denial that fishery management policy failed. Such analysis may also focus attention on testing, acceptance, and commitment to new, sustainable management practices.

Restoration strategies for California abalones need to vary by species because different factors caused the collapses and surviving populations were left in differing circumstances. Species now as rare as white abalone will probably require more expensive and invasive actions than those needed to rebuild pink or red abalone, for which a few aggregations still persist. Species left with larger aggregations of large individuals distributed widely throughout the species range when the fishery was closed appear to have a better chance of survival and restoration than those reduced to a few solitary individuals in a few isolated locations.

Ecosystem-based restoration appears to be a promising strategy both to rebuild depleted populations (Tegner 1993) and to sustain them in the future (Karpov et al. 1998). Abalones are interdependent parts of biological communities: they are not separate and apart from the sea urchins, kelp, lobsters, and fishes in their environment. Advances in ecology during the past 40 years have revealed the critical importance of competition and predation in sustaining biological communities (Paine 1966; Dayton and Tegner 1984). In the 1940s and 1950s, scientists and fishery managers hypothesized that selected species could be reduced without substantially affecting the rest of the ecosystem, but failed to rigorously test the concept. Modern ecology has falsified that hypothesis, as ecologists unraveled the complex interconnections of kelp forest and rocky reef ecosystems identifying myriad obligate relationships among species (Dayton et al. 1998). Early ignorance of the underlying principles of ecology and the critical importance of maintaining high abalone population density virtually doomed the traditional fishery management schemes used to regulate southern California abalone fisheries. We now know that species can not be effectively managed in isolation from other species or their environment and that we must seek ecologically based solutions for sustainability. The objectives of this paper are to describe the conditions of remnant abalone populations in southern California and to explore potential restoration and sustainable management strategies for three of these species.

White abalone restoration

Legal commercial fishing removed more than 280 tonnes of white abalone from the southern California population during the 1970s, while recreational divers removed an unknown additional amount. This reduced adult densities from 2 000–10 000 ha^{-1} in the 1970s to fewer than $100 \cdot ha^{-1}$ in the 1980s (Fig. 1) (Davis et al. 1996). That population density was below the critical minimum population density needed for successful reproduction and by the mid-1990s an average of one white abalone was found on each hectare of suitable habitat (Davis et al. 1998). The absence of large aggregations of white abalone shells on historic white abalone reefs, such as those observed near black abalone aggregations in the 1980s (Richards and Davis 1993), indicated that mass mortality in situ was not a cause of white abalone population collapse. The latest significant cohort of white abalone in southern California appears to have been spawned in the late 1960s. The virtual absence of small white abalone in the current population indicated that the late 1960s cohort would expire without replacement between 2005 and 2009 when the cohort reached the maximum age of 40 years (Tutschulte 1976). The paucity of fishery-independent information and fishery managers' apparent ignorance of the available fishery-dependent data obscured the population collapse and no warning was issued before the population reached critically low levels and densities. Without unprecedented human intervention to reverse the impact of earlier neglect, survival of this species is unlikely.

White abalone is the deepest-living of the California abalones. They are generally found at depths of 25–30 m and the center of their distribution in the 1970s was the California Channel Islands (Cox 1960; Tutschulte 1976). Beyond the range of normal SCUBA operations, restoration will require extensive submersible surveys at remote islands to find and collect brood stock. During seven days, Davis et al. (1998) found only nine live white abalone while searching 77 050 m^2 of historic white abalone habitat for 37 h at the Channel Islands with the research submersible DELTA.

Fig. 1. White abalone commercial landings (bars) and mean population densities (squares and line) in California 1965–1997 (after Davis et al. 1998).

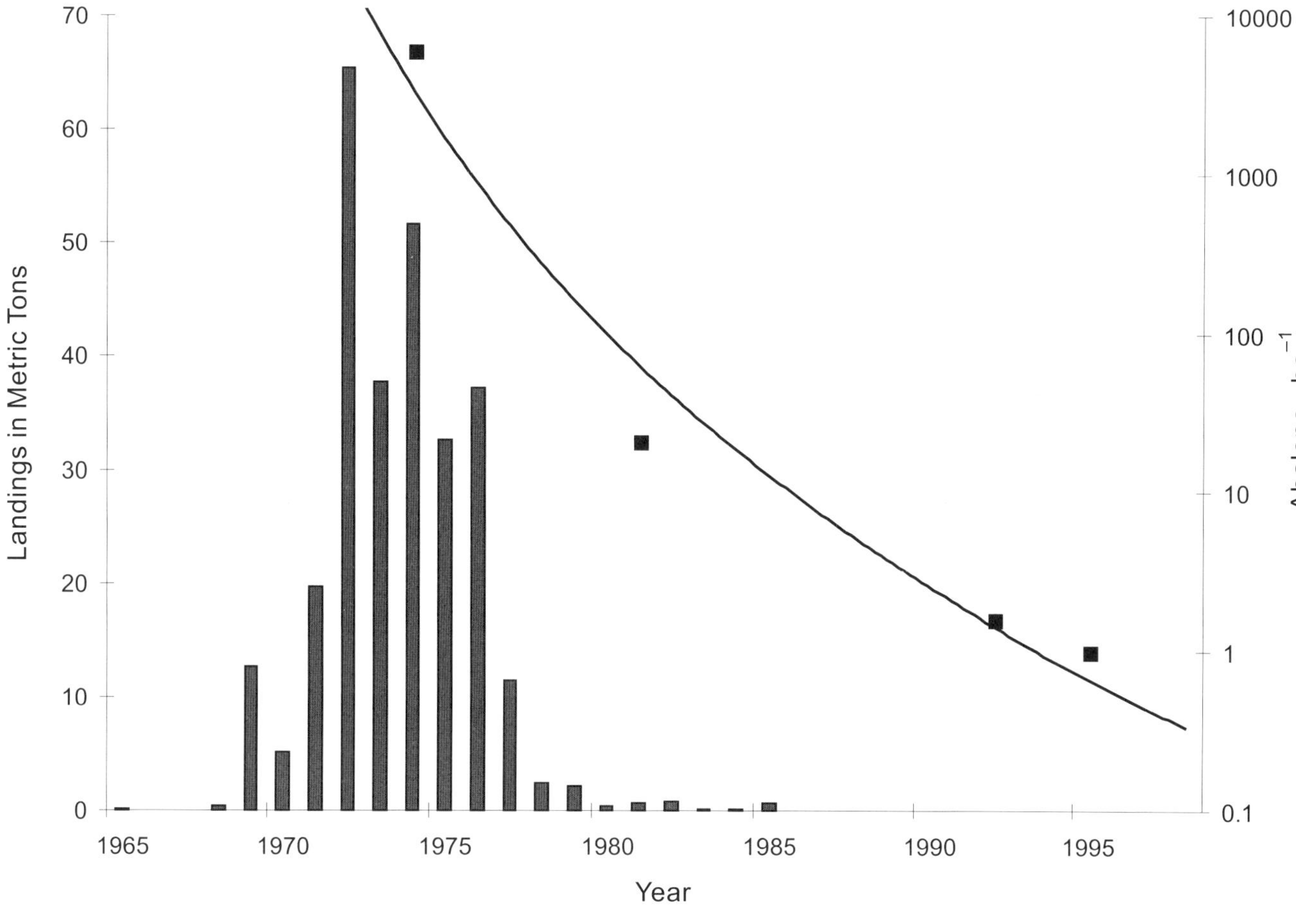

Population estimates indicated that fewer than 2000 white abalone existed in 1997 (Davis et al. 1998).

Two options for producing another generation of white abalone are apparent: (*i*) concentrate critical aggregations of wild adults on refugia reefs; and (*ii*) bring wild adults into culture facilities, spawn them, raise the next generation in captivity, and reintroduce them into the wild as adults. If sufficient numbers of survivors can be found and collected, these options could be exercised simultaneously, but both options will require collection of a large proportion of the wild population. The first option, wild husbandry, will require protecting refugia populations from poaching, monitoring their condition, and waiting hopefully for success. The second option, captive rearing, is more certain of producing another generation of white abalone, but still faces the difficulty of establishing cultured populations in refugia in the wild, something not yet accomplished in spite of vigorous attempts around the world (Tegner 2000). In addition, the minimum size of founding populations for sustainable wild abalone populations remains unknown. Shepherd and Brown (1993) estimated the minimum viable population size for wild abalone, *H. laevicauda,* is several thousand individuals.

Gathering a brood stock for multiple restoration efforts in culture facilities and in the wild will be expensive. For example, at an average cost of US$8000 per day and with the above average density, it will cost US$1.3 million for 154 days of submersible time and ship support to collect a brood stock of 200 white abalone. Several hundred adults will be required for each culturing facility or refugia reef to maintain genetic diversity of subsequent cohorts. Several independent culture facilities need to be used to reduce the risk of catastrophic loss from disease infestation, toxic chemical contamination, and (or) equipment failure.

Current proposals call for out-planting 1000 five-year-old adults (>10 cm shell length, SL) at each of 10 refugia every year for 10 years (Davis et al. 1998), which will cost more than US$10 million. Studies of natural recruitment and growth rates (Tutschulte and Connell 1981, 1988) indicated that an additional 10–15 years would be required for out-planted brood stock to repopulate surrounding areas with wild adult progeny.

Pink abalone restoration

Legal fishing reduced adult pink abalone populations below critical densities of $1000 \cdot ha^{-1}$ in the 1980s (Parker et al. 1992; Davis et al. 1992). Egg-per-recruit analysis (Tegner et al. 1989) failed to predict the impending population collapse, probably because of unpredictable stochastic influences and Allee effects in the wild. As the adult population declined, young-of-the-year (≤31 mm SL) recruitment became erratic and then failed at the northern edge of its range, i.e., the northern Channel Islands (Davis 1995). Unlike the white abalone situation, fishery-independent data warned of

Fig. 2. Pink abalone population densities in fished areas (dots and solid line) and in a 15-ha fishing-free reserve (triangles and dashed line) at Anacapa Island, Channel Islands National Park, California, 1983–1998.

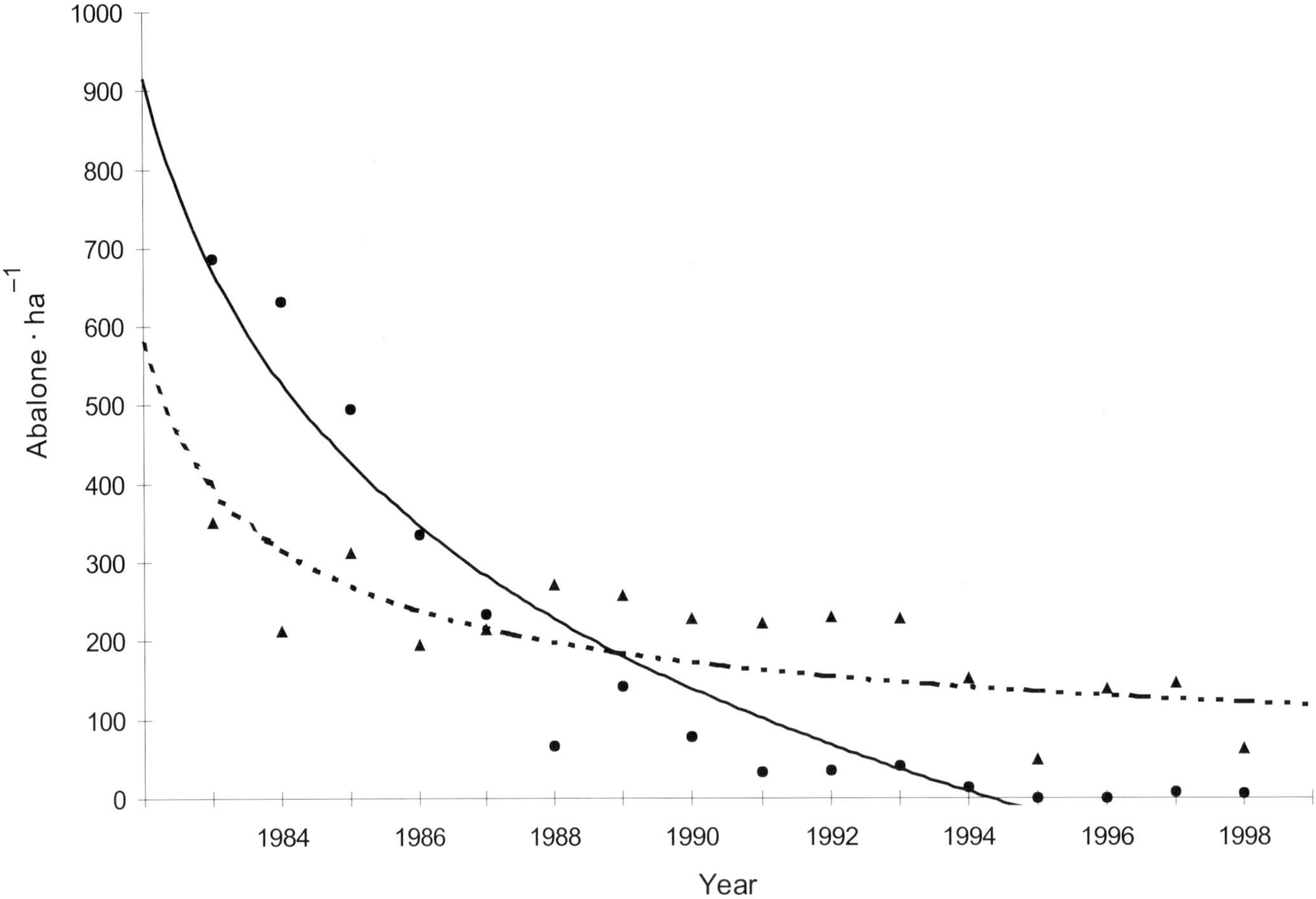

pink abalone population collapse (Davis et al. 1992). Eventually the reduced landings and declining number of areas that produced landings were interpreted as further evidence of unsustainability (Parker et al. 1992; Anonymous 1995). California closed both sport and commercial pink abalone fisheries in 1996, when some adult aggregations still occurred at remote islands near the geographic center of the species range. Nevertheless, adult densities fell below naturally sustainable densities and continued to dwindle even in the 15-ha fishing-free Anacapa Island Ecological Reserve (Fig. 2). For populations allowed to decline below critical densities, protection from fishing mortality alone was not sufficient to prevent collapse and reproductive failure.

Although population densities may be too low for effective reproduction, if aggregated, enough pink abalone may still exist in the wild to restore the southern California population without resorting to culturing in artificial environments. Simply waiting for natural recruitment from the sparsely distributed survivors to rebuild and spread the species throughout its historic range appeared futile, based on 20+ years of experience along the mainland coast in southern California (Tegner 1993).

Restoration could be accelerated by decades if adults were concentrated into refugia in the current range of the species along the mainland coast and northern Channel Islands. The National Park Service and the California Department of Fish and Game established an experimental pink abalone refugium in Channel Islands National Park in 1995 to explore this restoration strategy. Adding 600 adults collected from surrounding reefs to a small (0.18 ha) kelp forest restored historic adult densities ($3600 \cdot ha^{-1}$). The translocation disrupted spawning during the 1995–96 season and a major El Niño event reduced food and generally disrupted the 1996–97 season. Marginal recruitment, measured against a background of none for several years, appeared to have occurred in 1998. To conclude this effort has been successful is premature, but the adult density was still in excess of $2000 \cdot ha^{-1}$ in 1998 and estimated to be falling at about 15% per year. Translocations in replicate refugia and many years of monitoring adult demographics and juvenile recruitment will be needed before this strategy is adequately tested.

Black abalone restoration

Black abalone present yet another restoration situation. Remnants of a once widespread, high-density (commonly $>100 \cdot m^{-2}$) population appeared to flourish on offshore islands and isolated mainland sites north of Santa Barbara in the 1970s (Duros 1987; Parker et al. 1992). A fatal disease of unknown etiology called withering syndrome (Haaker et al. 1992) rapidly reduced most populations to <1% of former densities, beginning in 1985 (Altstatt et al. 1996). As adult densities fell, juvenile recruitment became erratic and then failed when adult densities became less than 50% of former levels (15–20 $\cdot$ m^{-2}) (Richards and Davis 1993). This recruitment failure was similar to the stock-recruitment relationship described for *H. laevicauda* in South Australia (Shepherd and Brown 1993). The few *H. cracherodii* survivors tended to be large individuals and were apparently resistant to withering syndrome, an observation also confirmed in laboratories (Carolyn Friedman, California Fish & Game, P.O. Box 247, Bodega Bay, CA 94923, personal communication). Fishery independent data near the epicenter of the epidemic gave an early warning of population collapse and California closed the fisheries in 1993 statewide to protect *H. cracherodii* survivors in areas remote from the disease outbreak.

Survivor densities are exceptionally low in many areas that recently supported huge numbers of black abalone. Thirty-minute searches of prime habitat (several hundred square metres) now yield an occasional single individual at sites in Channel Islands National Park that contained mean black abalone densities of 27–74 $\cdot$ m^{-2} in the early 1980s (Richards and Davis 1993; Daniel Richards, Channel Islands National Park, 1901 Spinnaker Drive, Ventura, CA 93001, personal communication). Waiting for natural recovery of such a severely depleted population seems hopeless, given other experience with depleted abalone populations in southern California (Tegner 1993, 2000).

Identifying the cause(s) of withering syndrome would be helpful in guiding black abalone restoration but may not be essential to complete before restoration begins. Given current rudimentary knowledge of marine invertebrate pathogens, many years may be required to identify the infectious agents responsible for withering syndrome in black abalone. However, when disease-resistant brood stock can be identified from symptomology, captive breeding and rearing of the next generation could begin restoring black abalone to once productive areas along the mainland and remote islands. Just as with white abalone, reintroducing a laboratory-raised generation to the wild faces considerable uncertainty. If sufficient numbers of disease-resistant individuals can be located in the wild, they could be concentrated in refugia. Alternatively, experimental translocations of wild adults into sites historically occupied by black abalone, such as San Diego County, where black abalone were extirpated before withering syndrome appeared in the 1980s, could escape the disease and restore a viable population.

Discussion

Depleted abalone species will each require a slightly different strategy for population rebuilding. If white abalone brood stock can be found, captive spawning and rearing to adulthood, with later re-establishment of brood stocks in refugia, appears to be the most promising option. Concentrating wild adults in refugia and monitoring recruitment may also yield good results as it did with green abalone at Palos Verdes (Tegner 1993), but I think the number of white abalone brood stock left in the wild is too small to recommend leaving them in the ocean. Pink abalone appear to be numerous enough to justify concentrating wild brood stock from low-density scattered populations in refugia and monitoring recruitment, without captive spawning and rearing. Characterizing the cause of withering syndrome in black abalone and identifying and finding sufficient disease-resistant brood are the highest priority restoration actions for this species. Population monitoring is needed to allow necessary adjustments in any strategy.

A few guiding principles for abalone restoration emerge from these case studies. In contrast to fishery-dependent data that chronicle past conditions, fishery-independent population monitoring can provide essential early warnings of impending population collapses and the subsequent cascading ecosystem dysfunction (Dayton et al. 1998). By identifying problems early when they can be addressed more easily, fishery-independent monitoring that includes other key ecosystem species can also reduce costs and increase certainty of sustaining populations and restoration success (Davis 2000). The true costs of conservation failure are frequently discounted when fishery allocation decisions are made (Safina 1997). There seems to be little accountability for public expenses incurred to rebuild depleted populations and for the impacts of social and ecological disruption, due to neglect of the precautionary approach in managing exploitation of vulnerable species, such as California abalones. How the burdens of the additional public expense of periodic population rebuilding and the risks of extinction from the current failed strategy are borne will set an important precedent. It may also influence how the precautionary principle is applied to management of newer fisheries that have replaced abalone in California, such as the red sea urchin, *Strongylocentrotus franciscanus*. The successful refugia-based red abalone fishery in northern California (Karpov et al.1998), the Palos Verdes translocation experiment with green abalone (Tegner 1993), and preliminary results of the experimental pink abalone refugia reported here, all indicate that management strategies that reserve portions of populations in ecologically sustainable refugia would cost less, cause less social disruption, and conserve resources better than the traditional species-based strategies employed in southern California.

References

Altstatt, J.M., Ambrose, R.F., Engle, J.M., Haaker, P.L., Lafferty, K.D., and Raimondi, P.T. 1996. Recent declines of black abalone *Haliotis cracherodii* on the mainland coast of central California. Mar. Ecol. Prog. Ser. **142:** 185–192.

Anonymous. 1995. Pink, green, and white abalone fishery closure. California Resources Agency, Department of Fish and Game, Environmental Document. 1416 Ninth Street, Suite 1311, Sacramento, Calif. 95814.

Cox, K.W. 1960. Review of the abalone of California. Calif. Fish Game, **46**: 381–406.

Davis, G.E. 1995. Recruitment of juvenile abalone (*Haliotis* spp.) measured in artificial habitats. Mar. Freshw. Res. **46**: 549–554.

Davis, G.E. 2000. General ecological monitoring program design, implementation, and applications: a cases study from Channel Islands National Park, California. *In* Measuring and monitoring biodiversity for conservation science and adaptive management. *Edited by* J.K. Reaser and F. Dallmeier. Smithsonian Institution, Washington, DC. (In press).

Davis, G.E., Haaker, P.L., and Richards, D.V. 1996. Status and trends of white abalone at the California Channel Islands. Trans. Am. Fish. Soc. **125**: 42–48.

Davis, G.E., Haaker, P.L., and Richards, D.V. 1998. The perilous condition of white abalone, *Haliotis sorenseni* Bartsch 1940. J. Shellfish Res. **17:** 871–876.

Davis, G.E., Richards, D.V., Haaker, P.L., and Parker, D.O. 1992. Abalone population declines and fishery management in southern California. *In* Abalone of the world: biology, fisheries, and culture. *Edited by* S.A. Shepherd, M.J. Tegner, and S.A. Guzmán del Próo. Fishing News Books, Blackwell Science Ltd., Oxford, U.K. pp. 237–249.

Dayton, P.K., and Tegner, M.J. 1984. The importance of scale in community ecology: a kelp forest example with terrestrial analogs. *In* Novel approaches to interactive systems. *Edited by* P.W. Price, C.N. Slobodchikoff, and W.S. Gaud. John Wiley & Sons, Inc., New York. pp. 457–481.

Dayton, P.K., Tegner, M.J., Edwards, P.B., and Riser, K.L. 1998. Siding baselines, ghosts, and reduced expectations in kelp forest communities. Ecol. Appl. **8:** 309–322.

Duros, W.J. 1987. Stacking behavior of an intertidal abalone: an adaptive response or a consequence of space limitation? J. Exp. Mar. Biol. Ecol. **108**: 1–14.

Haaker, P.L., Richards, D.V., Friedman, C.S., Davis, G.E., and Parker, D.O. 1992. Mass mortality and withering syndrome in black abalone, *Haliotis cracherodii*, in California. *In* Abalone of the world: biology, fisheries, and culture. *Edited by* S.A. Shepherd, M.J. Tegner, and S.A. Guzmán del Próo. Fishing News Books, Blackwell Science Ltd., Oxford, U.K. pp. 214–224.

Karpov, K.A., Haaker, P.L., Albin, D., Taniguichi, I.K., and Kushner, D. 1998. The red abalone, *Haliotis rufescens*, in California: importance of a depth refuge to abalone management. J. Shellfish Res. **17:** 863–870.

Karpov, K.A., Haaker P.L., Taniguchi, I., and Rogers-Bennett, L. 2000. Serial depletion and collapse of the California abalone (*Haliotis* spp.) fishery. *In* Workshop on Rebuilding Abalone Stocks in British Columbia. *Edited by* A. Campbell. Can. Spec. Publ. Fish. Aquat. Sci. 130. pp. 11–24.

Paine, R.T. 1966. Food web diversity and species diversity. Am. Nat. **100**: 65–75.

Parker, D.O., Haaker, P.L., and Togstad, H.A. 1992. Case histories for three species of California abalone, *Haliotis corrugata*, *H. fulgens*, and *H. cracherodii*. *In* Abalone of the world: biology, fisheries, and culture. *Edited by* S.A. Shepherd, M.J. Tegner, and S.A. Guzmán del Próo. Fishing News Books, Blackwell Science Ltd., Oxford, U.K. pp. 384–394.

Richards, D.V., and Davis, G.E. 1993. Early warnings of modern population collapse of black abalone, *Haliotis cracherodii*, Leach 1814 on the California Channel Islands. J. Shellfish Res. **12:** 189–194.

Safina, C. 1997. Song for the blue ocean. Henry Holt and Co., Inc., New York. 458 p.

Shepherd, S.A., and Brown, L.D. 1993. What is an abalone stock: implications for the role of refugia in conservation. Can. J. Fish. Aquat. Sci. **50**: 2001–2009.

Tegner, M.J. 1993. Southern California abalone: can stocks be rebuilt using marine harvest refugia? Can. J. Fish. Aquat. Sci. **50**: 2010–2018.

Tegner, M.J. 2000. Abalone (*Haliotis* spp.) enhancement in California: what we've learned and where do we go from here. *In* Workshop on Rebuilding Abalone Stocks in British Columbia. *Edited by* A. Campbell. Can. Spec. Publ. Fish. Aquat. Sci. 130. pp. 61–71.

Tegner, M.J., Breen, P.A., and Lennert, C.E. 1989. Population biology of red abalones, *Haliotis rufescens*, in southern California and management of the red and pink, *H. corrugata*, abalone fisheries. Fish. Bull. (U.S.), **87**: 313–339.

Tutschulte, T.C. 1976. The comparative ecology of three sympatric abalones. Doctoral dissertation. University of California, San Diego, Calif.

Tutschulte, T.C., and Connell, J.H. 1981. Reproductive biology of three species of abalones (*Haliotis*) in southern California. Veliger, **23**: 195–206.

Tutschulte, T.C., and Connell, J.H. 1988. Growth of three species of abalones (*Haliotis*) in southern California. Veliger, **31**: 204–213.

Wilson, P., and Wheeler, D.P. 1997. California's ocean resources: an agenda for the future. The Resources Agency of California, 1416 Ninth Street, Suite 1311, Sacramento, Calif. 95814.

Marine protected areas and their relevance to abalone (*Haliotis kamtschatkana*) conservation in British Columbia

Glen S. Jamieson

Abstract: Biological data suggest that the absolute abundance of adult abalone (*Haliotis kamtschatkana*) in close proximity to each other affects spawning success and that the geographic range of larval dispersal is limited. Individual concentrations of abalone may thus be largely self-sustaining and potentially vulnerable to local extirpation if adult mortality rate is too high. A species-specific threshold of spawning stock density may exist below which the risk of recruitment failure is high. Many northern abalone concentrations may have been reduced below this threshold. The threat of regional abalone extirpation should be reduced if patches of relatively high abalone density can be maintained and illegal harvest prevented. The current total ban on abalone fishing in British Columbia is proving difficult to enforce and abalone poaching remains a major concern. Achieving protection of abalone may be possible in Marine Protected Areas where a total diving prohibition can be enforced. The optimum size and spatial distribution of such protected areas can only be evaluated after appropriate objectives required for maintaining viable abalone populations are identified.

Résumé : Les données biologiques suggèrent que l'abondance absolue des ormeaux (*Haliotis kamtschatkana*)) adultes vivant à proximité les uns des autres affecte leur réussite de reproduction et que l'aire de dispersion géographique des larves est limitée. Les concentrations d'ormeaux pourraient donc être en bonne partie autonomes, et risquent de disparaître localement si le taux de mortalité est trop élevé. Il existe peut-être un seuil de densité des stocks de reproducteurs spécifique à l'espèce au-dessous duquel le risque d'échec du recrutement est élevé. Dans un grand nombre de concentrations d'ormeaux nordiques, la densité a peut-être été réduite au-dessous de ce seuil. Si l'on parvient à préserver les endroits où la densité d'ormeaux est relativement élevée et à empêcher les récoltes illégales, la menace d'extinction de l'espèce à l'échelle locale devrait diminuer. L'interdiction complète de la récolte d'ormeaux en Colombie-Britannique semble difficile à faire respecter, et le braconnage est encore très présent. Il serait possible d'assurer la protection des ormeaux dans les zones marines protégées où l'on pourrait appliquer une interdiction complète de plongée. La dimension optimale et la distribution spatiale des zones marines protégées ne pourront être évaluées qu'après l'établissement des objectifs nécessaires au maintien de populations d'ormeaux viables.

[Traduit par la Rédaction]

Introduction

As of 1997, 124 marine protected areas (mpas) had been legislated in British Columbia (B.C.), but fishing has also been permitted in most of these areas (Jamieson and Lessard 2000). British Columbia and Canada are now in the process of harmonizing areas that will be recognized jointly as protected (Anonymous 1998). Whether fishing of any type occurs will be determined on a site-by-site basis and will not necessarily be limited. There are also plans to emphasis "ecosystem" management in Canada, rather than simply to continue to manage each species in isolation of other species. The most relevant co-existing species here are sea urchins, particularly the red sea urchin (*Stronglyocentrotus franciscanus*), which are often the most dominant herbivores in the shallow subtidal. The fishery for red sea urchins, with 110 licensed vessels and a landing of 6501.6 t in 1999 (DFO 1999), occurs in all areas of B.C. where northern abalone, *Haliotis kamtschatkana* (hence only referred to as abalone), occur.

Over the past 15 years, a series of Canadian stock assessment documents (e.g., Farlinger and Campbell 1992; Winther et al. 1995; Thomas and Campbell 1996; Cripps and Campbell 1998; Campbell et al. 1998) have documented the gradual decline in abundance of abalone in B.C. For 13 years, this species supported a significant commercial fishery but all fishing was terminated in 1990 because abundance indices were low and there was no evidence of forthcoming significant new recruitment. Abundance of this species continues to decline, with increased concern that local populations may be being extirpated (Campbell et al. 1998; Jamieson 1999). In 1999, the Committee on the Status of Endangered Wildlife in Canada (COSEWIC) declared *H. kamtschatkana* a threatened species in Canada.

In this paper, I review the history of fishing restrictions in existing legislated mpas in B.C. and consider the relevance

G.S. Jamieson. Pacific Biological Station, Fisheries and Oceans Canada, Nanaimo, BC V9R 5K6, Canada.
Correct citation: Jamieson, G.S. 2000. Marine protected areas and their relevance to abalone (*Haliotis kamtschatkana*) conservation in British Columbia. *In* Workshop on Rebuilding Abalone Stocks in British Columbia. *Edited by* A. Campbell. Can. Spec. Publ. Fish. Aquat. Sci. 130. pp. 139–147.

and potential implications of mpas in general to abalone. With all harvest of abalone banned in B.C. since 1990, theoretically the whole province of B.C. is already a mpa for this species, although as will be shown, substantial illegal harvesting persists. I also suggest general functional differences between protected areas in tropical versus temperate waters and review connectivity considerations relative to the spatial size and distribution of site-specific mpas, focused specifically towards abalone abundance enhancement.

History of mpas in B.C.

Historical First Nation resource management

Although coastal peoples indigenous to B.C. had no centralized system of resource management, resource management was nevertheless practised. With limited opportunity for food transport, it was critical to the survival of these hunter–fisher–gather societies that natural food resources be exploited in a manner that guaranteed their long-term sustainability (Turner 1997). Close observation of life cycles and seasonality of plants and animals allowed the use of ecological indicators to determine appropriate harvesting times and anticipated levels of resource abundance. Diversity was a key to sustainable resource use. Abalone were used for food, but the shells were also used for decoration and currency (Turner 1997). Virtually all areas of the land, water, and shoreline fell under the specific "ownership" of one individual, usually a hereditary chief, who while being a recipient of the benefits of using the area, also had the responsibility of maintaining, sustaining, and sharing the resources equitably among other community members. Every productive area was closely observed and maintained to ensure resource sustainability. Because there appears to be no documentation that Canadian indigenous peoples dove for abalone, most harvested live abalone probably came from the intertidal or shallow subtidal zones. Presumably this limited their ability to exploit abalone concentrations extensively, as abalone concentrations, at least during the 1980s, were mostly centred in abundance around 0–10 m depth (Sloan and Breen 1988). There is no data to suggest that abalone spatial distribution, in contrast to absolute abundance, changed as a result of recent harvesting and so historically, deeper (>1 m) waters would have functionally represented a natural abalone refugia, or reserve, from First Nations harvesting.

Historically, the major predator of adult abalone was the sea otter, *Enhydra lutris,* not humans, as abalone are a preferred prey (Johnson 1982; Estes and Van Blaricom 1985). In areas where sea otters have been reintroduced in B.C., almost all surviving abalone are found only in rock crevices, while outside the range of sea otters, live abalone are abundant outside crevices on open rock faces (Hines and Pearse 1982; Watson 1993). However, native people hunted sea otters as well as abalone and sea otters likely avoided the immediate areas around native villages (Russ Jones, Haida Fisheries Program, Skidegate, B.C., personal communication). Coastal human habitations were historically relatively abundant along the coast, as shown by the many coastal middens today, and this may have created many refuges where abalone could have occurred in at least modest abundance subtidally. Sea otter and indigenous human populations were both devastated almost simultaneously in the early 1800s by the arrival of Europeans, the former by hunting and the latter by diseases such as smallpox. With the removal of sea otters and release of predation pressure, the potentially numerous, albeit likely small and geographically isolated, abalone populations that might have existed along the coast may have gradually expanded in both size and number. Also, there may have been occasional significant large natural recruitment events, which in the absence of extensive predation, allowed an accumulation of abalone stocks. The increase in abalone abundance helped to support the recent legal and illegal commercial abalone fisheries.

The current situation

Of the 124 existing legislated mpas in Canadian Pacific waters, 116 have been designated by B.C. (Jamieson and Lessard 2000) (Table 1). The federal Government of Canada is the senior level of government and has the mandate to manage renewable fishery resources, and so federal Fisheries and Oceans Canada (DFO) regulations supersede provincial regulations. There are no Marine Protected Areas (MPAs, words capitalized, because this is the specific name designated for marine protected areas under the Oceans Act) currently established in B.C. (two inshore candidate MPAs were established in 1998 at Race Rocks and Gabriola Pass, but these have yet to be approved through legislation), and few Fisheries Act fishing restrictions are specifically associated with any other legislated mpa. At the present, then, fishing can and does occur in most areas in B.C. that are recognized as "protected" (Jamieson and Lessard 2000). "Protected" simply means that some natural resources are managed and that some human activities may be regulated. Restrictions may be for activities (e.g., ocean dumping, dredging, nonrenewable resource exploitation, etc.) other than fishing.

The proposed mpa strategy

In August 1998, Canada and B.C. jointly released a discussion paper on a mpa strategy for Canada's Pacific coast (Anonymous 1998), which summarized the designations for existing or potential mpas and the mandates of the Acts under which they were designated (Table 1).

Until January 1997, the DFO had no mechanism other than fishery closures under the Fisheries Act to protect resources; however, with passage of the Oceans Act, Canada now has the authority to establish mpas. Other federal legislation can establish other mpas, such as National Parks (Parks Act) and National Wildlife Areas (Canada Wildlife Act). British Columbia and Canada are now beginning the process of harmonizing their regulations for areas both governments feel need renewable resource protection designation. Specific fishing regulations in any mpa will be determined on a site-by-site basis and will not necessarily prohibit fishing for any species.

The strategy proposes to establish a network of mpas through integrated management (IM) processes using existing legislative mpa designations. With IM, the marine environment will effectively be zoned, with specified areas designated for various marine uses, including aquaculture, protection, industry (e.g., log booming), and fisheries. The

Table 1: Federal and British Columbia statutory powers for protecting marine areas and the numbers of each marine protected area type in 1997 (in brackets) (total = 124).

Agency	Legislative tools	Designations	Mandate
Fisheries and Oceans Canada (Federal) (0)	*Oceans Act*	Marine Protected Areas (0)	To protect and conserve • fisheries resources, including marine mammals and their habitats • endangered or threatened species and their habitats • unique habitats • areas of high biodiversity or biological productivity • areas for scientific and research purposes
	Fisheries Act	Fishery closures	Conservation mandate to manage and regulate fisheries, conserve and protect fish, protect fish habitat, and prevent pollution of waters frequented by fish
Environment Canada (Federal) (7)	*Canada Wildlife Act*	National Wildlife Areas (2) Marine Wildlife Areas (0)	To protect and conserve marine areas that are nationally or internationally significant for all wildlife but focusing on migratory marine birds
	Migratory Birds Convention Act	Migratory Bird Sanctuaries (5)	To protect coastal and marine habitats being used heavily by birds for breeding, feeding, migration, and overwintering
Parks Canada (Federal) (1)	*National Parks Act* Proposed *National Marine Conservation Areas Act*	National Parks (1) Proposed National Marine Conservation Areas (0)	To protect and conserve for all time marine conservation areas of Canadian significance that are representative of the 5 natural marine regions identified on the Pacific coast of Canada and to encourage public understanding, appreciation, and enjoyment
Ministry of Environment, Lands and Parks (Provincial) (116)	*Ecological Reserve Act*	Ecological Reserves (15)	To protect • representative examples of B.C.'s marine environment • rare, endangered, or sensitive species or habitats • unique, outstanding, or special features • areas for scientific research and marine awareness
	Park Act	Provincial Parks (81)	To protect representative examples of marine diversity, recreational and cultural heritage special natural, cultural heritage, and recreational features To serve a variety of outdoor recreation functions including • enhancing major tourism travel routes • providing attractions for outdoor holiday destinations
	Wildlife Act	Wildlife Management Areas (4)	To conserve and manage areas of importance to fish and wildlife and to protect endangered or threatened species and their habitats, whether resident or migratory, of regional, national, or global significance
	Land Act	Wildlife Reserves (15)	To reserve Crown land for specified periods for a variety of reasons, including the protection of sensitive habitat or habitat important for the conservation or well-being of species
	Environment and Land Use Act	"Protected Areas" (1)	To protect • representative examples of marine diversity, recreational and cultural heritage • special natural, cultural heritage, and recreational features

strategy will achieve an expanded network of mpas for coastal B.C. by 2010. Because of the large geographical area involved, the overall coastal area has been divided into six regions, which will undergo IM planning.

Where an IM planning process is not contemplated in the near term, independent mpa proposals may also be considered. Interim management guidelines may be applied to proposed or candidate mpas where guidelines are demonstrated to be necessary to protect specific marine resources, habitats, or values that may be under threat until coastal planning is completed. Because abalone are now a recognized "threatened" species (COSEWIC 1999), mpas whose objectives are specifically associated with abalone stock rebuilding could be justifiable in the short term.

Mpas in temperate waters

What does "protection" mean?

Protection can apply to both habitat and specific renewable resources. While many protected areas primarily focus on habitat or substrate features (e.g., deserts, canyons, bays, etc.), many others focus on protecting biological structure (e.g., forests, coral reefs, etc.). In tropical waters and on land, protected areas have often been established to protect long-lived, structurally important species (e.g., corals, climax forest trees, etc.). Community biodiversity is often particularly high in such environments and these communities are readily visible and aesthetically interesting to many people, which means that support for their protection is likely to be popular. This does not mean that biological communities with shorter-lived species, such as grasslands, wetlands, etc., also are often not protected, but communities structured around longer-lived species have often been identified as particularly important to conserve. With the general absence of long-lived structural species in temperate waters (with the exception of deep-water corals), it is less obvious to many people what exactly should be protected and why. A particular body of water may not appear as unique biologically from the water surface since one can not differentiate most underwater habitats visually, except where surface and floating kelp beds are visible. Early global attempts at marine conservation showed this confusion and lack of understanding of marine biological systems. For example, one of the key criteria used for delimitation of early Japanese marine reserves was to define the seaward boundary as the furthest viewing point from land (Dutton 1992). Equally restrictive approaches have characterized the selection of most Australian marine and estuarine protected areas (Dutton and Saenger 1994). Although Australian selection criteria were largely based on the classification schema proposed by Ray (1976), they also identify socio–political factors as important. Kriwoken (1993) described the first comprehensive identification of potential marine reserves in Tasmania as being restricted to areas adjacent to existing terrestrial parks. This was rationalized on legal establishment convenience, easier enforcement, and easier control of terrestrial impacts. Regardless of justification, the net effect was described by Kriwoken (1993) as "islands of management in a sea of mismanagement."

Dutton and Saenger (1994) also commented on the early Australian failure to recognize the diversity of marine systems that may be of ecological significance and in particular to distinguish between the extremes of estuarine/near coastal and oceanic regimes. The danger here is that allocation and management approaches being applied may not be appropriate to the nature of the system being protected and the human impact on it. Associated with this consideration is the matter of scale and the possible fragmentation of management responsibilities for protection. As in the terrestrial sense, there may be limitations associated with small, isolated protected areas. However, as Jamieson and Levings (1998) pointed out, for pelagic species or species with meroplanktonic larvae, such issues are typically of lesser concern than in the terrestrial environment. However, the issue of poor co-ordination and planning around the often complex problems of resource use and allocation can be a major one. This problem is also difficult to ameliorate (Dutton et al. 1994), often partly because of bureaucratic delays and the inflexibility in dealing with dynamic, evolving problems.

Although mpas may not necessarily restrict any fisheries, some restrictions in every mpa generally prevent significant habitat modification (e.g., no ocean dumping or dredging). These restrictions should protect the productive potential of a habitat, even if adults of targeted species are not present in harvestable abundances, because the natural physical substrate of the habitat would then be maintained. Natural succession may still change community structure over time, but abalone, as an indigenous species, are presumably adapted to such events. The conservation of an unperturbed substrate is particularly important in temperate waters, which as indicated above, are less likely to have long-lived structural species present. In any specific year, settlement of a particular species may or may not be abundant at a specific site because of the vagaries of the currents dispersing planktonic larvae; however, there is always the potential that a year-class will be substantial if an undisturbed substrate exists.

Permitted human activities in mpas may also indirectly affect the biological structural integrity of ecosystems in a manner initially unanticipated by managers, with perhaps the most significant being marine community perturbation because of fishing. While physical habitat destruction is often quite discernible, at least in shallow waters, humans have a long history of greatly affecting a region's productivity and biodiversity by fishing and here the consequences may not be as obvious. Some fishing activities, such as trawling and longlining, can be quite destructive (Breeze et al. 1997; Macdonald et al. 1996) to biological communities that either constitute (e.g., corals, macrophyte beds) or depend on spatial structure (e.g., burrows, etc.) in the marine environment. Our awareness of this destruction and ability to measure the biological consequences is often impeded by topographic complexity and water depth. Abalone generally inhabit rocky areas where habitat destruction by any fishing method is minimal, but as was shown with extirpation of sea otters, benthic nearshore community structure was nevertheless influenced by renewable resource harvesting, i.e., the hunting of sea otters.

The issue of identifying sites and deciding which particular marine areas most merit protection is a significant challenge (Levings and Jamieson 1999). Evaluating marine ecosystem characteristics, particularly in a comparative sense, may be relatively time consuming and costly if relevant species aren't always present to indicate areas of important productivity. Many sedentary species such as bivalves and other benthic species are cryptic and motile species such as many fishes are typically transient, often on a seasonal basis. Species' occurrences are often influenced by physical parameters such as exposure, substrate type, and currents, and less obviously, although not necessarily less importantly, by biological community structure. For this reason, physical topography rather than community biology has been used to date to delineate virtually all actual or potential mpa boundaries.

Connectivity issues

Jamieson and Levings (1998) reviewed connectivity issues in the marine environment and concluded that the need for corridors in a terrestrial sense is not matched by the same need in the aquatic environment. Corridor issues affecting the community structure of mpas exist in the marine and aquatic environment, but are relatively few compared to the terrestrial realm. Options for "shaping" the distribution of organisms using corridors are also fewer because of the "open nature" of oceans and the dynamics of natural (Roberts 1997) and human-influenced marine connections. Abalone, like most marine benthic invertebrates, have adults that move relatively little and meroplanktonic larvae that are their primary dispersal stage. However, the duration of the abalone planktonic stage is only a few days long and the larvae likely have behavior that limits the range of their dispersal. The eggs are heavier than water, and while juveniles <10 mm shell length (SL) are hard to find, they seem to occur mostly on bedrock or boulders encrusted with red coralline algae at 5–15 m depth (Sloan and Breen 1988).

St. Mary et al. (2000) discussed the implications of large-scale spatial segregation among life stages of a species with respect to optimizing issues relating to mpa network design. They concluded that for species with a large geographic range explicit integration of biological and socio–economic models would be required to optimize a design. This would be the case for most marine species, but abalone in general probably have small-scale spatial segregation between life stages. Modest-sized (e.g., 2–10 km^2), no-abalone-harvest mpas may therefore protect all the life stages produced from a particular abalone concentration. However, specific information on northern abalone larval dispersal and subsequent settlement patterns, such as would be obtained from deployment of larval abalone settlement collectors around an isolated abalone concentration, is needed to confirm this likelihood. This suggests that the primary rationale for a network of "no-take" abalone mpas may not be to facilitate fairly rapid large-scale rebuilding of abalone abundance, but rather to preserve the nuclei of concentrations for a possible long-term, more gradual expansion of areas of relatively high abalone abundance. This might occur if small no-harvest mpas were placed a "biologically significant" distance (optimal value presently unknown) apart in a local network and if fishery management could ensure no illegal fishing. The primary rationale for a network of mpas and the nature of the network will be determined by the specific management objectives established for abalone.

Application of mpas to abalone population maintenance or stock rebuilding

Abalone biology

Reproduction

Abalone are dioecious, i.e., individuals are of only one sex and spawning is via the release of gametes into the water, with external fertilization. Initiation of spawning by one individual typically triggers spawning by near-by individuals. Fertilization rate is therefore maximized when abalone concentrate in abundance during spawning, and as with all broadcast spawners, there is concern that adult density may often be too low to ensure acceptable fertilization rates (Levitan and Sewell 1998). The minimal population size for a viable population is unknown. Abalone can have ripe gonads year round (Quayle 1971), but significant periods of spawning activity may occur over the course of the main spawning season, April–August, with some rebuilding of gonad size between spawnings (Sloan and Breen 1988).

Larval dispersal

We currently know nothing about the dispersal of larvae from northern abalone populations but this species probably has larval dispersal characteristics similar to other haliotids. Olson (1984) and Calderwood (1985) reported that larval duration in hatcheries for northern abalone ranged from 4 to 8 days at 14–10°C, respectively. In the absence of specific data, northern abalone dispersal is assumed to be similar to *H. rubra* (Prince et al. 1988), which is limited in geographic extent.

Growth and maturity

Abalone cannot be aged from shell characteristics (Fournier and Breen 1983). However, by tagging individuals in the wild and monitoring size change over time, age at legal size was estimated at about 6–8 years (Qualye 1971; Breen 1986) and annual growth was about 5–6 mm SL per year at recruitment (Sloan and Breen 1988). Average longevity is not known but growth appears to slow down with age and individuals have been hypothesized to live up to 50 years (Breen 1980). While 50% of abalone were estimated to be mature at 55 mm SL, the importance of large mature female abalone in contributing to total population fecundity had probably been underestimated until recently (Campbell et al. 1992).

Spatial distribution

Although abalone occur throughout the outer coastal areas of B.C., even when abundant, their spatial distribution was quite contagious rather than evenly distributed. Many small, relatively isolated concentrations exist because of local topography and the availability of hard substrates, with often no possibility of adult abalone movement among them. Abalone are mobile but adults likely move only over a range of a few hundreds of metres at most during their lives. Where abalone occur, they may further concentrate at times of spawning, which can make them particularly vulnerable to predation or harvest by humans. Unfortunately, precise locations of historic abalone concentrations, reported in the early years of the fishery after logbook completion was required, remain unanalyzed. Thus, precise geo-referenced locations of historical abalone concentrations are currently largely unavailable.

Conservation issues

Fishing mortality

Legal fishing: *Haliotis kamtschatkana* is the only invertebrate species for which all fishing in B.C. is currently totally banned. The Fisheries Act provides as much protection as possible under existing legislation on the basis of specific recognition of the necessity for this species' conservation. First Nations' food, social, and ceremonial harvests have precedence in any fishing plan, so if fishing is once again permitted after stocks rebuild, First Nation fishing will be the first to be permitted. "No take" mpa designations may not exclude First Nation harvests if fishing is permitted.

Campbell (1997) considered criteria that might be used to reopen the northern abalone fishery in B.C. if stocks rebuild. He concluded that fishery-reopening criteria could not be fully defined due to a lack of data on the frequency and patch size of adult abalone concentrations required to maintain sufficient recruitment for a healthy population. At very low population densities, Allee effects, i.e., depressed per capita survivorship or fecundity as populations become small (Lande 1987; Hastings 1991; Quinn et al. 1993), may become important.

Illegal fishing: A major problem has been illegal harvesting (Farlinger and Thomas 1989; Farlinger 1990). Coincident with a reduction in the legal quota for abalone, recreational SCUBA diving and dive fisheries for other species were expanding, increasing the numbers of divers encountering abalone. Knowledge of abalone concentrations became more widespread. Reduced quotas for abalone increased the price per kilogram in the late 1980s and current abalone prices in B.C. are estimated to range from \$55 to \$110 kg^{-1} (B. Jubinville, 60 Front St., Nanaimo, B.C., personal communication), making abalone increasingly attractive for poachers to harvest. The DFO fishery officers try to enforce the law but the problem seems to be similar to that of the illegal drug trade: for some people, value and potential gain appear sufficiently high to override the risk of being caught. The result is that illegal fishing is continuing.

Natural mortality

In the absence of sea otters: Sloan and Breen (1988) list the abalone predators in B.C. currently most likely impacting on abalone abundance. These include *Octopus dofleini*, cabezon (*Scoroaenichthys marmoratus*), wolf eels (*Anarrhishthys ocellatus*), rock crab (*Cancer productus*), and the sea star (*Pycnopodia helianthoides*). Little is known about the mortality rates of juveniles. Thermal mortality occurs at temperatures constantly >16°C. (Paul and Paul 1981). Emmett and Jamieson (1988) tagged pre-recruit abalone and estimated that natural mortality (M) ranged from 0.48 to 2.51 at four sites. Recovery of dead shells gave a minimum estimate of M = 0.15–0.27. Estimates of M for various abalone species mostly range from 0.2 to 0.4 (Sloan and Breen 1988) but M is likely to be variable over both time and space.

With sea otters: Following their reestablishment at a few isolated locations during 1969–1972, the range and abundance of sea otters is still expanding in B.C. (average of 18–20% increase in number per year on the west coast of Vancouver Island (J. Watson, Malaspina University College, Nanaimo, B.C., personal communication). The population expanded from 89 animals introduced between 1969 and 1972 to over 1522 in 1995 (Watson et al. 1997). Ultimately, unless extensively harvested in certain localities, sea otters can be expected to regain all of their original range, which includes most, if not all, habitats presently occupied by abalone. Abalone can coexist with sea otters but at a relatively low density as cryptic individuals (Watson and Smith 1996).

Enhancement

The current depressed state of natural abalone populations and the high price of abalone have generated considerable interest in the opportunities of abalone aquaculture for both wild abalone population enhancement and intensive product production. Culture technology for abalone production has been developed and, in the 1980s, Pacific Trident Ltd. attempted northern abalone aquaculture in Saanich Inlet and at Sooke, Vancouver Island. The production of thousands of juveniles was achieved, but before the economics of grow-out could be rigorously tested, extremely high mortality of abalone occurred in the first few months after settlement due to a previously undescribed protozan parasite, *Labyrinthuloides haliotidis* (Bower 1987*a*, 1987*b*). Adult abalone were not noticeably affected by the parasite. How these facilities became infected is unknown. Studies to ascertain the prevalence of this parasite in the wild have not been conducted.

Although the release of cultured abalone juveniles into the wild, where they can be exploited later, has been practised for years in Japan, the procedure has failed to enhance wild populations of abalone (Seki and Sano 1998). Landings have declined through most of the time period of abalone release. At best, enhancement of wild populations may have only slowed the rate of decline. Recapture studies (Seki and Sano 1998) suggested a direct relationship between survival of cultured abalone after release into the wild and the size at which abalone were released. Since culture is a relatively expensive operation, maximum economic return may be best achieved when cultured abalone are held until harvest in a controlled environment, thereby eliminating any loss from either predation or being unable to find all the abalone in the wild that were initially released. However, the idea of releasing cultured stock is also being considered in B.C. as a conservation approach only, at least in the short term, with juveniles to be released to enhance currently unexploited wild abalone populations. Release in either existing or future mpas may then be likely. Clearly defining why enhancement of wild populations is being attempted is therefore important, i.e., strictly for species conservation or to harvest the released individual abalone at a later date. Enhancement may be justifiable for conservation reasons and to rationalize mpas in specific locations but enhancement is probably not justifiable for the economic sustenance of a wild harvest fishery.

Conservation reasons for release of either cultured abalone or transplanted wild abalone into specific locations include the need to lower the extinction probability of the species by having as many distinct metapopulations of the species as possible, to lessen demographic stochasticity (chance variation in the number of births and deaths in a population with time-variant vital rates: Fogarty 1993), and to stem possible inbreeding depression. However, the latter is only achieved if cultured or transplanted abalone include a diversity of genetic stocks. There is presently no information about differences in genetic diversity among different abalone stocks.

Mpa options

Mpa objectives relevant to abalone

Abalone conservation

Jamieson (1999) summarized the continued decline in abalone populations in B.C. and expressed concern that, if this trend was not reversed, the risk that abalone might be extirpated from at least portions of the coast would be high. Campbell et al. (1998) documented the increasing number of abalone index survey locations over time in which no abalone were being found and this might be indicative of local abalone population collapses. Effectively conserving the integrity of abalone concentrations that remain seems increasingly important, because these will be the primary source of recruits if abalone abundance is to rebuild in the near future towards the recent historic level. I emphasize "effectively" here, because as with all abalone fisheries presently closed, in theory the whole coast now is functionally a no-abalone-harvest mpa. That abalone populations are not rebuilding suggests that while this may be true in theory, in practice, either recruitment may no longer be sufficient to sustain populations, possibly as a result of the Allee effect, or exploitation through poaching must still be of sufficient magnitude to prevent many concentrations from rebuilding. Tegner (1993) documented a similar problem in California: specific areas were closed to abalone fishing but divers exploiting other species were still permitted access. Abalone populations did not rebuild and Tegner suggested this was apparently because of poaching. This did not mean that the area had to be "no-take" for all exploited species. Since subtidal abalone can only be harvested by diving, all diving should have been prohibited (Tegner 1993); hook and line fishing could still have occurred.

Representative biodiversity conservation

The objective for a number of Canadian mpas, i.e., National Marine Conservation Areas, Ecological Reserves, and Provincial Parks, is to sustain the biological ecosystem representative of that particular area. No-take areas, which may comprise the whole or only a part of a specific mpa, could act as reference points, helping to maintain a community structure as undisturbed as possible, where natural community dynamics could be studied and hence provide baseline data for use in trying to explain changes in surrounding exploited areas. Fishing for some species may be permitted in specific areas outside the no-take areas, as long as representative examples of the pristine biological community are sustained in the overall area. Since abalone are an integral part of the natural community, sustaining their continued presence would likely be an objective in mpas. Managers would be given the required authority to control human activities negatively impacting the continued presence of important species, including abalone.

Logistic issues and monitoring needs

Regulation enforcement

The main issue with respect to the efficacy of mpas as a management tool is to ensure that all illegal activity, such as poaching, is prevented, even if in only a few small, relatively discrete areas where abalone naturally occur in abundance. A major problem with enforcement of existing regulations in B.C. is that the coastline is long [38 516 km (Jamieson and Lessard 2000)], with only a relatively few coastal communities present in most outer coast areas. Abalone often occur in remote, isolated locations and fishery officers can not be everywhere at once. The most convenient fishery enforcement location is at the points of landing, i.e., the few ports at which legal catches for most species are brought ashore. Because the daily weight of abalone that can be harvested by a diver, whether legally or illegally, is relatively small, abalone can effectively be landed almost anywhere on the coast that a vessel can access. Abalone can even be caged under water, attached to submerged buoys, then quickly hauled up later at a more convenient time for transport to shore. Because all abalone landings therefore can not be effectively monitored, the only management option left to prevent fishing is to try to eliminate diving of any type, excluding diving with approved guides or wardens for surveillance or research purposes, in at least those areas where continuous surveillance of the water surface can not be provided. Even a few hours of fishing effort can substantially impact a small abalone concentration and Tegner (1993) emphasized the importance of removing all opportunities for poaching. Prevention of all potentially exploitative diving, though, demands that guardians always be present.

The only abalone population in B.C. known to never have been subjected to commercial fishing is in the immediate vicinity of William's Head Penitentiary near Victoria, B.C. There, penitentiary guards have prevented nearshore access to both fishers and poachers in an effort to minimize opportunity for inmate escape, allowing abalone abundance and average size to remain relatively high (Wallace 1997, 1999). Population characteristics are presumably what other virgin populations resembled before commercial fishing began.

Sustained local enforcement should be achieved if a majority of residents in a local community support the objective of abalone conservation and are willing to collectively assume the responsibility of guardianship, because no single individual can always be around to keep watch over a specific resource. Coastal communities are widely scattered and relatively few in number (most may be First Nations), so there may be only limited opportunities geographically to achieve this type of management even if most coastal communities support this approach. The development of co-management with the involvement of First Nations is also an option to help establish improved surveillance. Charges for guided entry to mpas, where diving would otherwise be prohibited, could help provide additional surveillance and possibly some enforcement operating funds. As noted above, an absence of subtial abalone harvest likely occurred in the past before European settlement.

Mpa size, number, and spatial distribution

Community discussion is important to consider the vision for the future of abalone, given that the establishment of some mpas specifically to conserve abalone may be likely. Potential vision options include the following: (*i*) assuring the continued presence of at least a modest number of spatially isolated abalone concentrations; (*ii*) letting abalone abundance remain at current low levels because of an inability to effectively limit poaching, with all fisheries closed indefinitely, and perhaps having intensive abalone culture; or (*iii*) trying to initiate new, hopefully effective management approaches as soon as possible allowing abalone populations to rebuild substantially over most of the coast.

With respect to (*i*), analysis to determine the appropriate sizes of abalone concentrations, the number of concentrations, and the spatial pattern of concentrations that should persist should consider both biological (e.g., alleviation of demographic stochasticity, lowering of extinction rate in the sense of equilibrium theory, and the stemming of inbreeding depression) and social factors (e.g., enforcement concerns, community interest, etc.). Such an analysis is beyond the scope of this paper and would likely require study of the larval dispersal characteristics of this abalone species.

In conclusion, where we proceed from here depends on our collective ability to agree on a vision. Strong scientific advice for the minimum size of an effective mpa and the spatial distribution of a network of protected areas does not currently exist for abalone, so any initiatives described above will require experimental management. The successes or failures of the various adaptive management initiatives will have to be closely monitored in a timely manner.

References

Anonymous. 1998. Marine protected areas: a strategy for Canada's Pacific Coast. Department of Fisheries and Oceans, Vancouver, B.C. and the British Columbia Land Use Coordination Office, Victoria, B.C. 28 p.

Bower, S.M. 1987*a*. *Labyrinthuloides haliotidis* n. sp. (Protoza: Labyrinthormorpha), a pathogenic parasite of small juvenile abalone in a British Columbia mariculture facility. Can. J. Zool. **65**: 1996–2007.

Bower, S.M. 1987*b*. Pathogenicity and host specificity of *Labyrinthuloides haliotidis* (Protoza: Labyrinthormorpha), a parasite of juvenile abalone. Can. J. Zool. **65**: 2008–2012.

Breen, P.A. 1980. Muscled mollusc: the northern abalone. Diver, **6**: 26–28.

Breen, P.A. 1986. Management of the British Columbia fishery for northern abalone (*Haliotis kamtschatkana*). *In* North Pacific Workshop on Invertebrate Stock Assessment and Management of Invertebrates. *Edited by* G.S. Jamieson and N. Bourne. Can. Spec. Publ. Fish. Aquat. Sci. 92. pp. 300–312.

Breeze, H., Davis, D.S., Butler, M., and Kostylev, V. 1997. Distribution and status of deep-sea corals off Nova Scotia. Marine Issues Committee Spec. Publ. 1, Ecology Action Centre, Halifax, Nova Scotia. 58 p.

Calderwood, G. 1985. Technological advances in the field of abalone mariculture. *In* Final Rep. (1982–1985) prepared by Pacific Trident Mariculture Ltd., Victoria, B.C. Department of Supply and Services, Ottawa, Canada. No. 053B. FP 501-2-005. 125 p.

Campbell, A. 1997. Possible criteria for reopening the northern abalone (*Haliotis kamtschatkana*) fishery in British Columbia. Can. Stock Assess. Sec. Res. Doc. 97/64. Fisheries and Oceans Canada, 200 Kent Street, Ont. K1A 0E6, Canada. 47 p.

Campbell, A., Manley, I., and Carolsfeld, W. 1992. Size at maturity and fecundity of the abalone, *Haliotis kamtschatkana*, in northern British Columbia. Can. Manuscr. Rep. Fish. Aquat. Sci. 2169. pp. 47–65.

Campbell. A., Winther, I., Adkins, B., Brouwer, D., and Miller, D. 1998. Survey of the northern abalone (*Haliotis kamtschatkana*) in the central coast of British Columbia, May 1997. Fisheries and Oceans Canada. Can. Stock Assess. Sec. Res. Doc. 98/99. Fisheries and Oceans Canada, 200 Kent Street, Ont. K1A 0E6, Canada. 28 p.

Committee on the Status of Endangered Wildlife in Canada (COSEWIC). 1999. A listing of all species considered by COSEWIC in 1999. Available as Internet document URL: http://www.cosewic.gc.ca/COSEWIC/QueryResults.cfm?DList=1999.

Cripps, K., and Campbell, A. 1998. Survey of abalone populations at Dallain Point and Higgins Pass, Central Coast of British Columbia, 1995–96. Can. Manuscr. Rep. Fish. Aquat. Sci. 2445. 31 p.

DFO (Fisheries and Oceans Canada). 1999. Red sea urchin. Fisheries and Oceans Canada. Science Stock Status Rep. C6-09 (1999). 3 p.

Dutton, I.M. 1992. National parks and nature conservation in Japan. Aust. Parks Recreation, **28**: 28–39.

Dutton I., and Saegner, P. 1994. Expanding the horizon(s) of marine conservation: the challenge of integrated coastal management. *In* Marine protected areas and biosphere reserves: 'towards a new paradigm.' *Edited by* D.J. Brunckhorst. Australian Nature Conservtion Agency Publ., Canberra. pp. 18–27. (Also available as Internet document URL: http://www.environment.gov.au/portfolio/anca/mpa/idutton.html).

Dutton, I.M., Saenger, P., Perry, T., Luker G., and Worboys, G.L. 1994. An integrated approach to management of coastal aquatic resources — a case study from Jervis Bay, Australia. Aquat. Conserv. **4**: 57–74.

Emmett, B., and Jamieson, G.S. 1988. An experimental transplant of northern abalone, *Haliotis kamtschatkana,* in Barkley Sound, British Columbia. Fish. Bull. **87**: 95–104.

Estes, J.A., and Van Blaricom, G.R. 1985. Sea otters and shellfisheries. *In* Marine mammals and fisheries. *Edited by* J.R. Beddington, R.J.H. Beverton, and D.M. Lavigne. George Allen and Unwin, London. pp. 187–235.

Farlinger, S. 1990. Review of the biological basis for management of the British Columbia abalone fishery. Can. Manuscr. Rep. Fish. Aquat. Sci. 2099. pp. 41–51.

Farlinger, S., and Campbell, A. 1992. Fisheries management and biology of northern abalone, *Haliotis kamtschatkana*, in the northeast Pacific. *In* Abalone of the world: biology, fisheries and culture. *Edited by* S.A. Shepherd, M.J. Tegner, and S.A. Guzmán del Próo. Fishing News Books, Blackwell Science Ltd., Oxford, U.K. pp. 395–406.

Farlinger, S., and Thomas, G. 1989. Review of the abalone fishery in British Columbia. PSARC Doc. I88–10. *Summarized in* Can. Manuscr. Rep. Fish. Aquat. Sci. 2020. pp. 138.

Fogarty, M.J. 1993. Recruitment distributions revisited. Can. J. Fish. Aquat. Sci. **50**: 2723–2728.

Fournier, D.A., and Breen, P.A. 1983. Estimation of abalone mortality rates with growth analysis. Trans. Am. Fish. Soc. **112**: 403–411.

Hastings, A. 1991. Structure models of metapopulation dynamics. *In* Metapopulation dynamics: empirical and theoretical investigations. *Edited by* M. Gilpin and I. Hansk. Academic Press, London. pp. 57–71.

Hines, A.H., and Pearse, J.S. 1982. Abalone, shells and sea otters: dynamics of prey populations in central California. Ecology, **63**: 1547–1560.

Jamieson, G.S. 1999. Review of status of northern, or pinto, abalone, *Haliotis kamtschatkana*, in Canada. Fisheries and Oceans Canada. Can. Stock Assess. Sec. Res. Doc. 99/190. Fisheries and Oceans Canada, 200 Kent Street, Ont. K1A 0E6, Canada. 22 p.

Jamieson, G.S., and Lessard, J. 2000. Marine Protected Areas and fishery closures in British Columbia. Can. Spec. Publ. Fish. Aquat. Sci. 131. In press.

Jamieson, G.S., and Levings, C.D. 1998. Marine Protected Areas: connectivity considerations. *In* Linking protected areas with working landscapes conserving biodiversity. *Edited by* N.W.P. Munro and J.H.M. Willison. Science and Management of Protected Areas Assoc., Wolfville, N.S., Canada. pp. 151–160.

Johnson, A.M. 1982. Status of Alaska sea otter populations and developing conflicts with fisheries. Trans. N. Am. Wildl. Nat. Resour. Conf., Washington, D.C.: Wildlife Management Institute, Washington, D.C. pp. 293–299.

Kriwoken, L. 1993. Islands of management in a sea of mismanagement. *In* Protection of marine and estuarine areas — a challenge for Australians. *Edited by* A. Ivanovici, D. Tarte, and M. Olson. Proceedings of the Fourth Fenner Conference on the Environment, Canberra, 9–11 October, 1991. Int. Union Conserv. Nat. Nat. Resour. and Australian National Parks and Wildlife Service, Canberra, Australia. pp. 67–72.

Lande, R. 1987. Extinction thresholds in demographic models of territorial populations. Am. Nat. **116**: 463–479.

Levings, C.D., and Jamieson, G.S. 1999. Evaluation of criteria for creating MPAs in the Pacific region: a proposed semi-quantitative scheme. Fisheries and Oceans Canada. Can. Stock Assess. Sec. Res. Doc. 99/210. 30 p.

Levitan, D.R., and Sewell, M.A. 1998. Fertilization success in free-spawning marine invertebrates: review of the evidence and fisheries implications. *In* Proceedings of the North Pacific Symposium on Invertebrate Stock Assessment and Management. *Edited by.* G.S. Jamieson and A. Campbell. Can. Spec. Publ. Fish. Aquat. Sci. 125. pp. 59–164.

MacDonald, D.S., Little, M., Eno, N.C., and Hiscock, K. 1996. Disturbance of benthic species by fishing activities: a sensitivity index. Aquat. Conserv.: Mar. Freshw. Ecosyst. **6**: 257–268

Olson, S. 1984. Shellfish enhancement report. NOAA, NMFS, Final Rep. Project 1-144-R. Washington State Dep. Fisheries, Olympia, Wash. 85 p. (Mimeograph.)

Paul, A.J., and Paul, J.M. 1981. Temperature and growth of maturing *Haliotis kamtschatkana* Jonas. Veliger, **23**: 321–324.

Prince, J.D., Sellers, T.L., Ford, W.B., and Talbot, S.R. 1988. Confirmation of a relationship between the localized abundance of breeding stock and recruitment for *Haliotis rubra* Leach (Molluscs: Gastropoda). J. Exp. Mar. Biol. Ecol. **122**: 91–104.

Quayle, D.B. 1971. Growth, morphometry and breeding in the British Columbia abalone (*Haliotis kamtschatkana* Jonas). Fish. Res. Board Can. Tech. Rep. 279. 84 p.

Quinn, J.F., Wing, S.R., and Botsford, L.W. 1993. Harvest refugia in marine invertebrate fisheries: models and applications in the red sea urchin, *Stronglyocentrotus franciscanus*. Am. Zool. **33**: 537–550.

Ray, G.C. 1976. Critical marine habitats: definition, description, criteria and guidelines for identification and management. *In* Proceedings of the International Conference on Marine Parks and Reserves. Tokyo, 12–14 May 1975. Int. Union Conserv. Nat. Nat. Resour. Publ. New Ser. **37**: 15–59.

Roberts. C.M. 1997. Connectivity and management of Caribbean coral reefs. Science (Washington, D.C.), **278**: 1454–1457.

St. Mary, C.M., Osenberg, C.W., Frazer, T.K., and Lindberg, W.J. 2000. Stage structure, density dependence, and the efficacy of marine reserves. Bull. Mar. Sci. (In press.)

Seki, T., and Sano, M. 1998. An ecological basis for the restoration of abalone populations. Bull. Tohoku Nat. Fish. Res. Inst. **60**: 23–40. (Abstract in English).

Sloan, N.A., and Breen, P.A. 1988. Northern abalone, *Haliotis kamtschatkana*, in British Columbia: fisheries and synopsis of life history information. Can. Spec. Publ. Fish. Aquat. Sci. 103. 46 p.

Tegner, M.J. 1993. Southern California abalones: can stocks be rebuilt using marine harvest refugia. Can. J. Fish. Aquat. Sci. **50**: 2010–2018.

Thomas, G., and Campbell, A. 1996. Abalone resurvey in Aristazabal island, the Estevan Group and Banks Island, June 1993. Can. Tech. Rep. Fish. Aquat. Sci. 2089. pp. 97–109.

Turner, N. 1997. Traditional ecological knowledge. *In* The rainforests of home: profile of a North American bioregion. *Edited by* P.K. Schoonmaker, B. von Hagen, and E.C. Wolf. Island Press, Washington, D.C. pp. 275–299.

Wallace, S. 1997. The role of marine reserves in the management of non-migratory species in coastal British Columbia. *In* The design and monitoring of marine reserves. University of British Columbia, Vancouver, B.C. Fisheries Centre Res. Rep. 5. pp. 24–25.

Wallace, S.S. 1999. Evaluating three forms of marine reserves on northern abalone (*Haliotis kamtschatkana*) populations in British Columbia, Canada. Conserv. Biol. **13**: 882–887.

Watson, J.C. 1993. The effects of sea otter (*Enhydra lutris*) foraging on shallow rocky communities off northwestern Vancouver Island, British Columbia. Ph.D. dissertation, University of California, Santa Cruz, Calif.

Watson, J.C., and Smith, T.G. 1996. The effects of sea otters on invertebrate fisheries in British Columbia: a review. Can. Tech. Rep. Fish. Aquat. Sci. 2089. pp. 262–303.

Watson, J.C., Ellis, G.M., Smith T.G., and Ford, J.K.B. 1997. Updated status of the sea otter, *Enhydra lutris*, in Canada. Can. Field-Nat. **111**: 277–286.

Winther, I., Campbell, A., Thomas, G.A., Adkins, B.E., and Clapp, B.G. 1995. Abalone resurvey in the southeast Queen Charlotte Islands, 1994. Can. Manuscr. Rep. Fish. Aquat. Sci. 2273. 46 p.

Summary of Rebuilding Strategy

Summary of a strategy for rebuilding abalone stocks in British Columbia

J. Gardner, J. Griggs, and A. Campbell

Introduction

The strategy to rebuild northern abalone (*Haliotis kamtschatkana*) stocks in British Columbia (B.C.) is based on discussions during a four-day workshop held at the Coast Bastion Hotel in Nanaimo, B.C., February 23–26, 1999. Working group participants had community, First Nations, aquaculture, recreational, commercial, and conservation interests. Government and university representatives, including First Nations, scientists, and management specialists made presentations during the first two days of the workshop.

Historically, northern abalone provided a traditional fishery for First Nations and the species supported a commercial dive fishery from the 1970s until the late 1980s. Fisheries and Oceans Canada (DFO) closed the abalone fishery in December 1990 because of conservation concerns about the decline in abalone densities (Adkins 2000; Campbell 2000). The plan was to review the closure after five years, but surveys showed, nine years later, that stocks continued to decline. Northern abalone was now the only species with a total harvesting closure on the B.C. coast. This led to the need to determine how to better encourage northern abalone stocks to rebuild.

Experience in other countries shows that B.C. is not the only jurisdiction facing this challenge. Abalone fisheries have closed in Alaska and California (Karpov et al. 2000; Woodby et al. 2000) and in one case face the prospect of species extinction (Davis et al. 1998). Stocks in South Africa have been overharvested (Tarr 2000). While some progress is being made in stock restoration worldwide (e.g., Tegner 2000; Shepherd et al. 2000; Seki and Taniguchi 2000), poaching continues to threaten rebuilding efforts in many places, including B.C. (e.g., Daniels and Floren 1998; Jubinville 2000; Tarr 2000).

To begin meeting the rebuilding challenge, workshop participants came together to clarify roles, expectations, and shared interests; to examine abalone biology and ecology; to learn about rehabilitation efforts in other places; and to develop this strategy. The purpose of this paper is to summarize the main components of the strategy which were generated by facilitated working groups in the last two days of the workshop. The strategy forms the first step in a comprehensive plan for rebuilding northern abalone stocks in B.C.

J. Gardner and J. Griggs. Dovetail Consulting Inc., Suite 105–2590 Granville Street, Vancouver, BC V6H 3H1, Canada.
A. Campbell. Fisheries and Oceans Canada, Pacific Biological Station, Nanaimo, BC V9R 5K6, Canada.
Correct citation: Gardner, J., Griggs, J., and Campbell, A. 2000. Summary of a strategy for rebuilding abalone stocks in British Columbia. *In* Workshop on Rebuilding Abalone Stocks in British Columbia. *Edited by* A. Campbell. Can. Spec. Publ. Fish. Aquat. Sci. 130. pp. 151–155.

Goals and guiding principles for the strategy

The primary goal of the Abalone Rebuilding Strategy for British Columbia is to conserve the northern abalone resource through the co-operative efforts of all involved parties.

The secondary goals of the strategy are as follows:

1. halt the decline of wild stocks;
2. develop self-sufficient recruitment and ensure healthy meta-populations (including conservation of genetic diversity) of abalone throughout its range; and
3. support the rebuilding of stocks to defined, optimal levels while ensuring ecosystem health (ecological integrity) for present and future generations.

Fundamental to the strategy is the continued application of the hierarchy of priorities shaping the mandate of DFO:

1. conservation;
2. provision of access for First Nations' food, social, and ceremonial needs as defined in the Canadian Constitution under Section 35(1); and
3. provision of other opportunities for resource harvesting and use.

Additional guiding principles for the strategy are to

- take precautionary (e.g., FAO 1995) and ecosystem-based (e.g., Perry 1999) approaches;
- be adaptive by using pilot projects to experiment with different rebuilding methods;
- foster co-operation and local community involvement in a shared stewardship effort;
- develop a sound information and data base, including traditional local knowledge and international science;
- incorporate public education to build awareness of the abalone resource and abalone status; and
- balance costs and benefits fairly.

Key elements of the strategy

Prevention of illegal harvesting

Poaching is one of the most serious problems confronting efforts to rebuild abalone stocks. Actions required to prevent illegal harvesting include the following:

- ensure enforcement coverage along the B.C. coast, focusing on problem areas;

- combine necessary resources for enforcement activities, including help from First Nations fishery "guardian" officers;
- ensure that penalties and publicity are sufficient to deter illegal harvesting of abalone;
- encourage community involvement in looking for and reporting on violations;
- make the illegal harvest, sale, and (or) consumption of abalone socially unacceptable through education of the public, especially children, fishermen, processors, retailers, and consumers;
- create an abalone DNA bank in co-operation with other countries, so that identification of abalone species (and their geographic origin) can support effective enforcement;
- pass legislation allowing enforcement staff to prosecute people for the possession of illegal product from other countries; and
- co-operate over the control of the resource to share information relevant to enforcement and to ensure that "legal" farmed abalone can be distinguished from "illegal" wild abalone.

Stock assessment

Stock assessment should help in evaluating progress towards the goals of the rebuilding strategy by estimating changes in abalone population sizes over time and providing a monitoring and feedback tool to determine if the rebuilding strategy is effective. Stock assessment should also incorporate an understanding of population and community ecology to support an ecosystem-based approach to management.

Priorities for stock assessment are as follows:

- complete genetics work to allow for species identification and the identification of local abalone stocks to support enforcement efforts and identify meta-populations of abalone;
- complete baseline surveys to identify suitable pilot study sites; and
- communicate stock assessment information to various target audiences, including public, user groups, enforcement, and politicians.

Actions for stock assessment include the following:

- compilation and review of existing information and knowledge;
- use of quantitative surveys and mapping to monitor changes over time in abalone population sizes and biological attributes (e.g., recruitment, density, patch size, and patch frequency), and key ecosystem components, regime shifts, and other physical parameters of habitat;
- interpretation of survey data to understand abalone stock dynamics (e.g., stock–recruitment relationships) and physical factors such as oceanographic currents and habitat types in relation to abalone patches, and to identify abalone refugia; and
- making the results of stock assessment available to the science community and the public.

Abalone aquaculture

Abalone aquaculture or farming generally involves a private sector farming product with business in the lead role, although communities and co-operatives also can be involved. Abalone aquaculture can be a useful rebuilding tool by providing a safe haven for rearing abalone out of the wild environment. Farming and enhancement can be undertaken in parallel with respect to learning about the biology of the abalone species, monitoring, and technology development. Through farming, co-operation of the private and public sector can combine resources for rebuilding. Aquaculture will not replace natural production; the 40 or so abalone farms around the world have not proven this to be an easy enterprise (McCormick 2000).

Effective development of abalone aquaculture in B.C. depends on the following approaches.

- An experimental approach should be taken to ensure that we learn from aquaculture and enhancement projects through research, monitoring, and evaluation. The risk of failure must be taken into account.
- Aquaculture must not be conducted to the detriment of natural stocks and we need to ensure the protection of the existing resource through enhancement and enforcement parallel to aquaculture.
- Communities should be involved in aquaculture and enhancement.
- Meaningful consultation with the First Nations on the coast and the approval of First Nations for aquaculture initiatives in their traditional territories are essential.
- Commercial requirements for a viable abalone aquaculture enterprise include the creation of a favorable economic climate attractive to investors and providing certainty in areas such as access to brood stock, tax incentives, and government policies.
- Partnerships or at least collaboration with government are necessary to achieve the goals of both farmers and enhancement.
- Communities and the general public need assurances of environmental sustainability and good regulation of the aquaculture industry. DFO and provincial agencies should work with First Nations and aquaculturists on written collaborative agreements as a prerequisite for the development of aquaculture.

Stock restoration

The goal for abalone stock restoration in B.C. is to rebuild abalone stocks to reach self-sustaining stock levels for viable populations in B.C. and, where possible, reach maximum sustainable natural/historic stock levels. Note that the minimum self-sustaining stock level is biologically determined, whereas the maximum attainable level that could be built above the minimum level is, in part, socially or politically determined.

Priority actions for restoration include

- prioritization of locations for stock restoration work, which should include selection criteria, such as the ease in monitoring and enforcing closed fishing areas and the probability of success in reaching the goals;
- development of long-term goals, specific strategies, or interventions to reach local target levels with local communities; and
- restoration of stocks through stock interventions, habitat interventions, and community activities.

Natural mechanisms may rebuild the abalone population if the poaching problem is removed, in which case, more costly interventions may not be needed. However, if the poaching problem is addressed but natural rebuilding does not occur, the following potential interventions may be considered.

- *Stock interventions* could include larval seeding, juvenile out-planting, and aggregating adults to increase spawning densities. Considerations should include collecting and replanting brood stock or offspring in the same area [to maintain genetic diversity and minimize the risk of disease (Bower 2000; Withler 2000)] and develop tags to differentiate and track wild and out-planted individuals. Research is needed to improve understanding of density-dependent growth and survival and larval/juvenile behavior.
- *Habitat interventions* could include restoring habitat, removing or excluding competing or predator species, and (or) establishing marine protected areas or experimental closed areas. The time frames may determine what kinds of interventions to choose (e.g., if establishing "formal" marine protected areas take a long time, experimental closed areas could be considered).
- *Community activities* could include "neighborhood" or "coast watch" and stewardship programs for community involvement.

A strategic approach to stock restoration requires the following.

- The fishery should be kept closed during the rebuilding process.
- Specific interventions and research should be prioritized in terms of economic and biological feasibility, while the vision should be long-term (i.e., >10 years). The stock manipulation options should be considered with respect to distance from hatchery, area to be enhanced, cost, and time to recovery.
- Local community support is critical to the success of the restoration efforts.
- Because this is an international problem, we need to work with our neighbors and other international stakeholders.

Ecosystem-based management

Fundamental to an ecosystem-based approach are the following actions or considerations:

- Management of human activities and use of our knowledge of habitat and ecosystems are necessary to achieve the goal of sustainable populations.
- Humans should be recognized as one source among many other impacts. Others include introduced species, recovery of reintroduced species, such as sea otters (Watson 2000), and climate change affecting water temperature. The interrelationships of these factors and their cumulative effects need to be considered.
- We should strive for more regionalized management and monitoring of the abalone resource with local involvement, including First Nations communities.
- A system of marine protected areas or experimental closed areas throughout the coast large enough to hold self-sustaining abalone populations should be established. Refugia-based management holds significant potential as an alternative to managing commercial fishing effort by openings, size restrictions, quotas, and licensing (Davis 2000; Jamieson 2000). Refugia or marine protected areas adhere to the precautionary approach in a situation where uncertainty around abalone biology and ecosystem complexity are both high. If properly designed with regard to patch size, recruitment, and growth rates, marine protected areas could act as nurseries for abalone recovery.
- Research to build the knowledge base on ecosystem-based management for abalone, including the effects of sea otters and sea urchins and variability in ocean conditions on abalone should be undertaken.

Strategy implementation

The following actions are critical to the success of strategy implementation.

- Broad ownership of, and buy-in to the strategy from First Nations, communities, and other interest groups should be sought.
- Public awareness of the issues and challenges associated with the biology and management of abalone should be raised.
- The initial "hard choices" for management of abalone should be made and communicated. Action needs to be taken to remedy the "crisis" situation that faces abalone in B.C. Hard political and social decisions need to be made and communicated, including allowing no commercial fishing of northern abalone and establishing sustainable funding mechanisms for management of abalone involving all (i.e., First Nations, commercial, and recreational) users. How to improve management of abalone, allow brood stock harvest for aquaculture, and manage sea otter populations needs to be decided.
- Information should be made central to management. Traditional local ecological knowledge and scientific information need to be incorporated into an information base and made accessible to communities and First Nations. Scientific efforts need to be linked with communities to build involvement, understanding, and acceptance. Biological reference points need to be clearly identified. Effectively protected refugia need to be created to help in assessing alternative adaptive management actions.
- Ecosystem components should be included in management approaches. Although not all the information on an ecosystem may be available, ecosystem-based management of abalone should be considered, e.g., ecological influence of predator (e.g., sea otter, sea star, human) and competitor (e.g., sea urchin) populations. Associated challenges are as follows: addressing "zonal containment" of predators such as sea otters (e.g., use of lethal or nonlethal methods of removal); identifying "valued" ecosystem components and how to achieve a desired balance among abalone, sea otters, and humans; identifying "heritage harvest" areas and refugia; and fostering co-ordinated "international management" (e.g., involving market countries co-operating in poaching interventions).

Pilot projects are key to "taking action" and building community support for management of abalone. Care should be taken when initiating pilot projects to ensure they are part of the overall strategy for abalone management and help to meet the objectives for the rebuilding strategy.

Given the importance of abalone to First Nations culture and communities and the loss of access to the resource for perhaps an entire generation, First Nations peoples need to be involved in the development and execution of the strategy. Consultations and ongoing co-operation must continue on a government-to-government basis. A rebuilding strategy will not be effective without the commitment of First Nations because of their key role as stewards of the resource, especially in parts of the coast populated only by First Nations communities.

A co-operative effort is essential to the success of the rebuilding strategy. The highest level at which co-operation has to occur is international. "Pieces of the rebuilding puzzle" can continue to be drawn together from experience around the world. Here in B.C., to move the strategy forward, we need to identify what current programs (e.g., Marine Protected Area Strategy and Integrated Coastal Zone Management) can contribute.

DFO will continue to be the appropriate lead organization for the Rebuilding Strategy. The agency needs to be supported and advised in this role by a group representing those interested in abalone. Options for co-ordinating bodies for strategy implementation include an Abalone Advisory Committee, Abalone Research Team(s), an Abalone Action Team or Committee, an Abalone Council of British Columbia, or Regional Management Boards.

A detailed critical path for strategy implementation, with budgets, timelines, and assigned responsibilities has to be developed. Moving forward over the next two years, some implementation activities could happen concurrently, shortening the time frame required for sequential implementation of initiatives. However, projects should not be launched without sufficient knowledge, or to the detriment of a grassroots approach. Everyone must acknowledge that the duration of commitment required to rebuild abalone stocks will be long, in the order of decades.

Monitoring and evaluation are required to determine whether the strategy is being implemented in the manner intended and the objectives are being achieved. Monitoring should assess stocks as well as the "human" elements of the strategy (e.g., level of community buy-in and participation). Milestones should be established to act as stepping stones for progress towards the end result.

Acknowledgments

We thank D. Knight (Devon Knight Events), D. Duffy and P. Howe (Praxis Pacific), and K. Needham and C. Rankin (SALASAN Associates Inc.) for assisting with facilitation of working groups; V. Eccleston, C. Hand, G. Horne, I. Perry, J. Rogers, and R. Worsfold for note taking; B. Adkins and R. Mylchreest for providing input on the goals and guiding principles for the workshop strategy; and all the workshop participants for their enthusiastic participation. Funding to Dovetail Consulting Inc. to facilitate the workshop was provided by Fisheries and Oceans Canada.

References

Adkins, B.E. 2000. The British Columbia fishery for northern abalone, *Haliotis kamtschatkana*: management from inception to closure and beyond [Abstract]. *In* Workshop on Rebuilding Abalone Stocks in British Columbia. *Edited by* A. Campbell. Can. Spec. Publ. Fish. Aquat. Sci. 130. pp. 51.

Bower, S.M. 2000. Infectious diseases of abalone (*Haliotis* spp.) and risks associated with transplantation. *In* Workshop on Rebuilding Abalone Stocks in British Columbia. *Edited by* A. Campbell. Can. Spec. Publ. Fish. Aquat. Sci. 130. pp. 111–122.

Campbell, A. 2000. Review of northern abalone, *Haliotis kamtschatkana*, stock status in British Columbia. *In* Workshop on Rebuilding Abalone Stocks in British Columbia. *Edited by* A. Campbell. Can. Spec. Publ. Fish. Aquat. Sci. 130. pp. 41–50.

Daniels, R., and Floren, R. 1998. Poaching pressures on northern California's abalone fishery. J. Shellfish Res. **17**: 859–862.

Davis, E.G., Haaker, P.L., and Richards, D.V. 1998. The perilous condition of the white abalone *Haliotis sorenseni*, Bartsch, 1940. J. Shellfish Res. **17**: 871–875.

Davis, E.G. 2000. Refugia-based strategies to restore and sustain depleted abalone (*Haliotis* spp.) populations in southern California. *In* Workshop on Rebuilding Abalone Stocks in British Columbia. *Edited by* A. Campbell. Can. Spec. Publ. Fish. Aquat. Sci. 130. pp. 133–138.

FAO. 1995. Precautionary approach to fisheries. Part 1: Guidelines on the precautionary approach to capture fisheries and species introductions. FAO Fish. Tech. Pap. 350, Part 1. 47 p.

Jamieson, G.S. 2000. Review of marine protected areas in temperate waters and their relevance to abalone (*Haliotis* spp.) in British Columbia. *In* Workshop on Rebuilding Abalone Stocks in British Columbia. *Edited by* A. Campbell. Can. Spec. Publ. Fish. Aquat. Sci. 130. pp. 139–147.

Jubinville, B. 2000. Enforcing the fishery closure for northern (pinto) abalone (*Haliotis kamtschatkana*) in British Columbia [Abstract]. *In* Workshop on Rebuilding Abalone Stocks in British Columbia. *Edited by* A. Campbell. Can. Spec. Publ. Fish. Aquat. Sci. 130. pp. 52.

Karpov, K.A., Haaker, P., Taniguchi, I., and Rogers-Bennett, L. 2000. Serial depletion and the collapse of the California abalone (*Haliotis* spp.) fishery. *In* Workshop on Rebuilding Abalone Stocks in British Columbia. *Edited by* A. Campbell. Can. Spec. Publ. Fish. Aquat. Sci. 130. pp. 11–24.

McCormick, T.B. 2000. Abalone (*Haliotis* spp.) aquaculture: farming status and stock enhancement potential. *In* Workshop on Rebuilding Abalone Stocks in British Columbia. *Edited by* A. Campbell. Can. Spec. Publ. Fish. Aquat. Sci. 130. pp. 55–60.

Perry, R.I. 1999. Scientific concepts for ecosystem-based management of marine invertebrates in Canada's Pacific Coast. Can. Stock Assessment Secretariat Res. Doc. 99/123. Department of Fisheries and Oceans, 200 Kent Street, Ottawa, ON K1A OE6, Canada.

Shepherd, S.A., Preece, P.A., and White, R.W.G. 2000. Tired nature's sweet restorer? Ecology of abalone stock enhancement in Australia. *In* Workshop on Rebuilding Abalone Stocks in British Columbia. *Edited by* A. Campbell. Can. Spec. Publ. Fish. Aquat. Sci. 130. pp. 84–97.

Seki, T., and Taniguchi, K. 2000. Rehabilitation of northern Japanese abalone, *Haliotis discus hannai*, populations by transplanting juveniles. *In* Workshop on Rebuilding Abalone Stocks in British Columbia. *Edited by* A. Campbell. Can. Spec. Publ. Fish. Aquat. Sci. 130. pp. 72–83.

Tarr, R.J.Q. 2000. The South African abalone (*Haliotis midae*) fishery: a decade of challenges and change. *In* Workshop on Rebuilding Abalone Stocks in British Columbia. *Edited by* A. Campbell. Can. Spec. Publ. Fish. Aquat. Sci. 130. pp. 32–40.

Tegner, M.J. 2000. Abalone (*Haliotis* spp.) enhancement in California: what we've learned and where we go from here. *In* Workshop on Rebuilding Abalone Stocks in British Columbia. *Edited by* A. Campbell. Can. Spec. Publ. Fish. Aquat. Sci. 130. pp. 61–71.

Watson, J.C. 2000. The effects of sea otters (*Enhydra lutris*) on abalone (*Haliotis* spp.) populations. *In* Workshop on Rebuilding Abalone Stocks in British Columbia. *Edited by* A. Campbell. Can. Spec. Publ. Fish. Aquat. Sci. 130. pp. 123–132.

Withler, R.E. 2000. Genetic tools for identification and conservation of exploited abalone (*Haliotis* spp.) species. *In* Workshop on Rebuilding Abalone Stocks in British Columbia. *Edited by* A. Campbell. Can. Spec. Publ. Fish. Aquat. Sci. 130. pp. 101–110.

Woodby, D., Larson, R., and Rumble, J. 2000. Decline of the Alaska abalone (*Haliotis* spp.) fishery and prospects for rebuilding the stock. *In* Workshop on Rebuilding Abalone Stocks in British Columbia. *Edited by* A. Campbell. Can. Spec. Publ. Fish. Aquat. Sci. 130. pp. 25–31.

Appendix 1. Program agenda

Day 1. Tuesday, February 23, 1999

Session One: Welcome and opening presentations

James Johnny, Chief Councillor, Nanaimo First Nation, Nanaimo, B.C.

Chris Dragseth, Director, Fisheries Management, Fisheries and Oceans Canada, Vancouver, B.C.

Julia Gardner and Julian Griggs, Workshop Facilitators, Dovetail Consulting, Vancouver, B.C.

First Nations' perspectives on abalone in B.C. (presentations and discussion).

Garry Reece: Tsimshian Nation abalone issues.

Russ Jones: Haida abalone issues.

Aquaculture perspectives on abalone in B.C. (presentation and discussion).

David Johnson: Aquaculture's stake in rebuilding northern abalone wild stock in British Columbia.

Discussion on other perspectives on abalone in B.C.

Session Two: International experience with the status of abalone fisheries in California, Alaska, and South Africa

Konstantin A. Karpov, Peter L. Haaker, Ian Taniguchi, and Laura Rogers-Bennett: Serial depletion and the collapse of the California abalone (*Haliotis* spp.) fishery.

Doug Woodby, Robert Larson, and Jan Rumble: Decline of the southeast Alaska abalone (*Haliotis* spp.) fishery and prospects for rebuilding the stock.

Rob Tarr: The South African abalone (*Haliotis midae*) fishery: a decade of challenges and change.

Discussion of international experience on status of abalone fisheries.

Session Three: Current state of the abalone resource in B.C.

Alan Campbell: Review of northern abalone, *Haliotis kamtschatkana*, stock status in British Columbia.

Bruce Adkins: The British Columbia fishery for northern abalone, *Haliotis kamtschatkana*: management from inception to closure and beyond.

Bryan Jubinville: Enforcing the fishery closure for northern (pinto) abalone (*Haliotis kamtschatkana*) in British Columbia.

Discussion of current status of abalone resource in B.C.

Day 2. Wednesday, February 24, 1999

Session Four: International experience in rebuilding abalone stocks

Thomas B. McCormick: Abalone (*Haliotis* spp.) aquaculture: present status and a stock enhancement tool.

Mia J. Tegner: Abalone (*Haliotis* spp.) enhancement in California: what we've learned and where we go from here.

Tetsuo Seki and Kazuya Taniguchi: Rehabilitation of northern Japanese abalone, *Haliotis discus hannai*, populations by transplanting juveniles.

Scoresby A. Shepherd, P.A. Preece, and R.G.W. White: Tired nature's sweet restorer? Ecology of abalone (*Haliotis* spp.) stock enhancement in Australia.

Discussion of international experience in rebuilding abalone stocks.

Session Five: Additional pieces of the rebuilding puzzle

Ruth Withler: Genetic tools for identification and conservation of exploited abalone (*Haliotis* spp.) species.

Susan M. Bower: Infectious diseases of abalone (*Haliotis* spp.) and risks associated with transplantation.

Jane C. Watson: The effects of sea otters (*Enhydra lutris*) on abalone (*Haliotis* spp.) populations.

Gary E. Davis: Refugia-based strategies to restore and sustain abalone (*Haliotis* spp.) populations in southern California.

Glen S. Jamieson: Marine protected areas and their relevance to abalone (*Haliotis* spp.) conservation in British Columbia.

Discussions on rebuilding abalone stocks.

Session Six: Discussion on framework, guiding principles, and goals for a rebuilding strategy

Draft of rebuilding strategy framework, guiding principles, and goals.

Day 3. Thursday, February 25, 1999

Review of rebuilding strategy framework, guiding principles, and goals.

Facilitated working groups to discuss and report on strategy components: prevention of illegal harvest, stock assessment, aquaculture, stock restoration, ecosystem-based management.

Day 4. Friday, February 26, 1999

Facilitators' summary of draft strategy components.

Self-facilitated groups to discuss shared interests.

Final group discussion on agreement and revisions of draft strategy components.

Pn 1000 1—

2

Appendix 2. The abalone song

Riding the "abalone wave" of enthusiasm, a group of songsters led participants in a tribute to the abalone at the close of the workshop. The verse, composed incrementally over a series of abalone workshops around the world and sung to the tune of "Yankee Doodle," follows.

OH! Some folks boast of quail on toast
Because they think it's tony
But I'm content to owe my rent
And live on abalone!

OH! Mission points a friendly joint
Where every crab's a crony
And true and kind you'll ever find
The clinging abalone

He wanders free beneath the sea
Where 'ere the coast is stony
He flaps his wings and madly sings
The plaintive abalone.

On Carmel Bay the people say
We feed the Lazzaroni
On Boston beans and fresh sardines
And tender abalone.

Some live on hope and some on dope
And some on alimony
But my tomcat, he lives on fat
And tender abalone.

OH! Some drink rain and some champagne
Or brandy by the pony
But I will try a little rye
With a dash of abalone.

OH! Some like ham and some like jam
And some like macaroni
But bring me in a pail of gin
And a tub of abalone

He hides in caves beneath the waves
His ancient patrimony
And so 'tis shown that faith alone
Reveals the abalone.

The more we take the more they make
In deep sea matrimony
Race suicide cannot abide
The fertile abalone.

I telegraph my better half
By Morse or by Marconi
But if the need arises for speed
I send an abalone.

They came to hear from far and near
To meet in Californee
Scoresby, David, Martin and Kit
To save the abalone.

We came to Canada to meet
Nanaimo was so homey
Where DFO, First Nations tried
To save the Abalone